Trends in Logic

Volume 39

For further volumes:
http://www.springer.com/series/6645

TRENDS IN LOGIC
Studia Logica Library

VOLUME 39

SCOPE OF THE SERIES

The book series Trends in Logic covers essentially the same areas as the journal Studia Logica, that is, contemporary formal logic and its applications and relations to other disciplines. The series aims at publishing monographs and thematically coherent volumes dealing with important developments in logic and presenting significant contributions to logical research.

The series is open to contributions devoted to topics ranging from algebraic logic, model theory, proof theory, philosophical logic, non-classical logic, and logic in computer science to mathematical linguistics and formal epistemology. However, this list is not exhaustive, moreover, the range of applications, comparisons and sources of inspiration is open and evolves over time.

Luiz Carlos Pereira · Edward Hermann Haeusler
Valeria de Paiva
Editors

Advances in Natural Deduction

A Celebration of Dag Prawitz's Work

Springer

Editors
Luiz Carlos Pereira
Departamento de Filosofia
Pontifícia Universidade Católica do Rio de Janeiro
Rio de Janeiro
Brazil

Edward Hermann Haeusler
Departamento de Informática
Pontifícia Universidade Católica do Rio de Janeiro
Rio de Janeiro
Brazil

Valeria de Paiva
School of Computer Science
University of Birmingham
Birmingham
UK

ISSN 1572-6126 ISSN 2212-7313 (electronic)
ISBN 978-94-007-7547-3 ISBN 978-94-007-7548-0 (eBook)
DOI 10.1007/978-94-007-7548-0
Springer Dordrecht Heidelberg New York London

Library of Congress Control Number: 2013955128

Printed on acid-free paper

Springer is part of Springer Science+Business Media (www.springer.com)

This book is dedicated to
Dag Prawitz

Preface

In 1926, Lukasiewicz proposed the following problem in his seminar at the University of Warsaw: to find a new formalization of classical predicate logic that could mirror the use of assumptions in real life mathematical arguments and reasoning. Jaskowski took up the task and formulated for the first time a system of Natural Deduction. The system was presented in 1927 at a meeting whose proceedings were published in 1929. The journal paper was published in 1934 under the title "On the Rules of Suppositions in Formal Logic".

In the same year Gentzen published his classical paper "Untersuchungen über das logische Schliessen" where we can find a different Natural Deduction system for both classical logic and intuitionistic logic. In Gentzen's Natural Deduction systems each logical connective has an *introduction* rule showing how to introduce an instance of the connective into a proof, and an *elimination* rule showing how to obtain a consequence of a sentence with that connective as main sign in a proof. These paired introduction and elimination rules provide a self-contained and intuitive explanation of the meaning of each connective. But these paired rules could also produce some "roundabouts" in proofs and Gentzen's *Hauptsatz* idea was that these roundabouts could be eliminated from proofs: Gentzen himself described how this idea could be carried out for intuitionistic predicate logic. But Gentzen's main goal was the consistency of classical arithmetic and for this reason the theoretical fate of natural deduction systems could have been sealed by Gentzen's prescient commentary:

> The Hauptsatz holds both for classical and for intuitionistic predicate logic. In order to be able to enunciate and prove the Hauptsatz in a convenient form, I had to provide a logical calculus especially suited to the purpose. For this the natural calculus proved unsuitable. For, although it already contains the properties essential to the validity of the Hauptsatz, it does so only with respect to its intuitionistic form, in view of the fact that the law of the excluded middle, as pointed out earlier, occupies a special position in relation to these properties.

As Gentzen said, although in the intuitionistic case the natural calculus had the desired properties for the formulation and proof of the Hauptsatz, classical logic formulated with the rule of double negation elimination or through the addition of the axiom of the excluded middle did not seem to possess the essential properties "for the validity of the Haupsatz". This was the main reason for the development of a new calculus, the *sequent calculus*, for which the Haupsatz was formulated and proved for both classical and intuitionistic logic. From the founding fathers (Gentzen and Jaskowski) to 1965, studies on natural deduction systems were basically restricted to (1) their use as a good deductive system for teaching logic, and (2) to problems related to the formulation of the rules for the quantifiers, especially to the rules corresponding to universal generalization and to existential instantiation. This situation changed dramatically 30 years later, when in 1965 Dag Prawitz and Andrés Raggio proved, independently of each other, the normalization theorem for classical first order logic.

This process of proof normalization, and its later extensions to higher order logics, was a breakthrough that paved the way for numerous fruitful applications of proof theory and logic to computer science. In particular, the Curry-Howard isomorphism showing the intimate connections between Natural Deduction and Type Theory, and especially proof normalization and type equivalence, laid the groundwork for the development of type systems in programming languages. This not only led to the creation of new programming languages, but also fed back into the development of new logics. The Extended Curry-Howard Correspondence furthered the connection between types, proofs, and programs to include morphisms in appropriate categories between programs and proofs. This algebraic dimension lent new mathematical insight into computational and syntactic issues in computer science, and stands to open up new areas of study in classical proof theory.

Natural Deduction is very dear to all of us involved with this book. The book grew out of the organization of a conference entitled simply "Natural Deduction" that took place in the Pontifical Catholic University of Rio de Janeiro (PUC-Rio) in 2001. The conference was conceived as a celebration of Dag Prawitz's work in logic, in particular of his work on proving normalization for logical systems and his careful philosophical discussion of the importance of proofs. Due to circumstances beyond our control, the project was shelved for several years, but as the possibility of resuming it presented itself, we jumped at the opportunity, as interdisciplinary and deep-meaning interactions between philosophers and computer scientists need to be fostered by all means at our disposal.

We have selected 12 papers associated with Natural Deduction, in an extended sense, as the theme of the conference and of this volume that we now briefly describe. The first contribution, by Schröder-Heister, reminds us that in 1979, Prawitz proposed a kind of operational completeness for the set of intuitionistic logical constants: he first defined abstract introduction and elimination schemes and claimed that any logical operator whose introduction and elimination rules were instances of these abstract schemes would be defined by the intuitionistic logical constants. This problem was solved by Peter Schröder-Heister who introduced for the first time higher-level natural deduction rules and in particular higher-level generalized elimination rules. In his contribution Schröder-Heister investigates some interesting relations between generalized higher-level elimination rules and standard-level generalized elimination rules and shows how some results for generalized higher-level elimination rules can produce new results in the standard-level, in particular a new left-introduction schema for implication can be defined from the interpretation of implication as rules.

Urban's paper re-addresses the problem of the correspondence between normalization steps in Natural Deduction and the corresponding reductions in Sequent Calculus, an issue going back to Kreisel's observations in 1971, heavily developed in Zucker's 1974 doctoral work. Returning to Zucker's investigation, Urban reproves Zucker's result that in the fragment of intuitionistic logic consisting of formulae build up from $\exists$ and $\forall$ only, every reduction sequence in natural deduction corresponds to a reduction sequence in the sequent calculus and vice versa. But this is proved in a much simpler way than Zucker's via terms associated to the sequent derivations, with the additional benefit that the cut-elimination procedure (from Urban 2000) is strongly normalising for all connectives, while the cut-elimination procedure in Zucker's work is not strongly normalising when $\bigvee$ and $\exists$ are included. Urban's work also shows that the negative result when $\bigvee$ and $\exists$ are included, is not because cut-elimination fails to be strongly normalising for these connectives, as asserted by Zucker, rather it is because there are cut-elimination reductions that do not correspond to any normalisation reduction.

Joinet's paper provides an insightful discussion on how logic changed in the twentieth century from a philosophical enterprise concerned with reasoning to a mathematical enterprise concerned with proofs, and more recently, to a computer science enterprise concerned with the dynamics of proofs. In particular he discusses the dynamic character of reasoning versus the static character of proofs; the referential dimension of reasoning versus the inferential nature of proofs and the indeterminacy of reasoning versus the correctness of proofs and reflects that this change of concerns (the metamorphosis in the title), especially the last one, from Mathematics to Computer Science, seen as a consequence of the Curry-Howard isomorphism, provides us with a *Computational Foundation for Logic*, which amazingly enough bring us back to a dynamic theory of reasoning as originally conceived by philosophers.

De Queiroz and de Oliveira's paper also discusses the origins of Natural Deduction, going back to Frege's introduction of variable-binding, and his idea of having terms representing incomplete "objects" whenever they contain free variables, as motivating their discussion of equality rules in first-order logic. De Queiroz and De Oliveira wish to see in first-order logic with equality a greater separation into two independent levels that they call the functional and the logical levels of the calculus. An analysis of how to keep these two levels of calculus independent, and yet harmonious, lead the authors to describe a system of Natural Deduction rules for propositional equality that sits between the extensional and the intensional versions of Martin-Loef's type theory. The paper contains a cogent discussion of why one would like to have a system which is a middle ground solution between the intensional and the extensional accounts of Martin-Loef's propositional equality. And then presents a system satisfying these required properties, via a notion of "rewrite reasons".

Legris' paper has a historical-conceptual nature. Its aim is to examine Paul Hertz logical theory and to show its influence and impact on Gentzen's logical systems. According to Legris, Hertz's work can be seen as an anticipation of a theory of proofs in the current sense. Legris also shows how Hertz's work can be considered as a kind of bridge that would connect traditional logic to Gentzen's systems. The paper concludes with an analysis of Hertz's philosophical ideas concerning the relation between the essence of logic and more general epistemological issues.

In the literature logics modelling the idea of "generally" have been considered as extensions of First-Order Logic. Deductive systems for these logics can be obtained either by a usual system for first-order plus a set of axioms introducing the quantifiers as a kind of defined symbol, or by tailoring specific inference rules to deal with these new quantifiers. The article by Vana, Veloso and Veloso describes how to develop good Natural Deduction systems for the quantifiers corresponding to "generally", "most" and "several". This article is a comprehensive tract on the techniques used to deal with quantification notions derived from filters and ultra-filters. The systems presented are shown to be normalizable and the article discusses some consequences of normalization.

Anyone using a theorem prover or a proof assistant wants to have confidence that the system implemented will not allow the derivation of false results. Theorem provers (and systems used in their implementation) must ensure that it is not possible to derive contradictions and that the implementation itself is correct with respect to the original system. Professor Seldin's article discusses clearly how the use of logical systems based on typed lambda-calculus are a most adequate basis for theorem provers when one wants to have confidence on the correctness of their implementation. Taking each of the many vertices presented in Barendregt's lambda-cube, his article discusses how consistency and confidence on a given implementation can be obtained by means of basic observations on the main meta-properties of the lambda-calculi considered. The article is also a nice step-by-step presentation of how to improve the expressive power of a theorem prover by following paths along the edges of the lambda-cube.

Very traditional philosophical questions concerning the nature of propositional logic are taken up in Chateaubriand's paper. Is logic a theory about something? Can one conceive of propositional logic as an ontological theory? What is the (epistemological) role of deductive systems? Do propositions have a structure in the same way that we can say that sentences have a structure? These questions become especially interesting and acute with respect to the connection between the material implication and the material conditional, a central topic in the development of natural deduction systems.

Since the work of Lambek, category theory has been used to provide semantics to substructural logics. This semantics is proof-theoretical, since proofs are incorporated in the models. Linear logic is a logic that interprets substructural implicational into intuitionistic implication by means of exponentials. There have been plenty of categorical models for providing semantics for linear logic. In this article, Valeria de Paiva compares several notions of categorical model of intuitionistic linear logic in the literature. Her article explains why choices can be made and what they amount to. Her conclusion is that one of the oldest and more complicated notions, namely linear categories are still the best way to describe models of linear logic, if one wants the correspondence between syntax and semantics to be as tight as possible.

Assertions and hypothesis are key notions in the Logic for Pragmatics introduced by Carlo Dalla Pozza and Claudio Garola. In his contribution to the present volume, Bellin reconsiders and revises some results that are based on the modal translation of his proposal for an intuitionistic logic for assertions and conjectures into classical S4: while assertions require epistemic necessity of the truth of the propositional content, doubts require the epistemic possibility of falsity and conjectures are justified by the possibility of epistemic necessity. In the final part of the paper, Bellin investigates the proof theory of a fragment of co-intuitionistic logic.

Anyone who has studied proof theory has already been exposed to strong normalization proofs. Compared with weak normalization, they are more complicated. They fall into to main proof patterns: (1) Using reducibility candidates, or; (2) Providing a direct proof that any sequence of reductions is terminating. The first case is known to be more elegant and more famous indeed. It was used in the case of stronger theories such as System F and Gödel's system T, for example. The second case is more informative in general, and has two forms: (1) Provide a special kind of reduction sequence that dominates any other reduction sequence and prove weak normalization for this (dominating) reduction sequence, or; (2) Prove directly by induction, using a tricky induction measure, that any reduction sequence is terminating. The first approach is also known as the worst-sequence approach, since the special sequence definition involved considers the longest possible reduction sequence for each term. Proofs of confluence can be combinatorial or semantical, in the case a stronger than termination/confluence inductive hypothesis is used. Zimmerman's article introduces a method of decomposition of reductions. He uses the terminology and main results from rewriting theory to show how one can decompose a reduction in such a way that its confluence and termination can be obtained from the termination and confluence of its constituents. This method is applied to Natural Deduction systems for Intuitionistic Logic and Intuitionistic Linear Logic. Zimmerman obtains termination and confluence in a rather elegant way by means of a general theory of reductions.

Forty years ago, Prawitz formulated a very interesting conjecture about the completeness of intuitionistic logic: are Gentzen's elimination rules for intuitionistic logic the strongest possible rules? This conjecture was formulated in the framework of a certain semantic approach to general proof theory. In his contribution to the present volume, Prawitz reconsiders this old conjecture and the very semantic approach that served as the base for the formulation of the conjecture. A revised notion of validity is defined and given that Gentzen's elimination rules are valid, the conjecture now assumes the form: are all valid inferences rules derivable in Gentzen's natural deduction system for intuitionistic logic?

Luiz Carlos Pereira
Valeria de Paiva
Edward Hermann Haeusler

Acknowledgments

The editors would like to thank the authors of the papers in this collection for their forbearance with the ups and downs of our editorial journey. The editors also would like to acknowledge their immense gratitude (thank you a thousand times!) to Ana Isabel de Azevedo Spinola Dias, Fernando Naufel do Amaral, Alexandre Rademaker and Bruno Lopes Vieira, in chronological order. These fantastic ex-students and young collaborators are one reason why this book finally reaches its published state. The editors also would like to express their deeply felt gratitude to Heinrich Wansing, the editor-in-chief of Trends in Logic, for his commitment to this project, especially we were feeling lost. Lastly, the editors would like to thank Dag Prawitz for being our intellectual mentor, our unending source of inspiration, a towering academic father figure and our friend.

Contents

Generalized Elimination Inferences, Higher-Level Rules, and the Implications-as-Rules Interpretation of the Sequent Calculus 1
Peter Schroeder-Heister

Revisiting Zucker's Work on the Correspondence Between Cut-Elimination and Normalisation 31
Christian Urban

Proofs, Reasoning and the Metamorphosis of Logic 51
Jean-Baptiste Joinet

Natural Deduction for Equality: The Missing Entity 63
Ruy J. G. B. de Queiroz and Anjolina G. de Oliveira

Paul Hertz's Systems of Propositions As a Proof-Theoretical Conception of Logic 93
Javier Legris

On the Structure of Natural Deduction Derivations for "Generally" 103
Leonardo B. Vana, Paulo A. S. Veloso and Sheila R. M. Veloso

Type Theories from Barendregt's Cube for Theorem Provers 129
Jonathan P. Seldin

What is Propositional Logic a Theory of, if Anything? 145
Oswaldo Chateaubriand

Categorical Semantics of Linear Logic for All 181
Valeria de Paiva

Assertions, Hypotheses, Conjectures, Expectations: Rough-Sets Semantics and Proof Theory 193
Gianluigi Bellin

Decomposition of Reduction 243
Ernst Zimmermann

An Approach to General Proof Theory and a Conjecture of a Kind of Completeness of Intuitionistic Logic Revisited 269
Dag Prawitz

Generalized Elimination Inferences, Higher-Level Rules, and the Implications-as-Rules Interpretation of the Sequent Calculus

Peter Schroeder-Heister

Abstract We investigate the significance of higher-level generalized elimination rules as proposed by the author and compare them with standard-level generalized elimination rules as proposed by Dyckhoff, Tennant, López-Escobar and von Plato. Many of the results established for natural deduction with higher-level rules such as normalization and the subformula principle immediately translate to the standard-level case. The sequent-style interpretation of higher-level natural deduction as proposed by Avron and by the author leads to a system with a weak rule of cut, which enjoys the subformula property. The interpretation of implications-as-rules motivates a different left-introduction schema for implication in the ordinary (standard-level) sequent calculus, which conceptually is more basic than the implication-left schema proposed by Gentzen. Corresponding to the result for the higher-level system, it enjoys the subformula property and cut elimination in a weak form.

This chapter was completed during a research stay at IHPST Paris supported by the Foundation Maison des Sciences de l'Homme, by the ESF research project "Dialogical Foundations of Semantics (DiFoS)" within the ESF-EUROCORES programme "LogICCC—Modelling Intelligent Interaction" (DFG Schr 275/15-1) and the French-German DFG-ANR project "Hypothetical Reasoning—Logical and Semantical Perspectives (HYPOTHESES)" (DFG Schr 275/16-1, 16-2).

P. Schroeder-Heister (✉)
Wilhelm-Schickard-Institut für Informatik, Universität Tübingen,
Sand 13, 72076 Tübingen, Germany
e-mail: psh@uni-tuebingen.de

L. C. Pereira et al. (eds.), *Advances in Natural Deduction*,
Trends in Logic 39, DOI: 10.1007/978-94-007-7548-0_1,

1 Generalized Higher-Level Elimination Rules

In [27, 28], generalized elimination rules for logical constants were proposed in order to obtain a general schema for introduction and elimination rules for propositional operators.[1] Given m introduction rules for an n-ary constant of propositional logic c

$$(c\mathrm{I})\quad \frac{\Delta_1(A_1,\ldots,A_n)}{c(A_1,\ldots,A_n)}\quad\cdots\quad\frac{\Delta_m(A_1,\ldots,A_n)}{c(A_1,\ldots,A_n)},$$

where the $\Delta_i(A_1,\ldots,A_n)$ are premisses structured in a certain way, the elimination rule is

$$(c\,\mathrm{E})\quad \frac{c(A_1,\ldots,A_n)\quad \begin{matrix}[\Delta_1(A_1,\ldots,A_n)]\\ C\end{matrix}\quad\cdots\quad\begin{matrix}[\Delta_m(A_1,\ldots,A_n)]\\ C\end{matrix}}{C},$$

where the brackets indicate the possibility of discharging the assumption structures mentioned. For conjunction and implication these rules specialize to the following:

$$(\wedge\,\mathrm{I})\ \frac{A\quad B}{A\wedge B}\qquad (\wedge\,\mathrm{E_{GEN}})\ \frac{A\wedge B\quad \begin{matrix}[A\quad B]\\ C\end{matrix}}{C}$$

$$(\rightarrow\mathrm{I})\ \frac{\begin{matrix}[A]\\ B\end{matrix}}{A\rightarrow B}\qquad (\rightarrow\mathrm{E_{HL}})\ \frac{A\rightarrow B\quad \begin{matrix}[A\Rightarrow B]\\ C\end{matrix}}{C}\ .$$

Here, $(\wedge\,\mathrm{I})$ and $(\rightarrow\mathrm{I})$ are the usual introduction rules for conjunction and implication, and $(\wedge\,\mathrm{E_{GEN}})$ and $(\rightarrow\mathrm{E_{HL}})$are their generalized elimination rules, following the pattern of $(c\,\mathrm{E})$. The index "HL" indicates that our generalized elimination schema for $\rightarrow$ is a rule of higher level, in contradistinction to generalized standard-level elimination rules, which will be discussed in Sect. 2.

The motivation for the generalized introduction and elimination rules $(c\,\mathrm{I})$ and $(c\,\mathrm{E})$ can be given in different ways: One is that the introduction rules represent a kind of 'definition' of c, and the elimination rule says that everything that follows from each defining condition $\Delta_i(A_1,\ldots,A_n)$ of $c(A_1,\ldots,A_n)$ follows from $c(A_1,\ldots,A_n)$ itself. This can be seen as corresponding to Gentzen's ([6] p. 189) view of introduction inferences in natural deduction as a sort of definition, or as applying an *inversion principle* or a principle of *definitional reflection* to the introduction rules (for Lorenzen's 'inversion principle' and the related principle of definitional reflection and its usage in proof-theoretic semantics, see [8, 32, 34, 35, 38]).

---

[1] In [29] this approach was extended to quantifiers, and in [30] to various other applications, including the sequent calculus, certain substructural logics (in particular relevant logic), and Martin-Löf's logic. In [7, 32] and in later publications it was extended to the realm of clausal-based reasoning in general. For a general description of the programme of proof-theoretic semantics see [38], and for a formal rendering of proof-theoretic harmony see [19, 40]

A slightly different idea (preferred by the author in [27, 28]) that focusses more on the elimination rules is that $c(A_1, \ldots, A_n)$ expresses the 'common content' of $\Delta_1(A_1, \ldots, A_n), \ldots, \Delta_m\ (A_1, \ldots, A_n)$, i.e., the set of consequences of $c(A_1, \ldots, A_n)$ is exactly the intersection of the sets of consequences of the $\Delta_i(A_1, \ldots, A_n)$ $(1 \leq i \leq n)$. Formally, both interpretations amount to the same, viz., the 'indirect' form of the elimination rule which generalizes the standard pattern of introduction and elimination rules for disjunction:

$$\frac{A}{A \vee B} \qquad \frac{B}{A \vee B} \qquad \frac{A \vee B \quad \begin{matrix}[A]\\ C\end{matrix} \quad \begin{matrix}[B]\\ C\end{matrix}}{C} \ .$$

In the generalized elimination rule (c E) the premiss structures occur in assumption position. In the case of $\to$ this means that the dependency of B from A must be represented as a form of assumption. Our proposal in [27, 28] was to read this dependency as a *rule*, since the fact that B has been derived from A can be naturally understood as the fact that the rule allowing one to pass over from A to B has been established. Conversely, assuming that the rule permitting to pass from A to B is available, is naturally understood as expressing that we can in fact derive B from A. Consequently, the minor premiss of generalized $\to$ elimination ($\to \mathrm{E_{HL}}$)depends on the rule $A \Rightarrow B$, which is discharged at the application of this elimination inference. ($\to \mathrm{E_{HL}}$)is therefore a rule of higher level, i.e., a rule that allows for a proper rule to be discharged, rather than for only formulas as (dischargeable) assumptions. Actually, the 'usual' $\to$ elimination rule, which is modus ponens, is a rule without any discharge of assumptions at all:

$$(\to \mathrm{E_{MP}})\ \frac{A \to B \quad A}{B},$$

as are the usual $\wedge$ elimination rules[2]:

[2] A generalized schema for elimination rules which contains ($\wedge \mathrm{E_{GEN}}$) as an instance was proposed by Prawitz in [25]. I have made this point clear in all relevant publications. In fact, a fundamental error in Prawitz's treatment of implication in his elimination schema was one of my motivations to develop the idea of rules of higher levels, another one being von Kutschera's [11] treatment of logical constants in terms of an iterated sequent arrow in what he very appropriately called *Gentzen semantics* (Ger. 'Gentzensemantik'). So I claim authorship for the general schema for elimination rules, but not for the particular idea of generalized conjunction elimination ($\wedge \mathrm{E_{GEN}}$). I mention this point as I have been frequently acknowledged as the author or ($\wedge \mathrm{E_{GEN}}$), without reference to its embedding into a higher-level framework. When the higher-level framework is mentioned, it is often not considered relevant, as transcending the means of expression of standard natural deduction. Although it extends natural deduction, the results obtained for the general higher-level case can easily be specialized to the case of generalized standard-level elimination rules for implication. In this sense, a direct normalization and subformula proof for standard-level systems with generalized rules is already contained in [27]. For a discussion of inversion in relation to Lorenzen's and Prawitz's semantical approaches see [16, 34, 35].

$$(\wedge\text{E})\ \frac{A\wedge B}{A} \quad \frac{A\wedge B}{B}.$$

This approach is formalized in a calculus of rules of higher levels: A rule of level 1 is a formula, and a rule of level $\ell + 1$ has the form $(R_1, \ldots, R_n) \Rightarrow$ A, where the premisses $R_1, \ldots, R_n$ are rules of maximum level ℓ and the conclusion A is a formula. A finite list of rules is denoted by a capital Greek letter, so in general a rule has the form $\Delta \Rightarrow A$.

A rule can be applied and discharged in the following way: A rule of level 1 works like an axiom, a rule of level 2 generates its conclusion from its premisses:

$$A_1, \ldots, A_n \Rightarrow B \quad \frac{A_1 \quad \ldots \quad A_n}{B}$$

and a rule of level ≥3 of the form

$$(\Delta_1 \Rightarrow A_1, \ \ldots \ , \Delta_n \Rightarrow A_n) \ \Rightarrow \ B$$

generates B from $A_1, \ldots, A_n$, whereby the assumptions $\Delta_1, \ldots, \Delta_n$ can be discharged at this application:

$$(\Delta_1 \Rightarrow A_1, \ \ldots \ , \Delta_n \Rightarrow A_n) \Rightarrow B \quad \frac{\begin{matrix}[\Delta_1] \\ A_1\end{matrix} \qquad \ldots \qquad \begin{matrix}[\Delta_n] \\ A_n\end{matrix}}{\text{B}} . \tag{1}$$

Applying $\Delta \Rightarrow A$ means at the same time assuming it, i.e., introducing it as an assumption except in the case where it is a basic rule (or 'primitive rule') such as ($\rightarrow$ I). Therefore, formally, applying ($\rightarrow$ I) means the same as applying the rule $(A \Rightarrow B) \ \Rightarrow \ A \rightarrow B$:

$$(A \Rightarrow B) \Rightarrow A \rightarrow B \ \frac{\begin{matrix}[A] \\ B\end{matrix}}{A \rightarrow B} . \tag{2}$$

However, as in this case ($\rightarrow$ I) is not introduced as an assumption and therefore not used as an *assumption rule*, but is basic (or 'primitive'), we write as usual

$$(\rightarrow \text{I}) \ \frac{\begin{matrix}[A] \\ B\end{matrix}}{A \rightarrow B} .$$

The counting of rule levels is according to the height of the tree it corresponds to. For example, in (2) the application of the level-3-rule $(A \Rightarrow B) \ \Rightarrow \ A \rightarrow B$ corresponds to a tree

$$\frac{\begin{matrix}[A] \\ B\end{matrix}}{A \rightarrow B}$$

of height 3. In this sense, the primitive rules ($\rightarrow$I) and ($\vee$E) are rules of level 3. Obviously, a rule discharging a rule of level 2, i.e., discharging a rule of the form $A_1, \ldots, A_n \Rightarrow B$, must be at least of level 4. The primitive rules used in standard natural deduction are therefore of maximum level 3. A rule of maximum level 3 is also called a *standard-level rule*, whereas a rule of level $\geq$4 is called a proper *higher-level rule*. Thus ($\rightarrow$E$_{\text{HL}}$)is a proper higher-level (viz., level-4) rule. Obviously, if ℓ is the level of an introduction rule for a logical constant c of the form (c I), then $\ell + 1$ is the level of the corresponding elimination rule of the form (c E). This raising of levels made it necessary to introduce the concept of higher-level rules and the idea of assuming and discharging rules. That this cannot be avoided, i.e. that elimination rules cannot always be 'flattened' (using a terminology proposed by Read [26]) has recently been formally established in [18, 19]. The idea of higher-level elimination rules can be generalized from the realm of logical constants to arbitrary clausal definitions in logic-programming style and therefore to inductive definitions, leading to a powerful principle of 'definitional reflection' that extends standard ways of dealing with clausal definitions (see [7, 8, 32]).

That modus ponens ($\rightarrow$E$_{\text{MP}}$) and ($\rightarrow$E$_{\text{HL}}$)are equivalent can be seen as follows: Suppose ($\rightarrow$E$_{\text{MP}}$) is a primitive rule. Suppose we have derived the premisses of ($\rightarrow$ E$_{HL}$), i.e. we have a derivation $\begin{matrix}\mathcal{D}_1\\ A\rightarrow B\end{matrix}$ of $A \rightarrow B$ and a derivation $\begin{matrix}A\Rightarrow B\\ \mathcal{D}_2\\ C\end{matrix}$ of C from the assumption rule $A \Rightarrow B$. If we replace every application of the assumption rule $A \Rightarrow B$ in $\mathcal{D}_2$

$$\begin{matrix}& \mathcal{D}_3\\ A \Rightarrow B & \dfrac{A}{B}\end{matrix}$$

with an application of ($\rightarrow$E$_{\text{MP}}$) using $\mathcal{D}_1$ as derivation of its major premiss

$$(\rightarrow\text{E}_{\text{MP}})\ \dfrac{\begin{matrix}\mathcal{D}_1\\ A\rightarrow B\end{matrix} \qquad \begin{matrix}\mathcal{D}_3\\ A\end{matrix}}{B}\ ,$$

then we obtain a derivation of C, i.e., of the conclusion of ($\rightarrow$E$_{\text{HL}}$). Note that this derivation of ($\rightarrow$E$_{\text{HL}}$)from ($\rightarrow$E$_{\text{MP}}$) works for every single rule instance, i.e., for A, B, C fixed, so we need not assume that ($\rightarrow$E$_{\text{HL}}$)and ($\rightarrow$E$_{\text{MP}}$) are rule schemata.

Conversely, suppose ($\rightarrow$E$_{\text{HL}}$)is a primitive rule. Suppose we have derived the premisses of ($\rightarrow$E$_{\text{MP}}$), i.e., we have derivations $\begin{matrix}\mathcal{D}_1\\ A\rightarrow B\end{matrix}$ and $\begin{matrix}\mathcal{D}_2\\ A\end{matrix}$ of $A \rightarrow B$ and A respectively. From $\mathcal{D}_2$, using the assumption rule $A \Rightarrow B$, we obtain

$$\begin{matrix}& \mathcal{D}_2\\ A \Rightarrow B & \dfrac{A}{B}\end{matrix}\ ,$$

from which, together with $\mathcal{D}_1$ we obtain by means of ($\rightarrow$E$_{\text{HL}}$)

$$\dfrac{\begin{matrix}\mathcal{D}_1 \\ A \to B\end{matrix} \qquad {}^1[A \Rightarrow B]\dfrac{\begin{matrix}\mathcal{D}_2 \\ A\end{matrix}}{B}}{B}\,1\ ,$$

which is a derivation of the conclusion B of ($\to E_{MP}$). However, it is important to note that unlike the other direction, the derivation of ($\to E_{MP}$) from ($\to E_{HL}$)does not work for every single rule instance, i.e., for A, B, C arbitrarily fixed. Rather, we must be able to substitute B for C. So what we have essentially proved is that ($\to E_{HL}$), as a *rule schema*, implies ($\to E_{MP}$).

This corresponds to the original idea of generalized elimination rules of the form (c E), where the C is implicitly universally quantified, expressing that *every* C that can be derived from each defining condition of $c(A_1, \dots, A_n)$ (for fixed $A_1, \dots, A_n$), can be derived from $c(A_1, \dots, A_n)$ itself.

We have an entirely similar situation when proving the equivalence of ($\wedge$E) and ($\wedge E_{GEN}$). For the direction from left to right, given derivations $\begin{matrix}\mathcal{D}_1 \\ A\wedge B\end{matrix}$ and $\begin{matrix}A\ B \\ \mathcal{D}_2 \\ C\end{matrix}$, we construct a derivation

$$\begin{matrix}\dfrac{\begin{matrix}\mathcal{D}_1 \\ A\wedge B\end{matrix}}{A} & \dfrac{\begin{matrix}\mathcal{D}_1 \\ A\wedge B\end{matrix}}{B} \\ \multicolumn{2}{c}{\mathcal{D}_2} \\ \multicolumn{2}{c}{C\ ,}\end{matrix}$$

whereas for the direction from right to left we instantiate C with A and B respectively, yielding derivations

$$\dfrac{A\wedge B \quad {}^1[A]}{A}\,1 \quad \text{and} \quad \dfrac{A\wedge B \quad {}^1[B]}{B}\,1\ ,$$

where $[A]$ represents a derivation of A from A and B, in which A is effectively discharged and B is vacuously discharged, and analogously with $[B]$.

A detailed proof of normalization and of the subformula property for higher-level natural deduction, for systems with and without the absurdity rule (the latter corresponding to minimal logic), and for a system with explicit denial can be found in [27].[3] Generalized elimination rules have also found their way into type-theoretical approaches such as Martin-Löf's type theory (see [15], p. i–iii), where they are particularly useful in the treatment of dependent products (see [5]), uniform approaches to the machine-assisted treatment of logical systems such as the Edinburgh logical framework (see [9]), and proof-editors and theorem provers such as Isabelle ([20]). For applications of this framework to relevance logic, logic programming and Martin-Löf-style logical systems see [30].

[3] Although stated as a result, the proof was omitted from the English journal publication [28], both for space constraints and because its character is that of an exercise.

In this chapter, we confine ourselves to minimal logic, or more precisely, to what corresponds to minimal logic in the higher-level framework, namely nonempty systems of rules and n-ary operators definable from them. Especially, we do not touch here the problem of dealing with empty systems of rules (which leads to intuitionistic absurdity) or the general problem of denial and negation. Also, concerning substructural issues, we just assume the standard provisions according to which lists of formulas or rules can be handled as sets, where appropriate. In particular, we do not discuss the issue of relevance, although the higher-level framework can be adapted to these fields (see [30, 34]). (Tennant's [42] approach to generalized elimination inferences is much inspired by the problems of relevant implication.) The vast area of using extended structural means of expression to describe or determine the inferential meaning of logical constants, which in particular leads to general frameworks for describing logics is also omitted from our discussion here. For this point see [44].

2 Generalized Standard-Level Elimination Rules

Dyckhoff [4] (in 1988), Tennant [41] (in 1992), López-Escobar [12] (in 1999), von Plato [17, 21, 22] (2001-2002) and Tennant [42] (in 2002) have independently (though, as is clear, with considerable temporal intervals) proposed a generalized form of the $\rightarrow$ elimination rule which is related to the left $\rightarrow$ introduction rule in sequent calculi. In our terminology it is of level 3 and thus a standard-level rule. We call it $(\rightarrow \mathrm{E}_{\mathrm{SL}})$[4]

$$(\rightarrow \mathrm{E}_{\mathrm{SL}})\ \frac{A \rightarrow B \qquad A \qquad \begin{array}{c}[B]\\ C\end{array}}{C}\ .$$

In these approaches the generalized elimination rule for conjunction is the same as before: $(\wedge \mathrm{E}_{\mathrm{GEN}})$, which means that it is a level-3-rule and thus one level higher than the $\wedge$ introduction rules. However, $(\rightarrow \mathrm{E}_{\mathrm{SL}})$ stays at the same level as $(\rightarrow \mathrm{I})$ (viz. level 3), so that no higher level is needed as far as the standard connectives are concerned. As these approaches are concerned with generalizing the elimination rules of the

---

[4] From the fact that none of these authors mentions the earlier one, we conclude that none of them was aware of the fact that $(\rightarrow \mathrm{E}_{\mathrm{SL}})$ had been discussed before. Dyckhoff's chapter appeared in a volume of a kind often called a 'grey' publication (similar to a technical report), which was difficult to notice and to get hold of, at least in pre-Internet times. It mentions $(\rightarrow \mathrm{E}_{\mathrm{SL}})$ neither in the title nor in the abstract nor in the display of the inference rules of natural deduction. The only sentence mentioning $(\rightarrow \mathrm{E}_{\mathrm{SL}})$ explicitly as a possible primitive rule for implication elimination occurs in the discussion of proof tactics for implication: 'this makes it clear that a possible form of the rule $\supset$_*elim* might be that from proofs of $A \supset B$, of A, and (assuming B) of C one can construct a proof of C. Such a form fits in better with the pattern for elimination rules now increasingly regarded as orthodox, and is clearer than the other possibility for $\supset$_*elim* advocated by Schroeder-Heister and Martin-Löf [...]' ([4], p. 55). Before the appearance of von Plato's papers, Dyckhoff never referred to this publication in connection with the particular form of $\rightarrow$ elimination, in order to make his idea visible.

standard connectives and not with a general schema for arbitrary n-ary connectives, they are content with $(\wedge E_{GEN})$ and $(\to E_{SL})$. $(\wedge E_{GEN})$ and $(\to E_{SL})$ are generalized elimination rules, as the indirect pattern of elimination found in $(\vee E)$ is carried over to conjunction and implication, in the sense that the right minor premiss C is repeated in the conclusion, even though, due to the presence of the minor premiss A, the uniformity inherent in the generalized higher-level rules is lost in $(\to E_{SL})$.

We do not discuss here the somewhat different conceptual intentions the above authors pursue when introducing $(\to E_{SL})$. We are solely interested in the form and power of this rule that we shall compare with $(\to E_{HL})$and with the higher-level approach in general. Concerning terminology, we use the term 'generalized elimination rules' as a generic term covering both the higher-level and the standard-level versions. Tennant [42] speaks of the 'parallel form' of elimination rules and of 'parallelized' natural deductions, von Plato [22] of 'general elimination rules.'

It can easily be shown that $(\to E_{MP})$ and $(\to E_{SL})$ are equivalent as primitive rules. Suppose $(\to E_{MP})$ is a primitive rule. Suppose we have derived the premisses of $(\to E_{SL})$, i.e., we have a derivation $\begin{array}{c}\mathcal{D}_1\\ A\to B\end{array}$ of $A\to B$, a derivation $\begin{array}{c}\mathcal{D}_2\\ A\end{array}$ of A and a derivation $\begin{array}{c}B\\ \mathcal{D}_3\\ C\end{array}$ of C from B. If we replace every occurrence of the assumption B in $\mathcal{D}_3$ with an application of $(\to E_{MP})$ using $\mathcal{D}_1$ and $\mathcal{D}_2$ as premiss derivations:

$$(\to E_{MP})\ \dfrac{\begin{array}{c}\mathcal{D}_1\\ A\to B\end{array} \qquad \begin{array}{c}\mathcal{D}_2\\ A\end{array}}{B},$$

we obtain a derivation of C, i.e., the conclusion of $(\to E_{SL})$. This derivation of $(\to E_{SL})$ from $(\to E_{MP})$ works for every single rule instance, i.e., for A, B, C fixed.

Conversely, suppose $(\to E_{SL})$ is a primitive rule. Suppose we have derived the premisses of $(\to E_{MP})$, i.e., we have derivations $\begin{array}{c}\mathcal{D}_1\\ A\to B\end{array}$ and $\begin{array}{c}\mathcal{D}_2\\ A\end{array}$ of $A\to B$ and A, respectively. Considering B to be a derivation of itself, by an application of $(\to E_{SL})$ we obtain

$$(\to E_{SL})\ \dfrac{\begin{array}{c}\mathcal{D}_1\\ A\to B\end{array} \qquad \begin{array}{c}\mathcal{D}_2\\ A\end{array} \qquad {}^1[B]}{B}\,1,$$

which is a derivation of the conclusion B of $(\to E_{MP})$. As before, it is important to note that unlike the other direction, the derivation of $(\to E_{MP})$ from $(\to E_{SL})$ does not work for every single rule instance, i.e., for A, B, C arbitrarily fixed. Rather, we must be able to substitute B for C. So what we have essentially proved is that $(\to E_{SL})$, as a *rule schema*, implies $(\to E_{MP})$.

This again corresponds to the idea of generalized elimination as an indirect schema, in which the C is implicitly quantified.

3 Comparison of the Higher Level with the Standard-Level Generalized → Elimination Rules

As rule schemata, or more precisely, with the letter C understood schematically, both ($\rightarrow E_{HL}$)and ($\rightarrow E_{SL}$) are equivalent to ($\rightarrow E_{MP}$), which implies that ($\rightarrow E_{HL}$)and ($\rightarrow E_{SL}$) are equivalent. If we compare ($\rightarrow E_{HL}$)and ($\rightarrow E_{SL}$) directly, i.e., not via ($\rightarrow E_{MP}$), then ($\rightarrow E_{HL}$)implies ($\rightarrow E_{SL}$) as a concrete rule, i.e., as an instance with A, B, C fixed, which can be seen as follows. Suppose we have derived the premisses of ($\rightarrow E_{SL}$), i.e., we have a derivation $\begin{matrix}\mathcal{D}_1\\ A\rightarrow B\end{matrix}$ of $A\rightarrow B$, a derivation $\begin{matrix}\mathcal{D}_2\\ A\end{matrix}$ of A and a derivation $\begin{matrix}B\\ \mathcal{D}_3\\ C\end{matrix}$ of C from B. If we replace every occurrence of the assumption B in $\mathcal{D}_3$ with the following application of the assumption rule $A \Rightarrow B$

$$A \Rightarrow B\ \frac{\begin{matrix}\mathcal{D}_2\\ A\end{matrix}}{B},$$

we obtain a derivation

$$\begin{matrix} A \Rightarrow B\ \dfrac{\begin{matrix}\mathcal{D}_2\\ A\end{matrix}}{B} \\ \mathcal{D}_3 \\ C \end{matrix}$$

of C from $A \Rightarrow B$. By application of ($\rightarrow E_{HL}$)we can discharge $A \Rightarrow B$:

$$(\rightarrow E_{HL})\ \frac{\begin{matrix}\mathcal{D}_1\\ A\rightarrow B\end{matrix} \qquad \begin{matrix} {}^1[A \Rightarrow B]\ \dfrac{\begin{matrix}\mathcal{D}_2\\ A\end{matrix}}{B} \\ \mathcal{D}_3 \\ C \end{matrix}}{C}\ 1$$

yielding the conclusion C of ($\rightarrow E_{SL}$).

For the converse direction, we have to keep C schematic, more precisely, we have to be able to specify it as B. Suppose we have derived the premisses of ($\rightarrow E_{HL}$), i.e., we have a derivation $\begin{matrix}\mathcal{D}_1\\ A\rightarrow B\end{matrix}$ of $A\rightarrow B$ and a derivation $\begin{matrix}A \Rightarrow B\\ \mathcal{D}_2\\ C\end{matrix}$ of C from the assumption rule $A \Rightarrow B$. Suppose $A \Rightarrow B$ is actually used as an assumption in $\mathcal{D}_2$ (otherwise $\mathcal{D}_2$ is already the desired derivation of C). If we replace every application of the assumption rule $A \Rightarrow B$ in $\mathcal{D}_2$

$$A \Rightarrow B \, \dfrac{\begin{matrix}\mathcal{D}_3\\ A\end{matrix}}{B}$$

with the following application of $(\rightarrow \mathrm{E}_{\mathrm{SL}})$

$$(\rightarrow \mathrm{E}_{\mathrm{SL}})\ \dfrac{\begin{matrix}\mathcal{D}_1\\ A \rightarrow B\end{matrix} \qquad \begin{matrix}\mathcal{D}_3\\ A\end{matrix} \qquad {}^1[B]}{B}\ 1 \ , \tag{3}$$

then we obtain a derivation of C, i.e., of the conclusion of $(\rightarrow \mathrm{E}_{\mathrm{HL}})$. The fact that in the application of $(\rightarrow \mathrm{E}_{\mathrm{SL}})$ in (3), by instantiating C to B we are using a trivialized version of $(\rightarrow \mathrm{E}_{\mathrm{SL}})$, which essentially is modus ponens $(\rightarrow \mathrm{E}_{\mathrm{MP}})$, cannot be avoided. It is tempting to consider replacing every application of the assumption rule $A \Rightarrow B$ in $\mathcal{D}_2$

$$\begin{matrix} A \Rightarrow B \, \dfrac{\begin{matrix}\mathcal{D}_3\\ A\end{matrix}}{B} \\ \mathcal{D}_4 \\ C \end{matrix} \tag{4}$$

with the following application of $(\rightarrow \mathrm{E}_{\mathrm{SL}})$:

$$(\rightarrow \mathrm{E}_{\mathrm{SL}})\ \dfrac{\begin{matrix}\mathcal{D}_1\\ A \rightarrow B\end{matrix} \qquad \begin{matrix}\mathcal{D}_3\\ A\end{matrix} \qquad \begin{matrix}{}^1[B]\\ \mathcal{D}_4\\ C\end{matrix}}{C}\ 1 \tag{5}$$

which would leave the C uninstantiated. However, this way is not viable as in (4) there may occur assumptions open in $\mathcal{D}_3$ but discharged at a rule application in $\mathcal{D}_4$. Such an assumption would remain open in (5), where the derivation (4) is split into two independent parts.

Therefore $(\rightarrow \mathrm{E}_{\mathrm{HL}})$and $(\rightarrow \mathrm{E}_{\mathrm{SL}})$ are equivalent with schematic C, but they are not equivalent as instances, i.e., for every particular C. In this sense $(\rightarrow \mathrm{E}_{\mathrm{HL}})$is stronger than $(\rightarrow \mathrm{E}_{\mathrm{SL}})$. However, one should not overestimate this difference in strength, as primitive rules of inference are normally understood as rule schemata.[5]

It is easy to see that the mutual translation between $(\rightarrow \mathrm{E}_{\mathrm{HL}})$and $(\rightarrow \mathrm{E}_{\mathrm{SL}})$ does not introduce new maximum formulas. Therefore from the normalization and subformula theorems proved in [27] for the generalized higher-level case we immediately obtain the corresponding theorems for generalized standard-level natural deduction. In fact, when specialized from the n-ary case to the case of the standard operators, this proof gives the same reductions as those found in Tennant's [41, 42], López-Escobar's [12], and Negri and von Plato's [17] work. This means there has been an earlier direct

[5] The way in which letters C are schematic, i.e., which propositions may be substituted for them, becomes important, if one investigates extensions of the language considered, i.e., in connection with the problem of uniqueness of connectives (see [3]).

proof of normalization for generalized natural deduction than the one given in 1992 by Tennant [41] (there for minimal logic and for Tennant's system of intuitionistic relevant logic), albeit in a more general setting.[6]

4 Comparison of the Expressive Power of Higher-Level with that of Standard-Level Rules

So far we have investigated the relationship between the generalized higher-level $\rightarrow$ elimination rule ($\rightarrow \mathrm{E_{HL}}$)and the generalized standard-level $\rightarrow$ elimination rule ($\rightarrow \mathrm{E_{SL}}$). We can carry over some of these results to a more general comparison which relates higher-level and standard-level rules independently of logical connectives. We may ask: Is it possible to express higher-level rules by means of standard-level rules, perhaps along the lines of the comparison of generalized $\rightarrow$ elimination rules in Sect. 3? Here we must again distinguish whether the rules compared are schematic with respect to some or all propositional letters or not.

The comparison of ($\rightarrow \mathrm{E_{HL}}$)and ($\rightarrow \mathrm{E_{SL}}$) might suggest that the rules

$$\frac{\begin{matrix}[A \Rightarrow B]\\ C\end{matrix}}{D} \qquad \text{and} \qquad \frac{A \quad \begin{matrix}[B]\\ C\end{matrix}}{D} \tag{6}$$

are equivalent. Corresponding to what was shown above, the left rule implies the right one for any fixed A, B, C, D: Given derivations $\begin{matrix}\mathcal{D}_1\\ A\end{matrix}$ and $\begin{matrix}B\\ \mathcal{D}_2\\ C\end{matrix}$ of the premisses of the right rule, we just need to replace each assumption B in $\mathcal{D}_2$ with a step

$$A \Rightarrow B \, \frac{\begin{matrix}\mathcal{D}_1\\ A\end{matrix}}{B}$$

in order to obtain a derivation of the premiss of the left rule. However, the converse direction is not valid: Already if in the left rule of (6) C does not depend on any assumption (i.e., the discharging of $A \Rightarrow B$ is vacuous), the right rule of (6) cannot be obtained, as there is no possibility of generating a derivation of A from the premiss derivations of the left rule. Therefore in (6) the higher-level rule is strictly stronger than the standard-level rule.

If we change D to C, which is more in the spirit of the generalized $\rightarrow$ elimination rule, do we then obtain the equivalence of

$$\frac{\begin{matrix}[A \Rightarrow B]\\ C\end{matrix}}{C} \qquad \text{and} \qquad \frac{A \quad \begin{matrix}[B]\\ C\end{matrix}}{C} \quad ? \tag{7}$$

6 López-Escobar [12] proves *strong* normalization. Other such proofs are given in [10] and [43].

As before, the left side implies the right one. However, the right side does not imply the left one, if C is fixed. Only if we allow for C to be substituted with B, which essentially means considering C to be schematic, it is the case that the right side of (7) implies the left side. By means of

$$\frac{A \quad {}^{1}[B]}{B}\,1,$$

which is the same as

$$\frac{A}{B},$$

we can eliminate every application of $A \Rightarrow B$ in a given derivation of C from $A \Rightarrow B$, yielding a derivation of C as required. As in the case described in Sect. 3, attempting to extract, from

$$\begin{array}{c} \mathcal{D}_1 \\ A \Rightarrow B \; \dfrac{A}{B} \\ \mathcal{D}_2 \\ C \end{array},$$

two independent derivations $\begin{array}{c}\mathcal{D}_1\\ A\end{array}$ and $\begin{array}{c}B\\ \mathcal{D}_2\\ C\end{array}$, is bound to fail, as in $\mathcal{D}_2$ assumptions open in $\mathcal{D}_1$ might have been discharged.

Of course, these are just examples showing that a translation of a higher-level rule into a standard-level rule is not possible according to the idea underlying $(\rightarrow E_{SL})$. However, this is not accidental: A reduction of higher-level to standard-level rules is not possible in general. For example, a higher-level rule of the form

$$\frac{\begin{array}{c}[A \Rightarrow B]\\ C\end{array}}{D},$$

which can be linearly written as $((A \Rightarrow B) \Rightarrow C) \Rightarrow D$, is never equivalent to a standard-level rule or to a finite set of standard-level rules. This follows from the general non-flattening results obtained by Olkhovikov and Schroeder-Heister in [19].

5 The Benefit of Higher-Level Rules: Strong Uniformity and Closure

In the following, we use this terminology: 'generalized$_{SL}$' stands for 'generalized higher-level', 'generalized$_{HL}$' for 'generalized standard level.' When we speak of the higher-level' or the 'standard-level' approach, we always mean the approaches with *generalized* higher-level or standard-level rules, respectively, and similarly when we speak of 'higher-level' or 'standard-level' natural deduction.

Both the higher-level and the standard-level generalized elimination rules satisfy the requirement that the elimination inferences for the standard connectives induced by the generalized form are equivalent to the 'common' elimination inferences, at least as inference schemata. In particular, modus ponens ($\rightarrow \mathrm{E}_{\mathrm{MP}}$) is equivalent to the generalized forms ($\rightarrow \mathrm{E}_{\mathrm{HL}}$)and ($\rightarrow \mathrm{E}_{\mathrm{SL}}$). A second requirement one should impose on a generalized form is its uniformity: The elimination inferences covered by it should follow a uniform pattern—achieving this is the main purpose of the generalization. The generalized$_{\mathrm{SL}}$ rules obviously satisfy this criterion. The generalized$_{\mathrm{HL}}$ rules satisfy it to the extent that they give an 'indirect' reading to the elimination inferences expressed by the schematic minor premiss and conclusion C. However, the standard-level implication rule ($\rightarrow \mathrm{E}_{\mathrm{SL}}$) is *hybrid* in so far as it has both the 'indirect' character expressed by the schematic C and the 'direct' character expressed by the minor premiss A which makes it a variant of modus ponens, if the derivation of C is trivialized to the derivation of B from B:

$$(\rightarrow \mathrm{E}_{\mathrm{SL}})\ \frac{A \rightarrow B \qquad A \qquad [B]}{B}\ .$$

This hybrid character is the price one pays for avoiding higher levels, which would be inevitable otherwise. Therefore, the generalized$_{\mathrm{SL}}$ eliminations are *uniform in a stronger sense* than the standard-level rules. There is a third criterion which only the generalized$_{\mathrm{SL}}$ rules satisfy and which is a *closure property*: Unlike the standard-level approach, the higher-level approach allows us to formulate a general schema for elimination rules for *all* introduction rules that are possible on the basis of *all* available means of expression. In particular, the means of expression used to formulate a given elimination rule must be admitted to formulate a new introduction rule. Putting it negatively: An elimination rule must not be of a form that is not suitable in principle as an introduction rule. In other words, for every list of introduction rules that can be formulated at all, a corresponding elimination rule is available.

Of course, by extending the means of expression, for example, by considering quantifiers or modal operators, one would be led to elimination rules of a different kind. However, every rule pattern used *in the given framework* should qualify to formulate introduction rules. That this is not the case with generalized$_{\mathrm{HL}}$ elimination rules in the manner of ($\rightarrow \mathrm{E}_{\mathrm{SL}}$) is seen by the following example. Consider, for example, the rule

$$\frac{A \qquad \begin{matrix}[B]\\ C\end{matrix}}{D}$$

which is a pattern used to formulate generalized$_{\mathrm{HL}}$ eliminations. Using this pattern as an introduction rule

$$(c_1\,\mathrm{I})\ \frac{A_1 \qquad \begin{matrix}[A_2]\\ A_3\end{matrix}}{c_1(A_1, A_2, A_3)}$$

gives us a ternary operator c_1, for which there is a generalized$_{\mathrm{SL}}$ elimination rule

$$(c_1\,\mathrm{E})\quad \frac{c_1(A_1, A_2, A_3) \qquad \begin{array}{c}[A_1 \quad A_2 \Rightarrow A_3]\\ C\end{array}}{C}\ ,$$

but no generalized elimination rule according to the standard-level pattern. This situation might be remedied by using two separate generalized$_{\mathrm{HL}}$ elimination rules according to the standard-level pattern:

$$(c_1\,\mathrm{E})'\quad \frac{c_1(A_1, A_2, A_3) \qquad \begin{array}{c}[A_1]\\ C\end{array}}{C} \qquad \frac{c_1(A_1, A_2, A_3) \qquad A_2 \qquad \begin{array}{c}[A_3]\\ C\end{array}}{C}\ .$$

However, this way out is not available if we consider the 4-ary connective c_2 with the introduction rules

$$(c_2\,\mathrm{I})\quad \frac{\begin{array}{c}[A_1]\\ A_2\end{array}}{c_2(A_1, A_2, A_3, A_4)} \qquad \frac{\begin{array}{c}[A_3]\\ A_4\end{array}}{c_2(A_1, A_2, A_3, A_4)}\ .$$

The corresponding higher-level elimination rule is

$$(c_2\,\mathrm{E})\quad \frac{c_2(A_1, A_2, A_3, A_n) \qquad \begin{array}{c}[A_1 \Rightarrow A_2]\\ C\end{array} \qquad \begin{array}{c}[A_3 \Rightarrow A_4]\\ C\end{array}}{C}\ .$$

This rule cannot be expressed along the lines of the standard-level rule ($\rightarrow \mathrm{E_{SL}}$). The natural way of attempting such a solution would be to propose the following elimination rule:

$$(c_2\,\mathrm{E})'\quad \frac{c_2(A_1, A_2, A_3, A_4) \qquad A_1 \quad \begin{array}{c}[A_2]\\ C\end{array} \qquad A_3 \quad \begin{array}{c}[A_4]\\ C\end{array}}{C}\ .$$

However, though it can be easily shown that $(c_2\,\mathrm{E})$ implies $(c_2\,\mathrm{E})'$ (the proof is similar to the demonstration that ($\rightarrow \mathrm{E_{HL}}$)implies ($\rightarrow \mathrm{E_{SL}}$) in Sect. 3), the converse does not hold. Even if in a given application of $(c_2\,\mathrm{E})$ there is no vacuous discharge of $A_1 \Rightarrow A_2$ or $A_3 \Rightarrow A_4$ (in which case the minor premiss A_1 or A_3 of $(c_2\,\mathrm{E})'$ cannot be generated), it may happen that in the derivation of the minor premisses C of $(c_2\,\mathrm{E})$ an assumption *above* $A_1 \Rightarrow A_2$ or $A_3 \Rightarrow A_4$ is discharged at a rule application *below* $A_1 \Rightarrow A_2$ or $A_3 \Rightarrow A_4$, respectively, such as the B in the following example of a derivation of $(B \rightarrow A_2)\vee(B \rightarrow A_4)$ from $c_2(A_1, A_2, A_3, A_4)$, $B \rightarrow A_1$ and $B \rightarrow A_3$:

$$
(c_2\,\mathrm{E})\;\dfrac{c_2(A_1,A_2,A_3,A_4)\qquad (\vee\mathrm{I})\,\dfrac{(\to\mathrm{I})\,\dfrac{{}^3[A_1\Rightarrow A_2]\;\dfrac{(\to\mathrm{E_{MP}})\,\dfrac{B\to A_1\qquad {}^1[B]}{A_1}}{A_2}}{B\to A_2}\,1}{(B\to A_2)\vee(B\to A_4)}\qquad (\vee\mathrm{I})\,\dfrac{(\to\mathrm{I})\,\dfrac{{}^3[A_3\Rightarrow A_4]\;\dfrac{(\to\mathrm{E_{MP}})\,\dfrac{B\to A_3\qquad {}^2[B]}{A_3}}{A_4}}{B\to A_4}\,2}{(B\to A_2)\vee(B\to A_4)}}{(B\to A_2)\vee(B\to A_4)}\,3\,.
$$

Since $(c_2\,\mathrm{E})$ is a generalized$_{\mathrm{SL}}$ rule, whereas $(c_2\,\mathrm{E})'$ is considered to be a generalized$_{\mathrm{HL}}$ rule, we have used applications of $(\to\mathrm{E_{MP}})$ (rather than applications of assumption rules), as they can easily be translated into either of these systems. The formula B occurs in both subderivations of the minor premisses in top position and thus *above* the assumptions $A_1\Rightarrow A_2$ and $A_3\Rightarrow A_4$, but is discharged at applications of $(\to\mathrm{I})$ *below* $A_1\Rightarrow A_2$ and $A_3\Rightarrow A_4$, so that we cannot split the subderivations of the minor premisses into two upper parts

$$
(\to\mathrm{E_{MP}})\,\dfrac{B\to A_1\qquad B}{A_1}\qquad\qquad (\to\mathrm{E_{MP}})\,\dfrac{B\to A_3\qquad B}{A_3}
$$

and two lower parts

$$
(\vee\mathrm{I})\,\dfrac{(\to\mathrm{I})\,\dfrac{A_2}{B\to A_2}}{(B\to A_2)\vee(B\to A_4)}\qquad\qquad (\vee\mathrm{I})\,\dfrac{(\to\mathrm{I})\,\dfrac{A_4}{B\to A_4}}{(B\to A_2)\vee(B\to A_4)}
$$

in order to obtain derivations of all four minor premisses of $(c_2\,\mathrm{E})'$.

This shows that the generalized$_{\mathrm{HL}}$ elimination rules do not cover the elimination rule necessary for such a simple connective as c_2, which has the meaning of $(A_1\to A_2)\vee(A_3\to A_2)$, although we can formulate its introduction rules in the standard-level framework. In this sense the generalized higher-level elimination rules are far more general than the standard-level ones[7]. A rigorous formal proof that a connective like c_2 (we used the ternary connective with the meaning $A_1\vee(A_2\to A_3)$) has no standard-level elimination rule, is given in [18].

As indicated above, even higher-level rules do not suffice to capture every propositional connective. As an example we need not consider modal connectives and the like, but the ternary connective c_3 with the meaning $A_1\to(A_2\vee A_3)$ suffices. However, the situation is entirely different from that of c_1 and c_2, as for c_3 we cannot even give introduction rules of the form $(c\,\mathrm{I})$ using the means of expression available. (In this sense c_3 resembles, e.g., a modal operator.) As soon as we have introduction rules according to the schema $(c\,\mathrm{I})$, we do have a corresponding generalized$_{\mathrm{SL}}$ elimination rule $(c\,\mathrm{E})$. This is not the case with generalized$_{\mathrm{HL}}$ rules.[8]

---

[7] In his discussion of generalized left inferences in his higher-level sequent framework, von Kutschera ([11], p. 15) gives an example similar to c_2 (namely the ternary connective with the meaning $(A_1\to A_2)\vee(A_3\to A_2)$) to show that the higher-level left rules cannot be expressed by lower-level left rules along the lines of the standard implication-left rule $(\to\mathrm{L})$ in the sequent calculus (which corresponds to $(\to\mathrm{E_{SL}})$, see Sect. 6)

[8] If we also consider operators *definable* in terms of others, i.e., if the premisses Δ_i of introduction rules $(c\,\mathrm{I})$ are allowed to contain operators c' which have already been given introduction

For this greater uniformity we pay the price of introducing higher levels. On the other hand, higher levels lead to considerable simplifications in other areas, notably in type theory (see [5]).

6 Generalized Elimination Rules, Sequent Calculus, and the Proudness Property

The motivations for using generalized$_{\text{HL}}$ elimination rules vary between the authors mentioned in Sect. 2. From the standpoint of proof-theoretic semantics, we [33, 36] proposed to use them in in a system called 'bidirectional natural deduction' that gives assumptions a proper standing. Another meaning-theoretical discussion of generalized$_{\text{HL}}$ elimination rules can be found in [16]. In what follows we are only concerned with the formal relationship of generalized$_{\text{HL}}$ elimination rules to left introduction rules in the sequent calculus along the lines advanced by von Plato and Tennant.

Generalized$_{\text{SL}}$ elimination rules give natural deduction elimination rules a form which is familiar from left introductions in sequent calculi.[9] Whereas the right introduction rules directly correspond to the introduction rules in natural deduction, this is not so in the case of the usual eliminations rules. The rule of $\vee$ elimination can be read as corresponding to the left introduction rule of $\vee$ in the sequent calculus

$$\frac{A \vee B \quad \begin{matrix}[A]\\ C\end{matrix} \quad \begin{matrix}[B]\\ C\end{matrix}}{C} \qquad \frac{\Gamma, A \vdash C \quad \Gamma, B \vdash C}{\Gamma, A \vee B \vdash C},$$

but the standard $\wedge$ and $\rightarrow$ eliminations cannot. However, for the generalized$_{\text{HL}}$ versions of these rules (for conjunction identical with the generalized$_{\text{SL}}$ version), this is indeed the case:

(Footnote 8 continued)
and elimination rules, then c_3 is, of course, trivially definable, with

$$\frac{A_1 \rightarrow (A_2 \vee A_3)}{c_3(A1, A_2, A_3)}$$

being its introduction rule. This may be considered a rationale to confine oneself, as in the standard-level approach, to the standard operators. Such an approach fails, of course, to tell anything about the distinguished character of the standard operators as being capable to express all possible operators based on rules of a certain form. This is a point in which the goals of the generalized$_{\text{SL}}$ and the generalized$_{\text{HL}}$ approaches fundamentally differ from one another. (In [27–29] operators definable from other operators in the higher-level framework are considered, in addition to those definable without reference to others.)

[9] For the symmetry in Gentzen's sequent calculus and its description in terms of definitional reflection see [2, 39].

$$\frac{A\wedge B \qquad \begin{matrix}[A\quad B]\\ C\end{matrix}}{C} \qquad\qquad \frac{\Gamma, A, B \vdash C}{\Gamma, A\wedge B\vdash C}$$

$$\frac{A\to B \qquad A \qquad \begin{matrix}[B]\\ C\end{matrix}}{C} \qquad\qquad \frac{\Gamma\vdash A \quad \Gamma, B\vdash C}{\Gamma, A\to B\vdash C}\ .$$

In this way a parallel between natural deduction and the sequent calculus is established.

The parallel between the sequent calculus and natural deduction goes even further in the standard-level approach. It can be shown that every normal derivation based on generalized$_{\mathrm{HL}}$ elimination rules can be transformed into one, in which major premisses only occur in top position. To see this, we have to observe that every formula C, which is a conclusion of a (generalized$_{\mathrm{HL}}$) elimination inference and at the same time major premiss of a (generalized$_{\mathrm{HL}}$) elimination inference can be eliminated, as the following example demonstrates, which shows that corresponding segments[10] of formulas of this kind are shortened:

$$(\to \mathrm{E}_{\mathrm{SL}})\ \dfrac{(\to \mathrm{E}_{\mathrm{SL}})\ \dfrac{\begin{matrix}\mathcal{D}_1\\ D\to E\end{matrix} \qquad \begin{matrix}\mathcal{D}_2\\ D\end{matrix} \qquad \begin{matrix}{}^1[E]\\ \mathcal{D}_3\\ A\to B\end{matrix}}{A\to B}\,1 \qquad \begin{matrix}\mathcal{D}_4\\ A\end{matrix} \qquad \begin{matrix}{}^2[B]\\ \mathcal{D}_5\\ C\end{matrix}}{C}\,2$$

reduces to

$$(\to \mathrm{E}_{\mathrm{SL}})\ \dfrac{\begin{matrix}\mathcal{D}_1\\ D\to E\end{matrix} \qquad \begin{matrix}\mathcal{D}_2\\ D\end{matrix} \qquad (\to \mathrm{E}_{\mathrm{SL}})\ \dfrac{\begin{matrix}{}^1[E]\\ \mathcal{D}_3\\ A\to B\end{matrix} \qquad \begin{matrix}\mathcal{D}_4\\ A\end{matrix} \qquad \begin{matrix}{}^2[B]\\ \mathcal{D}_5\\ C\end{matrix}}{C}\,2}{C}\,1\ .$$

Here it is assumed that below $A \to B$ there is no formula of the incriminated kind (in particular, C is not of that kind). Since maximum formulas (conclusions of introduction inferences being at the same time major premisses of (generalized$_{\mathrm{HL}}$) elimination inferences) are eliminated anyway, we can obtain a derivation of the required form.[11] This gives us a normal form theorem according to which every derivation based on generalized$_{\mathrm{HL}}$ elimination inferences can be transformed into a derivation with major premisses of elimination inferences standing always in top position. Following Tennant ([41], p.41), who speaks of a major premiss of an elimi-

---

[10] Following [23], a segment is a succession of formula occurrences of the same form C such that immediately succeeding occurrences are minor premiss C and conclusion C of a generalized$_{\mathrm{HL}}$ elimination step.

[11] What one essentially does here, is carrying out permutative reductions as known from [23]. Their general treatment, without assuming that segments are maximal (and thus start with the conclusion of an introduction inference), but only that they end with an elimination inference, was proposed by Martin-Löf (see [24], p. 253f.) For the higher-level case, these reductions are used in [27].

nation rules in top (i.e., assumption) position, as *standing proud*, we call this property the *proudness property*.

PROUDNESS PROPERTY OF GENERALIZED$_{\mathrm{SL}}$ NATURAL DEDUCTION:
Every derivation in generalized$_{\mathrm{SL}}$ natural deduction can be transformed into a derivation, in which every major premiss of an (generalized$_{\mathrm{SL}}$) elimination rule is an assumption.

A natural deduction derivation having the proudness property corresponds isomorphically (i.e., stepwise) to a cut-free derivation in the sequent calculus. Introduction of formulas in the sequent calculus on the left side of the sequent sign corresponds in generalized$_{\mathrm{HL}}$ natural deduction to introducing an assumption, which is the major premiss of an elimination rule.

The proudness property is not available without restriction in higher-level natural deduction. This is due to the fact that in the higher-level case we not only have *assumption formulas* but can also have *assumption rules*. Consider the following example of a derivation of B from $A \to B \wedge C$ and A:

$$(\to \mathrm{E_{HL}})\,\dfrac{A \to B\wedge C \qquad (\wedge \mathrm{E_{GEN}})\,\dfrac{{}^{1}[\mathrm{A} \Rightarrow \mathrm{B}\wedge\mathrm{C}]\,\dfrac{A}{B \wedge C} \qquad {}^{2}[B]}{B}\,2}{B}\,1 \;. \tag{8}$$

Another example is the following derivation of C from $A \to (B \to C)$, A and B:

$$(\to \mathrm{E_{HL}})\,\dfrac{A \to (B \to C) \qquad (\to \mathrm{E_{HL}})\,\dfrac{{}^{1}[A \Rightarrow (B \to C)]\,\dfrac{A}{B \to C} \qquad {}^{2}[B \Rightarrow C]\,\dfrac{B}{C}}{C}\,2}{C}\,1 \;. \tag{9}$$

In (8) the formula $B \wedge C$ is the conclusion of an assumption rule and at the same time the major premiss of an elimination rule, a situation which cannot be further reduced. Similarly, in (9) the formula $B \to C$ is the conclusion of an assumption rule and at the same time the major premiss of an elimination rule. However, if we weaken the notion of proudness to include this situation, then derivations in the higher-level approach satisfy it. We call a formula occurrence, which is a conclusion of the application of an assumption rule and at the same time major premiss of an elimination rule *weakly proud*. Then the following *weak proudness property* of higher-level natural deduction holds.

WEAK PROUDNESS PROPERTY OF GENERALIZED$_{\text{HL}}$ NATURAL DEDUCTION: *Every derivation in generalized$_{\text{HL}}$ natural deduction can be transformed into a derivation, in which every major premiss of an elimination rule is either an assumption or a conclusion of an assumption rule.*[12]

For example, the following derivation, in which $B \wedge C$ occurs both as a conclusion of an elimination inference and as the major premiss of another elimination inference, can be reduced to (8):

$$\dfrac{(\rightarrow \mathrm{E_{HL}})\,\dfrac{A \rightarrow B\wedge C \qquad {}^{1}[A \Rightarrow B\wedge C]\,\dfrac{A}{B\wedge C}}{B\wedge C}\,{}_{1} \qquad {}^{2}[B]}{B}\,{}_{2}\;(\wedge \mathrm{E_{GEN}}) \,.$$

Similarly, the following derivation, in which $B \rightarrow C$ occurs both as a conclusion of an elimination inference and as the major premiss of another elimination inference, can be reduced to (9):

$$\dfrac{(\rightarrow \mathrm{E_{HL}})\,\dfrac{A \rightarrow (B \rightarrow C) \qquad {}^{1}[A \Rightarrow (B\rightarrow C)]\,\dfrac{A}{B \rightarrow C}}{B \rightarrow C}\,{}_{1} \qquad {}^{2}[B \Rightarrow C]\,\dfrac{B}{C}}{C}\,{}_{2}\;(\rightarrow \mathrm{E_{HL}}) \,.$$

Weak proudness implies the subformula property. This is due to the fact that every formula C standing weakly proud is the subformula of an assumption rule. If this assumption rule is undischarged, then C is a subformula of an open assumption. Otherwise, this assumption rule is discharged at a higher-level introduction or elimination rule, in which case it is a subformula of the formula introduced or eliminated.[13] If one considers the subformula principle to be the fundamental corollary of normalization, the fact that for higher-level natural deduction we only have the weak but not the full proudness property, is no real disadvantage as compared to the standard-level (generalized) alternative.

7 A Sequent Calculus Variant of the Higher-Level Approach

Although intended for the purpose of uniform elimination rules in natural deduction, one might investigate how the generalized$_{\text{SL}}$ approach fits into a sequent-style framework and which form the weak proudness property then takes, in comparison to the proudness property of the standard-level approach, which immediately

---

[12] As an assumption formula can be viewed as the conclusion of a first-level assumption rule, the second alternative actually includes the first one.

[13] For a detailed proof of normalization and subformula property for higher-level natural deduction see [27].

corresponds to cut-free derivations. Some of the ideas and results presented here have been established in [1], others can be found in [30] and [7].[14]

We only consider the single-succedent (intuitionistic) variant. The generalized conjunction rule ($\wedge E_{GEN}$) simply translates into the $\wedge$ left rule

$$(\wedge \mathrm{L})\ \frac{\Gamma, A, B \vdash C}{\Gamma, A \wedge B \vdash C}\ ,$$

the only difference being that Γ may now stand for a set of higher-level rules rather than only for a set of formulas.[15] In order to frame the idea of assumption rules, we have to introduce a schema for the left introduction of a rule R as an assumption. We assume that R has the following general form:

$$(\Delta_1 \Rightarrow A_1, \ \ldots\ , \Delta_n \Rightarrow A_n) \ \Rightarrow \ B$$

which covers the limiting case $n = 0$ in which R is just the formula (= level-0-rule) B, and the cases in which some or all Δ_i, which are lists of rules, are empty, i.e., in which $\Delta_i \Rightarrow A_i$ is the same as A_i. Then the left introduction of a rule, which corresponds to using a rule as an assumption in generalized$_{\mathrm{SL}}$ natural deduction (1), proceeds as follows:

$$(\Rightarrow \mathrm{L})\ \frac{\Gamma, \Delta_1 \vdash A_1 \ \ \ldots \ \ \Gamma, \Delta_n \vdash A_n}{\Gamma, ((\Delta_1 \Rightarrow A_1, \ \ldots\ , \Delta_n \Rightarrow A_n) \ \Rightarrow \ B) \vdash B}\ ,$$

which covers as a limiting case:

$$(\Rightarrow \mathrm{L})^{\circ}\ \frac{\Gamma \vdash A}{\Gamma, (A \Rightarrow B) \vdash B}\ .$$

The right and left rules for implication are then the following:

$$(\rightarrow \mathrm{R})\ \frac{\Gamma, A \vdash B}{\Gamma \vdash A \rightarrow B} \qquad (\rightarrow \mathrm{L_{HL}})\ \frac{\Gamma}{, } - A \Rightarrow B \vdash C \Gamma, A \rightarrow B \vdash C\ .$$

The sequent calculus with higher-level rules, which results from the ordinary sequent calculus with cut by adding ($\Rightarrow$L), and by using ($\rightarrow$R) and ($\rightarrow \mathrm{L_{HL}}$) as rules for implication, will be called the *higher-level sequent calculus*. So we disregard here the feature that in addition to $\rightarrow$, $\wedge$ and $\vee$ we have right and left introduction rules for n-ary connectives and consider just the standard connectives, as this is the concern of the standard-level approach.

---

[14] Von Kutschera's [11] approach using an iteration of the sequent arrow should be mentioned as well, although it needs some reconstruction to fit into our framework.

[15] We can here neglect the difference between lists, multisets, and sets, as for simplicity we always assume that the usual structural rules of the intuitionistic sequent calculus (permutation, contraction, and thinning) are available—with the exception of cut, whose availability or non-availability as a primitive or admissible rule will always be explicitly stated.

Obviously, every derivation in the higher-level sequent calculus (with cut) can be translated into higher-level natural deduction, as the left-introduction rules are available as generalized$_{SL}$ inferences, and ($\Rightarrow$L) is available as the introduction of an assumption rule. Conversely, every derivation in higher-level natural deduction can be translated into the higher-level sequent calculus (with cut) along the lines described by Gentzen ([6], p. 422–424): Applications of introduction rules, of assumption rules, and of elimination rules with major premisses standing proud are homophonically translated into applications of right introduction rules, of ($\Rightarrow$L), and of left introduction rules, respectively. Only in the situation in which the major premiss of an elimination inference is not standing proud:

$$\text{E inference}\ \frac{\begin{matrix}\mathcal{D} \\ A\end{matrix} \quad \begin{matrix}\mathcal{D}_1 \\ C\end{matrix} \ \ldots \ \begin{matrix}\mathcal{D}_n \\ C\end{matrix}}{C},$$

we must apply cut, yielding

$$\text{Cut}\ \frac{\begin{matrix}\mathcal{D}' \\ \Gamma \vdash A\end{matrix} \qquad \text{L inference}\ \dfrac{\mathcal{D}'_1 \ \ldots \ \mathcal{D}'_m}{\Delta, A \vdash C}}{\Gamma, \Delta \vdash C},$$

where $\mathcal{D}'$, $\mathcal{D}'_1, \ldots, \mathcal{D}'_m$ are the sequent calculus translations of $\mathcal{D}, \mathcal{D}_1, \ldots, \mathcal{D}_m$. (This procedure also works for arbitrary n-ary connectives.)

The weak proudness property of higher-level natural deduction gives us immediately a *weak cut elimination theorem.*

WEAK CUT ELIMINATION FOR THE HIGHER- LEVEL SEQUENT CALCULUS: *Every derivation in the higher-level sequent calculus (with cut) can be transformed into a derivation, in which cut occurs only in the situation, where its left premiss is the conclusion of an application of an assumption rule, and the right premiss the conclusion of a left introduction rule for the cut formula, i.e., only in the following situation:*

$$(\text{Cut})\ \frac{(\Rightarrow\text{L})\ \dfrac{\vdots}{\Gamma \vdash A} \qquad (\text{L inference for } A)\ \dfrac{\vdots}{\Delta, A \vdash C}}{\Gamma, \Delta sseq C}.$$

As we have the subformula principle for higher-level natural deduction, it holds for the higher-level sequent calculus as well, if we only allow for cuts of the form described in the weak cut elimination theorem. Therefore cuts of this special form are harmless, although perhaps not most elegant.

That we do not have full cut elimination is demonstrated by the sequent-calculus translation of our example (8):

$$\dfrac{(\Rightarrow\text{L})\,\dfrac{A \vdash A}{A, (A \Rightarrow B \wedge C) \vdash B \wedge C} \qquad (\wedge\text{L})\,\dfrac{B \vdash B}{B \wedge C \vdash B}}{(\rightarrow\text{L}_{\text{HL}})\,\dfrac{A, (A \Rightarrow B \wedge C) \vdash B}{A, (A \rightarrow B \wedge C) \vdash B}}\,(\text{Cut}) \quad . \tag{10}$$

As there is no inference rule (apart from cut) which can generate $A, (A \Rightarrow B \wedge C) \vdash B$ (at least if A, B, and C are atomic and different from one another), cut is not eliminable. However, this application of cut is of the form permitted by the weak cut elimination theorem.

It might appear asymmetric at first glance that there is a left introduction rule ($\Rightarrow$L), but no right introduction rule for the rule arrow $\Rightarrow$. Why do we not have a right introduction rule for $\Rightarrow$ of the form

$$(\Rightarrow\text{R})\,\frac{\Gamma, A \vdash B}{\Gamma \vdash A \Rightarrow B}$$

and then introduce implication $\rightarrow$ on the right directly in terms of $\Rightarrow$, as it is done on the left:

$$(\rightarrow\text{R})'\,\frac{\Gamma \vdash A \Rightarrow B}{\Gamma \vdash A \rightarrow B} \quad ?$$

This asymmetry: only a left rule for $\Rightarrow$, is due to the fact that the rule arrow is not a logical constant in the genuine sense but a sign belonging to the structural apparatus, comparable to the comma. If we look at the comma and the rules for conjunction, we observe a similar phenomenon:

$$(\wedge\text{R})\,\frac{\Gamma \vdash A \quad \Gamma \vdash B}{\Gamma \vdash A \wedge B} \qquad (\wedge\text{L})\,\frac{\Gamma, A, B \vdash C}{\Gamma, A \wedge B \vdash C}.$$

When applying ($\wedge$R), we do not first introduce a comma on the right-hand side, which is conjunctively understood, in a way like

$$\frac{\Gamma \vdash A \quad \Gamma \vdash B}{\Gamma \vdash A, B},$$

and then introduce $\wedge$ in terms of the comma in a manner such as

$$\frac{\Gamma \vdash A, B}{\Gamma \vdash A \wedge B}.$$

Rather, we have a direct right-introduction rule ($\wedge$R) for conjunction, whereas on the left side, by means of ($\wedge$L), conjunction is reduced to the comma. This asymmetry is somewhat concealed by the fact that there is no formal left introduction rule for the comma in the sense in which there is a formal left introduction rule for the rule arrow $\Rightarrow$, as the comma is already there as a means to separate antecedent formulas (or antecedent rules). It nevertheless is a structural entity governed by rules which do not fit into the symmetric right-left-introduction schema. Analogously, the rule arrow must be looked upon as an enrichment of structural reasoning which essentially affects only the left side of the turnstile (in the intuitionistic framework).

The idea of higher-level rules, i.e., of rules as assumptions, is that we enrich our possibilities of formulating assumptions, in order to characterize logical operators as having the same consequences as certain assumption structures. Here $A \wedge B$ has the same consequences as the assumption structure (A, B), and $A \rightarrow B$ has the same consequences as the assumption structure $A \Rightarrow B$.

Although against the spirit of using rules as assumptions (or members of the antecedent), it is possible to express rules in terms of the standard operators. In fact, such a translation is used if we want to show that the standard operators suffice to express everything that can be expressed by means of higher-level rules, i.e., if we establish the completeness[16] of the standard operators (see [11, 27–30]). We translate the rule arrow $\Rightarrow$ by means of implication $\rightarrow$ and the comma by conjunction $\wedge$, so that, for example, a rule

$$(A, B \Rightarrow C), (E \Rightarrow F) \;\Rightarrow\; G$$

becomes the implication

$$((A \wedge B \rightarrow C) \wedge (E \rightarrow F)) \;\rightarrow\; G\,.$$

If we use this translation, then $(\rightarrow \mathrm{L_{HL}})$ becomes superfluous as premiss and conclusion are identical, and rule $(\Rightarrow \mathrm{L})$ becomes

$$\frac{\Gamma, \Delta_1 \vdash A_1, \;\ldots\; \Gamma, \Delta_n \vdash A_n}{\Gamma, ((\Delta_1 \rightarrow A_1) \wedge \;\ldots\; \wedge (\Delta_n \rightarrow A_n) \;\rightarrow\; B) \vdash B}\,. \tag{11}$$

This rule can be replaced with the simpler rule

$$(\rightarrow \mathrm{L})^{\circ} \; \frac{\Gamma \vdash A}{\Gamma, (A \rightarrow B) \vdash B}$$

which corresponds to $(\Rightarrow \mathrm{L})^{\circ}$. Given the premiss of (11), we simply need to use $(\rightarrow \mathrm{R})$ and $(\wedge \mathrm{R})$ to obtain

$$\Gamma \vdash (\Delta_1 \rightarrow A_1) \wedge \;\ldots\; \wedge (\Delta_n \rightarrow A_n)$$

from which by means of $(\rightarrow \mathrm{L})^{\circ}$ we obtain the conclusion of (11).

The result is a sequent calculus, in which the common $(\rightarrow \mathrm{L})$ rule

$$(\rightarrow \mathrm{L}) \; \frac{\Gamma \vdash A \quad \Gamma, B \vdash C}{\Gamma, A \rightarrow B \vdash C}$$

---

[16] This corresponds to functional completeness in the case of truth functions. In a further generalized setting this way of establishing completeness of systems of constants by translating structural operations into logical ones is used in [44] for many substructural logics.

is replaced with $(\to L)^\circ$. $(\to L)^\circ$ introduces the spirit of *rules* into a sequent calculus of the usual kind (i.e., without higher-level rules): From A we may pass over to B by using (= assuming) $A \to B$ understood as a rule which licenses this transition.

It is obvious that $(\to L)^\circ$ and $(\to L)$ are interderivable, if the letter C is understood schematically and, thus, can be replaced with B. We call the sequent calculus with $(\to L)^\circ$ as the left introduction rule for implication *the sequent calculus based on the implications-as-rules interpretation*, in short *rule-style sequent calculus* as opposed to the standard sequent calculus which has $(\to L)$ as left introduction rule. As it results by translation from the higher-level sequent calculus, we do not have cut elimination for this system. As a translation of (10), the following is a counterexample:

$$\text{(Cut)}\ \dfrac{(\to L)^\circ\ \dfrac{A \vdash A}{A, (A \to B \wedge C) \vdash B \wedge C} \qquad (\wedge L)\ \dfrac{B \vdash B}{B \wedge C \vdash B}}{A, (A \to B \wedge C) \vdash B}\ . \tag{12}$$

However, corresponding to the weak proudness property of higher-level natural deduction and the weak cut elimination theorem in the higher-level sequent calculus, we have a weak cut elimination theorem for the rule-style sequent calculus, which says that a situation such as (12) is essentially the only one where cuts must be admitted.

WEAK CUT ELIMINATION FOR THE RULE- STYLE SEQUENT CALCULUS:
Every derivation in the rule-style sequent calculus (with cut) can be transformed into a derivation, in which cut occurs only in the situation where its left premiss is the conclusion of $(\to L)^\circ$*, and where its right premiss results from introducing the cut formula in the last step, i.e., in the following situation:*

$$\text{(Cut)}\ \dfrac{(\to L)^\circ\ \dfrac{\vdots}{\Gamma \vdash A} \qquad \text{(L inference for } A)\ \dfrac{\vdots}{\Delta, A \vdash C}}{\Gamma, \Delta \vdash C}. \tag{13}$$

In fact, if we consider a purely implicational system with $(\to L)^\circ$ of the multi-ary form

$$\dfrac{\Gamma \vdash A_1 \quad \dots \quad \Gamma \vdash A_n}{\Gamma, A_1 \to (\dots (A_n \to B)\dots) \vdash B}$$

then we have full cut elimination, as remarked by Avron [1].[17] Analogously, the purely implicational natural deduction system with the following rule for implication

[17] Avron also remarks that the standard $(\to L)$ rule is a way of avoiding the multi-ary character of this rule, which cannot be effected by means of $(\to L)^\circ$ alone (if conjunction is not available). Negri and von Plato [17] (p. 184) mention the rule $(\to L)^\circ$ as a sequent calculus rule corresponding to modus ponens, followed by a counterexample to cut analogous to (12), which is based on implication only. This counterexample shows again that for cut elimination in the implicational system the multi-ary form of $(\to L)^\circ$ considered in [1] and the corresponding forms of rule introduction in the antecedent considered in [30] and [7] are really needed.

$$\frac{A_1 \to (\ldots (A_n \to B)\ldots) \to B \quad A_1 \quad \ldots \quad A_n}{B}$$

enjoys the full proudness property.

The rule-style sequent calculus satisfies the *subformula principle*, in spite of the weak form of cut which cannot be eliminated. This result is also carried over from the consideration of explicit higher-level rules. It is immediately plausible, too, as the cut formula A in (13) is contained in an implication $B \to A$ which is introduced by means of $(\to \mathrm{L})^\circ$ and therefore belongs to Γ.

8 Implications-as-Rules Versus Implications-as-Links[18]

We have seen that, when formalized as a sequent calculus, the interpretation of implications as rules yields a system with $(\to \mathrm{L})^\circ$ as the left introduction rule for implication. This system enjoys the subformula property, but only a weak form of cut elimination. Although the implications-as-rules view is very natural in the natural deduction framework, the corresponding rule-style sequent calculus might look strange at first glance, as one has become used to Gentzen's rule $(\to \mathrm{L})$ and to full cut elimination as a fundamental principle.

However, the alleged simplicity of $(\to \mathrm{L})$ is essentially a feature of *technical elegance*. If we want to have full cut elimination at any price in order to derive its corollaries such as the subformula property and other features with ease, then $(\to \mathrm{L})$ is the rule of choice. In fact, it were these technical considerations that led Gentzen to consider his sequent calculus. Unlike the calculus of natural deduction, for which Gentzen coined the term 'natural,' and for whose rules he gave a detailed philosophical motivation, the sequent calculus was not chosen by Gentzen for its philosophical plausibility, but merely for its suitability for the proof of the Hauptsatz (see [6], p. 191).

If we look at Gentzen's rule $(\to \mathrm{L})$ from a philosophical or conceptual point of view and compare it to the implications-as-rules view, then it loses some of its plausibility. Whereas the implications-as-rules interpretation gives a direct meaning to implication, as the notion of a rule is a very basic notion used to describe reasoning, and acts in general, from an elementary perspective, this does not hold for Gentzen's notion of implication as formalized in the sequent calculus. One can even argue that the feature of full cut elimination, which is distinctive of Gentzen's sequent calculus, is enforced by or at least embodied in his particular formulation of $(\to \mathrm{L})$, in contradistinction to $(\to \mathrm{L})^\circ$. Translated into natural deduction, an application $(\to \mathrm{L})^\circ$ can be displayed as follows:

$$A \to B \; \frac{\begin{array}{c}\mathcal{D}_1\\ A\end{array}}{B} \, . \tag{14}$$

[18] A more detailed exposition of the points made in this section can be found in [37].

As it represents the implications-as-rules interpretation, we write $A \to B$ as labelling the transition from A to B. The interpretation of implication underlying $(\to \mathrm{L})$ would then be displayed as

$$\begin{array}{c} \mathcal{D}_1 \\ A \to B \, \dfrac{A}{B} \\ \mathcal{D}_2 \\ C \;. \end{array} \tag{15}$$

Whereas in the first case, $A \to B$ just licenses to continue from A to B by extending $\mathcal{D}_1$, in the second case it *links* two derivations, namely the derivation $\mathcal{D}_1$ of A and the derivation $\mathcal{D}_2$ from B. In this sense we can speak of the *implications-as-links* interpretation as opposed to the *implications-as-rules* interpretation. Obviously, the implications-as-links interpretation adds to the implications-as-rules interpretation something that is expressed by the rule of cut, viz., the cut with cut formula B. Passing from (14) to (15) can be seen as requiring an implicit step which is expressed by the cut rule. So the fact that the implications-as-links interpretation leads to full cut elimination, is due to the fact that it embodies already some limited form of cut which in the implications-as-rules interpretation would have to be added separately.

That an implications-as-links view underlies Gentzen's sequent calculus is supported by the fact that even in systems, in which cut elimination fails, for example in systems with additional axioms or rules, a cut with the formula A can be enforced by adjoining $A \to A$ to the antecedent, as cut then then becomes an instance of $(\to \mathrm{L})$:

$$(\to \mathrm{L}) \; \frac{\Gamma \vdash A \quad \Gamma, A \vdash B}{\Gamma, A \to A \vdash B} \; .$$

From the standpoint of the implications-as-rules interpretation this effect is quite unplausible, as assuming the rule that allows one to pass over from A to A should be vacuous and have no effect whatsoever. The fact that adding $A \to A$ immediately enables cut with A shows that implication has received an interpretation which relates it to cut.

This argument not only shows that the implications-as-rules interpretation is more plausible than the implications-as-links interpretation, but also more elementary. As the implications-as-links interpretation adds certain features of cut to the implications-as-rules interpretation, it is, from the philosophical point of view, advisable to separate these two features, i.e., to use $(\to \mathrm{L})^\circ$ as the more elementary rule for implication, together with cut in a weakened form.

We should emphasize that these results apply to the intuitionistic case only. As soon as we consider the linear or classical case with more than one formula permitted in the succedent, different considerations apply which do not necessarily favor the implications-as-rules view, but might speak for the implications-as links view, due to the symmetry inherent in the multiple-succedent sequent calculus (see [39]).

However, in such systems implication has a different meaning, and, from a conceptual point, it can even be questioned if they contain a genuine notion of implication at all.

Concluding, the implications-as-links interpretation has substantial support from the simplicity of the underlying sequent calculus and its cut elimination feature, so for many technical considerations the implications-as-links interpretation is preferable. However, the implications-as-rules interpretation has the conceptual merit of carrying over the naturalness of natural deduction and the naturalness of the concept of a rule to the sequent calculus, including its natural extension with higher-level rules which allows for a very general treatment of logical constants. It motivates a sequent calculus which is not cut free, but needs a weak version of cut which does not obstruct the subformula principle.

References

1. Avron, A. (1990). Gentzenizing Schroeder-Heister's natural extension of natural deduction. *Notre Dame Journal of Formal Logic*, *31*, 127–135.
2. de Campos Sanz, W., & Piecha, T. (2009). Inversion by definitional reflection and the admissibility of logical rules. *Review of Symbolic Logic*, *2*, 550–569.
3. Došen, K., & Schroeder-Heister, P. (1988). Uniqueness, definability and interpolation. *Jounal of Symbolic Logic*, *53*, 554–570.
4. Dyckhoff, R. (1988). Implementing a simple proof assistant. *Workshop on Programming for Logic Teaching (Leeds, 6–8 July 1987): Proceedings 23/1988* (pp. 49–59). Centre for Theoretical Computer Science: University of Leeds.
5. Garner, R. (2009). On the strength of dependent products in the type theory of Martin-Löf. *Annals of Pure and Applied Logic*, *160*, 1–12.
6. Gentzen, G. (1934/35). Untersuchungen über das logische Schließen. *Mathematische Zeitschrift*, 39:176–210, 405–431 (English translation in: *The Collected Papers of Gerhard Gentzen* (ed. M. E. Szabo), Amsterdam: North Holland (1969), pp. 68–131).
7. Hallnäs, L., & Schroeder-Heister, P. (1990/91). A proof-theoretic approach to logic programming: I. Clauses as rules. II. Programs as definitions. *Journal of Logic and Computation*, *1*, 261–283, 635–660.
8. Hallnäs, L. & Schroeder-Heister, P. (2014). A survey of definitional reflection. In T. Piecha & P. Schroeder-Heister(Eds.), *Advances in Proof-Theoretic Semantics*, Heidelberg: Springer.
9. Harper, R., Honsell, F., & Plotkin, G. (1987). A framework for defining logics. *Journal of the Association for Computing Machinery*, *40*, 194–204.
10. Joachimski, F., & Matthes, R. (2003). Short proofs of normalization for the simply-typed λ-calculus, permutative conversions and Gödel's T. *Archive for Mathematical Logic*, *42*, 59–87.
11. von Kutschera, F. (1968). Die Vollständigkeit des Operatorensystems $\{\neg, \wedge, \vee, \supset\}$ für die intuitionistische Aussagenlogik im Rahmen der Gentzensemantik. *Archiv für mathematische Logik und Grundlagenforschung*, *11*, 3–16.
12. López-Escobar, E. G. K. (1999). Standardizing the N systems of Gentzen. In X. Caicedo & C. H. Montenegro (Eds.), *Models, Algebras and Proofs (Lecture Notes in Pure and Applied Mathematics Vol. 203)* (pp. 411–434). New York: Marcel Dekker.
13. Lorenzen, P. (1950). Konstruktive Begründung der Mathematik. *Mathematische Zeitschrift*, *53*, 162–202.
14. Lorenzen, P. (1955). *Einführung in die operative Logik und Mathematik* (2nd ed 1969). Berlin: Springer.
15. Martin-Löf, P. (1984). *Intuitionistic Type Theory*. Napoli: Bibliopolis.

16. Moriconi, E., & Tesconi, L. (2008). On inversion principles. *History and Philosophy of Logic*, *29*, 103–113.
17. Negri, S., & von Plato, J. (2001). *Structural Proof Theory*. Cambridge: Cambridge University Press.
18. Olkhovikov, G. K., & Schoeder-Heister, P. (2014). On flattening elimination rules. *Review of Symbolic Logic*, 7.
19. Olkhovikov, G. K., & Schoeder-Heister, P. (2014). Proof-theoretic harmony and the levels of rules: General non-flattening results. In E. Moriconi & L. Tesconi (Eds.), *Second Pisa Colloquim in Logic, Language and Epistemology*. ETS, Pisa.
20. Paulson, L. C. (1994). *Isabelle: A Generic Theorem Prover*. Berlin: Springer.
21. von Plato, J. (2000). A problem of normal form in natural deduction. *Mathematical Logic Quarterly*, *46*, 121–124.
22. von Plato, J. (2001). Natural deduction with general elimination rules. *Archive for Mathematical Logic*, *40*, 541–567.
23. Prawitz, D. (1965). *Natural Deduction: A Proof-Theoretical Study*. Almqvist & Wiksell (Reprinted Mineola NY: Dover Publ., 2006), Stockholm.
24. Prawitz, D. (1971). Ideas and results in proof theory. In J. E. Fenstad (Ed.), *Proceedings of the Second Scandinavian Logic Symposium (Oslo 1970)* (pp. 235–308). Amsterdam: North-Holland.
25. Prawitz, D. (1979). Proofs and the meaning and completeness of the logical constants. In J. Hintikka, I. Niiniluoto & E. Saarinen (Eds.), *Essays on Mathematical and Philosophical Logic: Proceedings of the Fourth Scandinavian Logic Symposium and the First Soviet-Finnish Logic Conference, Jyväskylä, Finland, June 29 - July 6, 1976* (pp. 25–40) (revised German translation 'Beweise und die Bedeutung und Vollständigkeit der logischen Konstanten', *Conceptus*, 16. (1982). *31–44)*. Dordrecht: Kluwer.
26. Read, S. (2014). General-elimination harmony and higher-level rules. In H. Wansing (Ed.), *Dag Prawitz on Proofs and Meaning*. Heidelberg: Springer.
27. Schroeder-Heister, P. (1981). *Untersuchungen zur regellogischen Deutung von Aussagenverknüpfungen*. Diss: Universität Bonn.
28. Schroeder-Heister, P. (1984). A natural extension of natural deduction. *Journal of Symbolic Logic*, *49*, 1284–1300.
29. Schroeder-Heister, P. (1984). Generalized rules for quantifiers and the completeness of the intuitionistic operators $\wedge, \vee, \rightarrow, \bot, \forall, \exists$. In M. M. Richter, E. Börger, W. Oberschelp, B. Schinzel & W. Thomas (Eds.) *Computation and Proof Theory. Proceedings of the Logic Colloquium held in Aachen, July 18–23, 1983, Part II. (Lecture Notes in Mathematics, Vol. 1104)*, (pp. 399–426). Berlin: Springer.
30. Schroeder-Heister, P. (1987). *Structural Frameworks with Higher-Level Rules*. Habilitationsschrift: Universität Konstanz.
31. Schroeder-Heister, P. (1991). Structural frameworks, substructural logics, and the role of elimination inferences. In G. Huet & G. Plotkin (Eds.) *Logical Frameworks* (pp. 385–403). Cambridge: Cambridge University Press.
32. Schroeder-Heister, P. (1993). Rules of definitional reflection. *Proceedings of the 8th Annual IEEE Symposium on Logic in Computer Science (Montreal 1993)* (pp. 222–232). Los Alamitos: IEEE Press.
33. Schroeder-Heister, P. (2004). On the notion of assumption in logical systems. In R. Bluhm & C. Nimtz (Eds.) *Selected Papers Contributed to the Sections of GAP5, Fifth International Congress of the Society for Analytical Philosophy, Bielefeld, 22–26 September 2003* (pp. 27–48). mentis (Online publication: http://www.gap5.de/proceedings), Paderborn.
34. Schroeder-Heister, P. (2007). Generalized definitional reflection and the inversion principle. *Logica Universalis*, *1*, 355–376.
35. Schroeder-Heister, P. (2008). Lorenzen's operative justification of intuitionistic logic. In M. van Atten, P. Boldini, M. Bourdeau & G. Heinzmann (Eds.), *One Hundred Years of Intuitionism (1907–2007): The Cerisy Conference*, (pp. 214–240) [References for whole volume: 391–416]. Birkhäuser, Basel.

36. Schroeder-Heister, P. (2009). Sequent calculi and bidirectional natural deduction: On the proper basis of proof-theoretic semantics. In M. Peliš (Ed.), *The Logica Yearbook 2008*. London: College Publications.
37. Schroeder-Heister, P. (2011). Implications-as-rules vs. implications-as-links: an alternative implication-left schema for the sequent calculus. *Journal of Philosophical Logic*, *40*, 95–101.
38. Schroeder-Heister, P. (2012). Proof-theoretic semantics. In E.N. Zalta, (Ed.), *Stanford Encyclopedia of Philosophy*.http://plato.stanford.edu.
39. Schroeder-Heister, P. (2013). Definitional reflection and Basic Logic. *Annals of Pure and Applied Logic* (Special issue, Festschrift 60th Birthday Giovanni Sambin), 164, 491–501.
40. Schroeder-Heister, P. (2014). Harmony in proof-theoretic semantics: A reductive analysis. In H. Wansing (Ed.), *Dag Prawitz on Proofs and Meaning*, Heidelberg: Springer.
41. Tennant, N. (1992). *Autologic*. Edinburgh: Edinburgh University Press.
42. Tennant, N. (2002). Ultimate normal forms for parallelized natural deductions. *Logic Journal of the IGPL*, *10*, 299–337.
43. Tesconi, L. (2004). *Strong Normalization Theorem for Natural Deduction with General Elimination Rules*. Ph. D. thesis: University of Pisa.
44. Wansing, H. (1993). *The Logic of Information Structures*. Berlin: Springer (Lecture Notes in Artificial Intelligence, Vol. 681).

Revisiting Zucker's Work on the Correspondence Between Cut-Elimination and Normalisation

Christian Urban

Abstract Zucker showed that in the fragment of intuitionistic logic whose formulae are build up from $\wedge$, $\supset$ and $\forall$ only, every reduction sequence in natural deduction corresponds to a reduction sequence in the sequent calculus and vice versa. Unfortunately, the technical machinery in Zucker's work is rather cumbersome and complicated. One contribution of this chapter is to greatly simplify his arguments. For example he defined a cut-elimination procedure modulo an equivalence relation; our cut-elimination procedure will be a simple term-rewriting system instead. Zucker also showed that the correspondence breaks down when the connectives $\vee$ or $\exists$ are included. We shall show that this negative result is not because cut-elimination fails to be strongly normalising for these connectives, as asserted by Zucker, rather it is because certain cut-elimination reductions do not correspond to any normalisation reduction.

Keywords Intuitionistic logic · Sequent calculus · Cut-elimination · Natural deduction · Normalisation

1 Introduction

Already Gentzen [7] presented a mapping from sequent proofs to natural deduction proofs, which however was later much improved by Prawitz [14]. According to this (many-to-one) mapping, instances of the cut-rule in the sequent calculus correspond to what is in natural deduction often called detours. From this observation arises the obvious question how cut-elimination, the process of eliminating all instances of the cut-rule from sequent proofs, and normalisation, the process of eliminating all detours from natural deduction proofs, correspond to each other under this mapping.

C. Urban (✉)
Department of Informatics, King's College London, London, UK
e-mail: christian.urban@kcl.ac.uk

L. C. Pereira et al. (eds.), *Advances in Natural Deduction*,
Trends in Logic 39, DOI: 10.1007/978-94-007-7548-0_2,

It seems Kreisel [10, p. 113] was the first who examined this "correspondence question" when he wrote:

> Consider…a calculus of sequents and a system of natural deduction. Inspection shows that …the normalization procedure for systems of natural deduction does not correspond to a particularly natural cut elimination procedure.

Indeed, if we employ Gentzen's method of eliminating cut-rules, that is eliminating innermost cut-rules first, then there is in general no correspondence with normalising natural deduction proofs. Consider for example the sequent proof

$$
\dfrac{
\dfrac{
\dfrac{
\dfrac{\overline{B\vdash B}\quad \overline{A\vdash A}}{B\supset A, B\vdash A}\;{\supset_L}
}{B\supset A\vdash B\supset A}\;{\supset_R}
\qquad
\dfrac{\overline{B\vdash B}\quad \overline{A\vdash A}}{B\supset A, B\vdash A}\;{\supset_L}
}{
\dfrac{B\supset A, B\vdash A}{B\vdash (B\supset A)\supset A}\;{\supset_R}
}\;\text{Cut}^{\bullet}
\qquad
\dfrac{
\dfrac{\dfrac{\overline{A\vdash A}}{A, B\vdash A}\;\text{Weak}_L}{A\vdash B\supset A}\;{\supset_R}
\quad \overline{A\vdash A}
}{(B\supset A)\supset A, A\vdash A}\;{\supset_L}
}{A, B\vdash A}\;\text{Cut}^{\star}
$$

which maps to the following natural deduction proof.

$$
\dfrac{
\dfrac{
\dfrac{
\dfrac{
\dfrac{\overline{B\supset A\vdash B\supset A}\quad \overline{B\vdash B}}{B\supset A, B\vdash A}\;{\supset_E}
}{B\supset A\vdash B\supset A}\;{\supset_I}^{\bullet}
\quad \overline{B\vdash B}
}{B, B\supset A\vdash A}\;{\supset_E}^{\bullet}
}{B\vdash (B\supset A)\supset A}\;{\supset_I}^{\star}
\qquad
\dfrac{\overline{A, B\vdash A}}{A\vdash B\supset A}\;{\supset_I}
}{A, B\vdash A}\;{\supset_E}^{\star}
$$

Note that the cut marked with a star maps to the starred detour (similarly the cut marked with a disc). Clearly we can remove the ⋆-detour using the reductions introduced in Prawitz [14]. In order to obtain a corresponding cut-elimination sequence, we have to reduce the ⋆-cut and move up the resulting smaller cuts, which is however *not* permitted by Gentzen's innermost strategy.

What Zucker [19] did is to remove the strategy and to introduce commuting conversions so that cuts can be eliminated independently from other cuts. A crucial point however is that cut-elimination needs to be strongly normalising, otherwise the correspondence fails: because Prawitz's reduction rules for natural deduction are strongly normalising, infinite cut-reduction sequences cannot be mapped to finite normalisation sequences. Under these circumstances, obtaining a strongly normalising cut-elimination procedure is rather tricky. As seen in the example above, the cut-elimination procedure has to allow cut-rules to pass over other cut-rules. For this consider the following cut-reduction, which allows a cut-rule (Suffix 2) to pass over another cut-rule (Suffix 1).

$$\dfrac{\ldots\vdash\ldots \qquad \dfrac{\ldots\vdash\ldots\quad\ldots\vdash\ldots}{\ldots\vdash\ldots}\,\text{Cut}_1}{\ldots\vdash\ldots}\,\text{Cut}_2 \qquad \dfrac{\ldots\vdash\ldots \qquad \dfrac{\ldots\vdash\ldots\quad\ldots\vdash\ldots}{\ldots\vdash\ldots}\,\text{Cut}_2}{\ldots\vdash\ldots}\,\text{Cut}_1$$

Clearly, this cut-reduction would immediately break strong normalisation because the reduct is again an instance of this reduction, and we can loop by constantly applying it. Zucker found a (rather *ad hoc*) way round this problem of loops: he considered only *proper reduction sequences*, that is reduction sequences which do not include repetitions; see Zucker [19, p. 93]. This restriction ensures that there are no infinite cut-reduction sequences in the $(\wedge, \supset, \forall)$-fragment of intuitionistic logic. In effect, he could give a positive answer to the correspondence question for this fragment (he also considered $\bot$, which we however shall omit). He established the following two facts ($|\_\_|$ stands for Prawitz's translation from sequent proofs to natural deduction proofs).

(i) Given a sequent proof M that reduces to N in one step. Then there is a (possibly empty) sequence of reductions starting from the natural deduction proof $|M|$ and ending with $|N|$.
(ii) Given a sequent proof M. If the natural deduction proof $|M|$ reduces to L, then there exist a sequent proof N such that $|N|$ is L and M reduces in one or more steps to N.

With these facts in place, Zucker could show that each reduction sequence in one system can be mapped to a reduction sequence in the other.

The restriction about proper, or non-repeating, reduction sequences fails, however, to ensure termination of cut-elimination when the connectives $\vee$ or $\exists$ are included. Figure 1 shows a simplified version of the non-terminating, non-repeating reduction sequence given by Zucker. Consequently, he gave a negative answer to the correspondence question for fragments that include $\vee$ or $\exists$. He wrote [19, p. 97]:

> We consider…a (natural) conversion rule in $\mathcal{S}$ for permuting cut with $\vee$L, which does not correspond to identity (or even a reduction) in $\mathcal{N}$. In fact, by use of this conversion rule, it turns out that one can define a non-termination, non-repeating reduction sequence in $\mathcal{S}$. It follows that for *any* set of conversion rules for $\mathcal{N}$ (with $\vee$) for which strong normalisation holds…the correspondence…between reductions in the two systems must fail.

In this paragraph $\mathcal{S}$ and $\mathcal{N}$ stand for Zucker's sequent calculus and natural deduction, respectively.

The main difference between Zucker's work and ours is that our cut-elimination procedure *is* strongly normalising for all connectives. Therefore we shall tackle the correspondence question again. In doing so we shall take care that the technical machinery is simpler than Zucker's. For example we shall not make use of an equivalence relation, which in his work includes Kleene-like permutation rules for cuts as well as rules for permuting contractions downwards in a sequent proof. Instead, we shall use the term-rewriting system introduced in Urban [17]; for a shorter account of this work see Urban and Bierman [18].

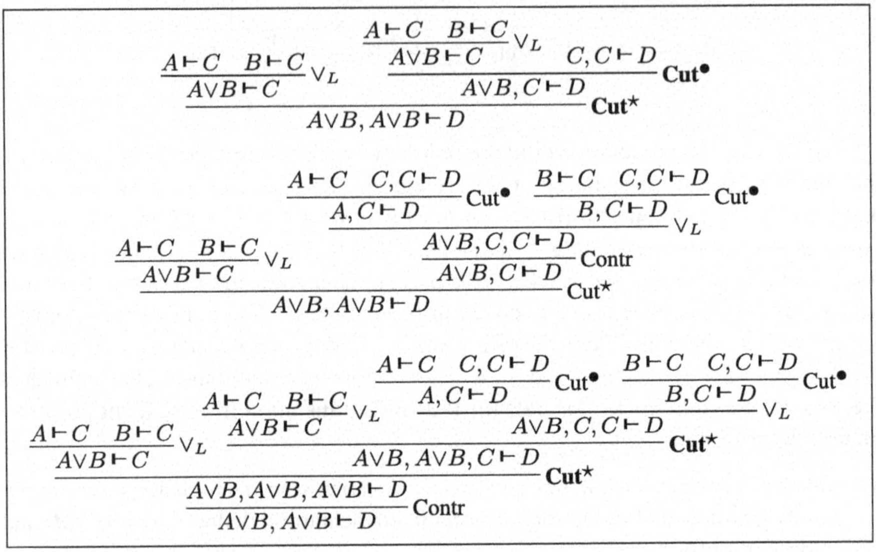

Fig. 1 A slight variant of the non-terminating reduction sequence given by Zucker; see Ungar [16]. In the first step, the cut marked with a disc is reduced creating two cuts as well as introducing a contraction. In the second step, the cut marked with a star is permuted with the contraction rule—again creating two cuts. Now the cuts written in bold face have the same shape as in the first proof, and we can repeat the reduction sequence infinitely many times

While the correspondence question seems to be of purely 'syntactic' interest, there are in fact a number of interesting issues. Natural deduction is the system of choice for any semantical investigation; to quote Girard et al. [8, p. 39]: 'In some sense, we should think of natural deductions as the true "proof" objects.' This is because in natural deduction (at least in some fragments) one can define a notion of equality of proof that is preserved under reduction. Consider now the correspondence question again: if cut-reductions do not match with normalisation reductions, then the sequent calculus might contain features that are not accounted for in natural deduction. This is particularly interesting, because in classical logic only sequent calculi seem to provide a framework for studying non-deterministic features; see Barbanera and Berardi [2], Urban [17] and Laird [11].

There are a number of earlier works that give an entirely positive answer to the correspondence question for all connectives. For example Pottinger [13] solved Zucker's problem with non-terminating reduction sequences by introducing normalisation-like reductions for the sequent calculus. Unfortunately, this work has a subtle defect: because Pottinger annotated sequent proofs with lambda-terms and then formalised cut-elimination as normalisation, the notion of normality for lambda-terms does *not* coincide with the notion of cut-freedom; see Pottinger [13, p. 323]. We shall avoid this problem by using proof annotations that encode precisely the structure of sequent proofs and also implement precisely the (standard) rules for cut-elimination.

More recently, Negri and von Plato [12] describe a natural deduction system with a particular formulation for the elimination rules—they are called general elimination rules. For this system a positive answer to the correspondence question can be given. However, we find it is questionable to make the natural deduction calculus to be 'sequent-calculus-like', meaning that many proofs are distinguished that differ only in the order of some inference rules. There is also work by Ungar [16] that modifies the natural deduction calculus for obtaining a positive answer to the correspondence question. His results build upon a very complicated, graph-like notion of natural deduction proofs, which is also not very convincing. Therefore, we feel it is prudent to reconsider the correspondence question.

The chapter is organised as follows: In Sects. 2 and 3 we shall describe a natural deduction system and a sequent calculus. In Sect. 4 we shall address the correspondence question for the $(\wedge, \supset, \forall)$-fragment of intuitionistic logic. An example will be given for why the correspondence breaks when $\vee$ is included. We shall conclude and suggest further work in Sect. 5.

2 Natural Deduction

For studying the correspondence question, Zucker introduced in [19] a minor variant of Gentzen's natural deduction system. We shall also use a slight variant, namely the system presented in (for example) Gallier [6]. It differs from Gentzen's formulation of natural deduction in that proofs are written in sequent style—that means open assumptions are recorded explicitly in every inference step—and that terms are annotated to proofs.

First we give the grammars for expressions and formulae (we use 'expression' instead of 'term' and moreover write expressions in a sans serif font in order to avoid confusions with proof annotations we shall introduce later).

$$\begin{aligned} \mathsf{t} &::= \mathsf{x} \mid \mathsf{f}\,\mathsf{t}_1 \ldots \mathsf{t}_n \\ B &::= A\,\mathsf{t}_1 \ldots \mathsf{t}_n \mid B{\wedge}B \mid B{\vee}B \mid B{\supset}B \mid \forall \mathsf{x}.B \mid \exists \mathsf{x}.B \end{aligned}$$

In this grammar x is taken from a set of variables, f from a set of functional symbols and A ranges over predicate symbols. As usual functional and predicate symbols have an arity (a natural number) assigned specifying the number of arguments. We shall often write x, y, z for variables and t for arbitrary expressions; square brackets (e.g., $[\mathsf{x}]$) will indicate that a variable becomes bound; and capture avoiding substitution for expressions and formulae is assumed as usual.

Next we define lambda-terms for annotating natural deduction proofs. Raw lambda-terms are defined by the grammar

p

$$
\begin{array}{rlll}
M, N, P ::= & x & \text{Var} & \\
\mid & \langle M, N \rangle & \text{And-I} & \\
\mid & \mathtt{fst}(M) \mid \mathtt{snd}(M) & \text{And-E}_i & i = 1, 2 \\
\mid & \mathtt{inl}(M) \mid \mathtt{inr}(M) & \text{Or-I}_i & i = 1, 2 \\
\mid & \mathtt{case}(P, \lambda x : B'.M, \lambda y : B''.N) & \text{Or-E} & \\
\mid & \lambda x : B.M & \text{Imp-I} & \\
\mid & M\,N & \text{Imp-E} & \\
\mid & \mathtt{inx}(\mathsf{t}, M) & \text{Exists-I} & \\
\mid & \mathtt{casex}(M, \lambda x{:}B.[\mathsf{y}]N) & \text{Exists-E} & \\
\mid & [\mathsf{y}]M & \text{Forall-I} & \\
\mid & M\,\mathsf{t} & \text{Forall-E} &
\end{array}
$$

where y and t are expressions, and the Bs are formulae. Let us briefly mention the conventions we shall assume for lambda-terms: free and bound variables are as usual; for brevity we shall often omit the formulae on binders and simply write $\lambda x.M$; a Barendregt-style naming convention for variables will always be observed and lambda-terms will be regarded as equivalent up to renaming of bound variables. A pleasing consequence of these conventions is that the notion of variable substitution for lambda-terms can be defined without worrying about clashes and capture of variables.

To annotate lambda-terms to natural deduction proofs, we shall have sequents of the form $\Gamma \triangleright M : B$ in which M is a lambda-term, B is a formula and Γ is a context—a set of (variable, formula)-pairs. We shall also employ some shorthand notation for contexts: rather than writing for example $\{(x, B), (y, C), (z, D)\}$, we shall write $x : B, y : C, z : D$. For contexts we have the convention that a context is *ill-formed*, if it contains more than one occurrence of a variable. For example, the context $x : B, x : C$ is not allowed.

In the sequel we are only interested in lambda-terms, M, for which there is a context Γ and a formula B such that $\Gamma \triangleright M : B$ is derivable given the inference rules shown in Fig. 2. The following set is defined to contain all those lambda-terms.

$$\mathcal{N} \stackrel{\text{def}}{=} \{\, M \mid \Gamma \triangleright M : B \text{ is derivable using the rules given in Fig. 2} \,\}$$

Apart from the initial sequent, the inference rules shown in Fig. 2 fall into two groups: introduction and elimination rules. A *detour* is a sequence of neighbouring inference rules beginning with an introduction rule introducing a formula that is eliminated by the last one in the sequence (necessarily an elimination rule). Detours in natural deduction proofs violate the *subformula property*; for a definition of this standard property see Troelstra and Schwichtenberg [15]. Normalisation is the process of stepwise eliminating detours from natural deduction proofs, converting them into normal forms for which the subformula property holds. Below we shall introduce term-rewriting rules, commonly referred to as *beta-reductions*, which eliminate a detour formed by an introduction rule and a neighbouring elimination rule.

$$\frac{}{x : B, \Gamma \triangleright x : B}$$

$$\frac{\Gamma \triangleright M : B \quad \Gamma \triangleright N : C}{\Gamma \triangleright \langle M, N \rangle : B \wedge C} \wedge_I \qquad \frac{\Gamma \triangleright M : B_1 \wedge B_2}{\Gamma \triangleright \mathtt{fst}(M) : B_1} \wedge_{E_1} \qquad \frac{\Gamma \triangleright M : B_1 \wedge B_2}{\Gamma \triangleright \mathtt{snd}(M) : B_2} \wedge_{E_2}$$

$$\frac{\Gamma \triangleright M : B_1}{\Gamma \triangleright \mathtt{inl}(M) : B_1 \vee B_2} \vee_{I_1} \qquad \frac{\Gamma \triangleright M : B_2}{\Gamma \triangleright \mathtt{inr}(M) : B_1 \vee B_2} \vee_{I_2}$$

$$\frac{\Gamma \triangleright P : B \vee C \quad x : B, \Gamma \triangleright M : D \quad y : C, \Gamma \triangleright N : D}{\Gamma \triangleright \mathtt{case}(P, \lambda x.M, \lambda y.N) : D} \vee_E$$

$$\frac{x : B, \Gamma \triangleright M : C}{\Gamma \triangleright \lambda x.M : B \supset C} \supset_I \qquad \frac{\Gamma \triangleright M : B \supset C \quad \Gamma \triangleright N : B}{\Gamma \triangleright M\, N : C} \supset_E$$

$$\frac{\Gamma \triangleright M : B(^{\mathsf{x}}/_{\mathsf{t}})}{\Gamma \triangleright \mathtt{inx}(\mathsf{t}, M) : \exists \mathsf{x}.B} \exists_I \qquad \frac{\Gamma \triangleright M : \exists \mathsf{x}.B \quad x : B(^{\mathsf{x}}/_{\mathsf{y}}), \Gamma \triangleright N : C}{\Gamma \triangleright \mathtt{casex}(M, \lambda x.[\mathsf{y}]N) : C} \exists_E$$

$$\frac{\Gamma \triangleright M : B(^{\mathsf{x}}/_{\mathsf{y}})}{\Gamma \triangleright [\mathsf{y}]M : \forall \mathsf{x}.B} \forall_I \qquad \frac{\Gamma \triangleright M : \forall \mathsf{x}.B}{\Gamma \triangleright M\, \mathsf{t} : B(^{\mathsf{x}}/_{\mathsf{t}})} \forall_E$$

Fig. 2 Term assignment for intuitionistic natural deduction proofs

$$\begin{aligned}
\mathtt{fst}(\langle M, N \rangle) &\xrightarrow{\beta} M \\
\mathtt{snd}(\langle M, N \rangle) &\xrightarrow{\beta} N \\
\mathtt{case}(\mathtt{inl}(P), \lambda x.M, \lambda y.N) &\xrightarrow{\beta} M(x/P) \\
\mathtt{case}(\mathtt{inr}(P), \lambda x.M, \lambda y.N) &\xrightarrow{\beta} N(y/P) \\
(\lambda x.M)\, N &\xrightarrow{\beta} M(x/N) \\
\mathtt{casex}(\mathtt{inx}(\mathsf{t}, M), \lambda x.[\mathsf{y}]N) &\xrightarrow{\beta} N(\mathsf{y}/t)(\mathsf{x}/M) \\
([\mathsf{y}]M)\, \mathsf{t} &\xrightarrow{\beta} M(\mathsf{y}/\mathsf{t})
\end{aligned}$$

In order to ensure that the normal forms satisfy the subformula property, further rewriting rules (we shall call them *gamma-reductions*) are needed for the term-constructors case and casex. A lucid explanation for why further rules are necessary is given in Girard et al. [8, p. 74]. We shall present the rewriting rules for case only (the rules for casex are similar).

$$\begin{aligned}
\mathtt{fst}(\mathtt{case}(P, \lambda x.M, \lambda y.N)) &\xrightarrow{\gamma} \mathtt{case}(P, \lambda x.\mathtt{fst}(M), \lambda y.\mathtt{fst}(N)) \\
\mathtt{snd}(\mathtt{case}(P, \lambda x.M, \lambda y.N)) &\xrightarrow{\gamma} \mathtt{case}(P, \lambda x.\mathtt{snd}(M), \lambda y.\mathtt{snd}(N)) \\
\mathtt{case}(P, \lambda x.M, \lambda y.N)\, Q &\xrightarrow{\gamma} \mathtt{case}(P, \lambda x.(M\, Q), \lambda y.(N\, Q)) \\
\mathtt{case}(P, \lambda x.M, \lambda y.N)\, \mathsf{t} &\xrightarrow{\gamma} \mathtt{case}(P, \lambda x.(M\, \mathsf{t}), \lambda y.(N\, \mathsf{t})) \\
\mathtt{case}(\mathtt{case}(P, \lambda x.M, \lambda y.N), \lambda u.S, \lambda v.T) &\xrightarrow{\gamma} \mathtt{case}(P, \lambda x.\mathtt{case}(M, \lambda u.S, \lambda v.T), \\
&\qquad \lambda y.\mathtt{case}(N, \lambda u.S, \lambda v.T)) \\
\mathtt{casex}(\mathtt{case}(P, \lambda x.M, \lambda y.N), \lambda z.[\mathsf{y}]S) &\xrightarrow{\gamma} \mathtt{case}(P, \lambda x.\mathtt{casex}(M, \lambda z.[\mathsf{y}]S), \\
&\qquad \lambda y.\mathtt{casex}(N, \lambda z.[\mathsf{y}]S))
\end{aligned}$$

We automatically assume that the reduction rules are closed under context formation, which is a standard convention in term rewriting. Properties of these reduction rules such as subject reduction, strong normalisation and confluence are well-known and therefore not further elaborated here.

3 Sequent Calculus

Zucker's chapter is not very concise, in part because he formalises a sequent calculus using the rather cumbersome notion of indexed formulae. A simpler formulation for an intuitionistic sequent calculus is given in Urban [17]; for an overviw of this work see Urban and Bierman [18]. The main idea is to formalise the sequent calculus in a similar fashion as natural deduction using proof-annotations. This will mean for example that contexts are sets, as in type-theory, and *not* multisets, as in Gentzen's LJ. A pleasing feature of 'contexts-as-sets' is that the structural rules are completely implicit in the form of the logical rules. In the remainder of this section, we shall repeat the basic definitions from Urban [17] in order to make this chapter self-contained.

Formulae and expressions are defined as in natural deduction. Sequents are of the form

$$\Gamma \triangleright M \triangleright a : B$$

where Γ is a context, $a : B$ is a labelled formula and M is a term. As in Sect. 2, contexts are sets of labelled formulae. However, in the sequent calculus we shall call the labels *names*; so contexts are sets of (name,formula)-pairs. Otherwise the same conventions for contexts apply as given in Sect. 2. The formula on the right-hand side of the sequent is labelled by a *co-name*. While co-names can be dispensed with when formalising an intuitionistic sequent calculus (they are necessary for classical logic), we shall *not* suppress them for reasons that will become clear later on when we define the operation of proof substitution.

The raw terms for annotating sequent proofs are given by the grammar:

$$
\begin{array}{rcll}
M, N & ::= & \mathsf{Ax}(x, a) & \text{Axiom} \\
& \mid & \mathsf{Cut}(\langle a{:}B\rangle M, (x{:}B)N) & \text{Cut} \\
& \mid & \mathsf{And}_R(\langle a{:}B'\rangle M, \langle b{:}B''\rangle N, c) & \text{And-R} \\
& \mid & \mathsf{And}_L^i((x{:}B)M, y) & \text{And-L}_i \quad i = 1, 2 \\
& \mid & \mathsf{Or}_R^i(\langle a{:}B\rangle M, b) & \text{Or-R}_i \quad i = 1, 2 \\
& \mid & \mathsf{Or}_L((x{:}B')M, (y{:}B'')N, z) & \text{Or-L} \\
& \mid & \mathsf{Imp}_R((x{:}B')\langle a{:}B''\rangle M, b) & \text{Imp-R} \\
& \mid & \mathsf{Imp}_L(\langle a{:}B'\rangle M, (x{:}B'')N, y) & \text{Imp-L} \\
& \mid & \mathsf{Exists}_R(\langle a{:}B\rangle M, \mathsf{t}, b) & \text{Exists-R} \\
& \mid & \mathsf{Exists}_L((x{:}B)[\mathsf{y}]M, y) & \text{Exists-L} \\
& \mid & \mathsf{Forall}_R(\langle a{:}B\rangle[\mathsf{y}]M, b) & \text{Forall-R} \\
& \mid & \mathsf{Forall}_L((x{:}B)M, \mathsf{t}, y) & \text{Forall-L}
\end{array}
$$

in which x, y, z are taken from a set of names; a, b, c from a set of co-names; y and t are expressions; and the Bs are formulae. We shall often write $\ldots, x, y, z$ for names, and $a, b, c, \ldots$ for co-names. Round brackets signify that a name becomes bound and angle brackets, that a co-name becomes bound. Again we have similar conventions as in Sect. 2: we shall omit the types on the bindings; regard terms as equal up to alpha-conversions and adopt a Barendregt-style convention for names and co-names. These conventions are standard in term rewriting. Notice however that names and co-names are not the same notions as a variable in lambda-terms: whilst a variable can be substituted with a lambda-term, a name or a co-name can only be "renamed". Rewriting a name x to y in a term M is written as $M[x \mapsto y]$, and similarly rewriting a co-name a to b is written as $M[a \mapsto b]$. The routine formalisation of these rewriting operations is omitted. For the terms defined above we have the relatively standard notions of free names and free co-names. Given a term, say M, its set of free names and free co-names is denoted by $FN(M)$ and $FC(M)$, respectively. Another useful notion is the following.

Definition 1 A term, M, introduces the name z or co-name c, if and only if M is of the form

for z:	$\mathsf{Ax}(z, c)$	for c:	$\mathsf{Ax}(z, c)$
	$\mathsf{And}^i_L(\langle x\rangle S, z)$		$\mathsf{And}_R(\langle a\rangle S, \langle b\rangle T, c)$
	$\mathsf{Or}_L(\langle x\rangle S, \langle y\rangle T, z)$		$\mathsf{Or}^i_R(\langle a\rangle S, c)$
	$\mathsf{Imp}_L(\langle a\rangle S, \langle x\rangle T, z)$		$\mathsf{Imp}_R(\langle x\rangle\langle a\rangle S, c)$
	$\mathsf{Exists}_L(\langle x\rangle[\mathsf{y}]S, z)$		$\mathsf{Exists}_R(\langle a\rangle S, \mathsf{t}, c)$
	$\mathsf{Forall}_L(\langle x\rangle S, \mathsf{t}, z)$		$\mathsf{Forall}_R(\langle a\rangle[\mathsf{y}]S, c)$

A term *freshly* introduces a name, if and only if none of its proper subterms introduces this name. In other words, the name must not be free in a proper subterm. Similarly for co-names. □

As we shall see later, this definition corresponds to the traditional notion of a main formula of an inference.

Again we shall be interested in only *well-typed* terms; this means those M for which there is a context Γ and a labelled formula $a : B$, such that $\Gamma \triangleright M \triangleright a : B$ holds given the inference rules in Fig. 3. The following set contains all such terms.

$$\mathcal{S} \stackrel{\text{def}}{=} \{ M \mid \Gamma \triangleright M \triangleright a : B \text{ is derivable using the rules given in Fig. 3} \}$$

Whilst the structural rules are *implicit* in our sequent calculus, i.e. the calculus has fewer inference rules, there are a number of subtleties concerning contexts. We assume for the commas in Fig. 3 the following conventions: a comma in a conclusion stands for set union and a comma in a premise stands for *disjoint* set union. Consider for example the $\wedge_{L_i}$-rule. This rule introduces the (name,formula)-pair $y : B_1 \wedge B_2$ in the conclusion, and consequently, y is a free name in $\mathsf{And}^i_L(\langle x\rangle M, y)$. However, y can already be free in the subterm M, in which case $y : B_1 \wedge B_2$ belongs to Γ. We refer to this as an *implicit contraction*. Hence the left-hand side of the conclusion of

$$\frac{}{x:B,\Gamma \triangleright \mathsf{Ax}(x,a) \triangleright a:B} \qquad \frac{\Gamma_1 \triangleright M \triangleright a:B \quad x:B,\Gamma_2 \triangleright N \triangleright b:C}{\Gamma_1,\Gamma_2 \triangleright \mathsf{Cut}(\langle a\rangle M,(x)N) \triangleright b:C}\,\mathsf{Cut}$$

$$\frac{x:B_i,\Gamma \triangleright M \triangleright a:C}{y:B_1\wedge B_2,\Gamma \triangleright \mathsf{And}^i_L((x)M,y) \triangleright a:C}\wedge_{L_i} \qquad \frac{\Gamma \triangleright M \triangleright a:B \quad \Gamma \triangleright N \triangleright b:C}{\Gamma \triangleright \mathsf{And}_R(\langle a\rangle M,\langle b\rangle N,c) \triangleright c:B\wedge C}\wedge_R$$

$$\frac{x:B,\Gamma \triangleright M \triangleright a:D \quad y:C,\Gamma \triangleright N \triangleright a:D}{z:B\vee C,\Gamma \triangleright \mathsf{Or}_L((x)M,(y)N,z) \triangleright a:D}\vee_L \qquad \frac{\Gamma \triangleright M \triangleright a:B_i}{\Gamma \triangleright \mathsf{Or}^i_R(\langle a\rangle M,b) \triangleright b:B_1\vee B_2}\vee_{R_i}$$

$$\frac{\Gamma \triangleright M \triangleright \Delta,a:B \quad x:C,\Gamma \triangleright N \triangleright b:D}{y:B\supset C,\Gamma \triangleright \mathsf{Imp}_L(\langle a\rangle M,(x)N,y) \triangleright b:D}\supset_L \qquad \frac{x:B,\Gamma \triangleright M \triangleright a:C}{\Gamma \triangleright \mathsf{Imp}_R((x)\langle a\rangle M,b) \triangleright b:B\supset C}\supset_R$$

$$\frac{x:B(^{\mathsf{x}}/_{\mathsf{y}}),\Gamma \triangleright M \triangleright a:C}{y:\exists \mathsf{x}.B,\Gamma \triangleright \mathsf{Exists}_L((x)[\mathsf{y}]M,y) \triangleright a:C}\exists_L \qquad \frac{\Gamma \triangleright M \triangleright a:B(^{\mathsf{x}}/_{\mathsf{t}})}{\Gamma \triangleright \mathsf{Exists}_R(\langle a\rangle M,\mathsf{t},b) \triangleright b:\exists \mathsf{x}.B}\exists_R$$

$$\frac{x:B(^{\mathsf{x}}/_{\mathsf{t}}),\Gamma \triangleright M \triangleright a:C}{y:\forall \mathsf{x}.B,\Gamma \triangleright \mathsf{Forall}_L((x)M,\mathsf{t},y) \triangleright a:C}\forall_L \qquad \frac{\Gamma \triangleright M \triangleright a:B(^{\mathsf{x}}/_{\mathsf{y}})}{\Gamma \triangleright \mathsf{Forall}_R(\langle a\rangle[\mathsf{y}]M,b) \triangleright b:\forall \mathsf{x}.B}\forall_R$$

Fig. 3 Term assignment for sequent proofs

$\wedge_{L_i}$ is of the form $y:B_1\wedge B_2 \oplus \Gamma$ where $\oplus$ denotes set union. Clearly, in the case that the term $\mathsf{And}^i_L((x)M, y)$ freshly introduces y, then this context is of the form $y{:}B_1\wedge B_2 \otimes \Gamma$ where $\otimes$ denotes *disjoint* set union. Note that $x{:}B_i$ cannot be part of the conclusion: x is intended to become bound in the term. Thus the context in the premise must be of the form $x{:}B_i \otimes \Gamma$.

There is one point worth mentioning in the cut-rule, because this is the only inference rules in our sequent calculus that does not share the contexts, but requires that two contexts are joined. Thus we take the cut-rule to be of the form

$$\frac{\Gamma_1 \triangleright M \triangleright a{:}B \quad x{:}B \otimes \Gamma_2 \triangleright N \triangleright b:C}{\Gamma_1 \oplus \Gamma_2 \triangleright \mathsf{Cut}(\langle a\rangle M,(x)N) \triangleright b:C}\ \mathrm{Cut}$$

In effect, this rule is only applicable, if it does not break the convention about ill-formed contexts, which can always be achieved by renaming some names appropriately. Notice however that we do not require that cut-rules have to be "fully" multiplicative: the Γ_is can share some formulae.

Next we focus on the term-rewriting rules. One reason for introducing terms is that they greatly simplify the formalisation of the cut-reduction rules, most notably the rules for commuting cuts. In LJ for example there are 360 different cases of commuting cuts! Using the notion of proof substitution, which we shall introduce below, the cut-reductions necessary for dealing with commuting cuts can be formalised in only 24 clauses. Clearly, this is an advantage of our use of terms.

Before we give the definition of the substitution, it is instructive to look at some examples. Commuting cuts need to permute to the places where the cut-formula is the main formula. At the level of terms this means the cuts need to be permuted to every subterm that introduces the cut-formula. This will be achieved with substitutions of the form

$$P(x{:}B/\langle a{:}B\rangle Q) \quad\text{and}\quad S(b{:}B/(y{:}B)T).$$

If a substitution is "next" to a term in which the cut-formula is introduced, the substitution becomes an instance of the Cut-term constructor. In the following, two examples we shall write $[\sigma]$ and $[\tau]$ for the substitutions $(c/(x)P)$ and $(x/\langle b\rangle Q)$, respectively (we do not write the type annotations in substitutions provided they are clear from the context).

$$\begin{aligned}\mathsf{And}_R(\langle a\rangle M, \langle b\rangle N, c)[\sigma] &= \mathsf{Cut}(\langle c\rangle \mathsf{And}_R(\langle a\rangle\, M, \langle b\rangle\, N, c), (x)P)\\ \mathsf{Imp}_L(\langle a\rangle M, (y)N, x)[\tau] &= \mathsf{Cut}(\langle b\rangle Q, (x)\mathsf{Imp}_L(\langle a\rangle\, M[\tau], (y)\, N[\tau], x))\end{aligned}$$

In the first term the formula labelled with c is the main formula and in the second the formula labelled with x is the main formula. So in both cases the substitutions "expand" to cuts, and in the second case, the substitution is also pushed inside the subterms. This is because there might be several occurrences of x (this name need not have been freshly introduced). An exception applies to axioms, where the substitution is defined differently, as shown below.

$$\begin{aligned}\mathsf{Ax}(x, a)(x/\langle b\rangle P) &= P[b \mapsto a]\\ \mathsf{Ax}(x, a)(a/(y)Q) &= Q[y \mapsto x]\end{aligned}$$

Recall that $P[b \mapsto a]$ stands for the term P in which every free occurrence of the co-name b is rewritten to a (similarly $Q[y \mapsto x]$). We are left with the cases where the name or co-name that is being substituted for is not a label of the main formula. In these cases, the substitutions are pushed inside the subterms or vanish in case of the axioms. Suppose the substitution $[\sigma]$ is *not* of the form $[z/\ldots]$ and $[a/\ldots]$, then we have the following clauses.

$$\begin{aligned}\mathsf{Or}_L((x)M, (y)N, z)[\sigma] &= \mathsf{Or}_L((x)\, M[\sigma], (y)\, N[\sigma], z)\\ \mathsf{Ax}(z, a)[\sigma] &= \mathsf{Ax}(z, a)\end{aligned}$$

Figure 4 gives the complete definition of proof substitution. We do not need to worry about inserting contraction rules when a term is duplicated, since our contexts are sets of labelled formulae and thus contractions are made implicitly. Another simplification is due to our use of the Barendregt-style naming convention, because we do not need to worry about possible capture of free names or co-names.

We are now in the position to formalise our cut-elimination procedure. We shall distinguish two kinds of cuts: *logical cuts* and *commuting cuts*. A cut of the form $\mathsf{Cut}(\langle a\rangle M, (x)N)$ is a logical cut provided M freshly introduces a and N freshly introduces x; otherwise the cut is a commuting cut. The corresponding reduction rules

$$
\begin{array}{rrcl}
1. & \mathsf{Ax}(x,c)({}^{c}/_{(y)}P) & \stackrel{\text{def}}{=} & P[y{\mapsto}x]\\
2. & \mathsf{Ax}(y,a)({}^{y}/_{\langle c\rangle}P) & \stackrel{\text{def}}{=} & P[c{\mapsto}a]\\
3. & \mathsf{And}_R(\langle a\rangle M,\langle b\rangle N,c)({}^{c}/_{(y)}P) & \stackrel{\text{def}}{=} & \mathsf{Cut}(\langle c\rangle\mathsf{And}_R(\langle a\rangle M,\langle b\rangle N,c),(y)P)\\
4. & \mathsf{Or}^i_R(\langle a\rangle M,c)({}^{c}/_{(y)}P) & \stackrel{\text{def}}{=} & \mathsf{Cut}(\langle c\rangle\mathsf{Or}^i_R(\langle a\rangle M,c),(y)P)\\
5. & \mathsf{Imp}_R((x)\langle a\rangle M,b)({}^{b}/_{(y)}P) & \stackrel{\text{def}}{=} & \mathsf{Cut}(\langle b\rangle\mathsf{Imp}_R((x)\langle a\rangle M,b),(y)P)\\
6. & \mathsf{Exists}_R(\langle a\rangle M,\mathsf{t},b)({}^{b}/_{(y)}P) & \stackrel{\text{def}}{=} & \mathsf{Cut}(\langle b\rangle\mathsf{Exists}_R(\langle a\rangle M,\mathsf{t},b),(y)P)\\
7. & \mathsf{Forall}_R(\langle x\rangle[\mathsf{y}]M,b)({}^{b}/_{(y)}P) & \stackrel{\text{def}}{=} & \mathsf{Cut}(\langle b\rangle\mathsf{Forall}_R(\langle a\rangle[\mathsf{y}]M,b),(y)P)\\
8. & \mathsf{And}^i_L((x)M,y)({}^{y}/_{\langle c\rangle}P) & \stackrel{\text{def}}{=} & \mathsf{Cut}(\langle c\rangle P,(y)\mathsf{And}^i_L((x)M({}^{y}/_{\langle c\rangle}P),y))\\
9. & \mathsf{Or}_L((x)M,(y)N,z)({}^{z}/_{\langle c\rangle}P) & \stackrel{\text{def}}{=} & \\
 & \multicolumn{3}{r}{\mathsf{Cut}(\langle c\rangle P,(z)\mathsf{Or}_L((x)M({}^{z}/_{\langle c\rangle}P),(y)N({}^{z}/_{\langle c\rangle}P),z))}\\
10. & \mathsf{Imp}_L(\langle a\rangle M,(x)N,y)({}^{y}/_{\langle c\rangle}P) & \stackrel{\text{def}}{=} & \\
 & \multicolumn{3}{r}{\mathsf{Cut}(\langle c\rangle P,(y)\mathsf{Imp}_L(\langle a\rangle M({}^{y}/_{\langle c\rangle}P),(x)N({}^{y}/_{\langle c\rangle}P),y))}\\
11. & \mathsf{Exists}_L((x)[\mathsf{y}]M,y)({}^{y}/_{\langle c\rangle}P) & \stackrel{\text{def}}{=} & \mathsf{Cut}(\langle c\rangle P,(y)\mathsf{Exists}_L((x)[\mathsf{y}]M({}^{y}/_{\langle c\rangle}P),y))\\
12. & \mathsf{Forall}_L((x)M,\mathsf{t},y)({}^{y}/_{\langle c\rangle}P) & \stackrel{\text{def}}{=} & \mathsf{Cut}(\langle c\rangle P,(y)\mathsf{Forall}_L((x)M({}^{y}/_{\langle c\rangle}P),\mathsf{t},y))
\end{array}
$$

Otherwise:

$$
\begin{array}{rrcl}
13. & \mathsf{Ax}(x,a)[\sigma] & \stackrel{\text{def}}{=} & \mathsf{Ax}(x,a)\\
14. & \mathsf{Cut}(\langle a\rangle M,(x)N)[\sigma] & \stackrel{\text{def}}{=} & \mathsf{Cut}(\langle a\rangle\ M[\sigma],(x)\ N[\sigma])\\
15. & \mathsf{And}_R(\langle a\rangle M,\langle b\rangle N,c)[\sigma] & \stackrel{\text{def}}{=} & \mathsf{And}_R(\langle a\rangle\ M[\sigma],\langle b\rangle\ N[\sigma],c)\\
16. & \mathsf{And}^i_L((x)M,y)[\sigma] & \stackrel{\text{def}}{=} & \mathsf{And}^i_L((x)\ M[\sigma],y)\\
17. & \mathsf{Or}^i_R(\langle a\rangle M,b)[\sigma] & \stackrel{\text{def}}{=} & \mathsf{Or}^i_R(\langle a\rangle\ M[\sigma],b)\\
18. & \mathsf{Or}_L((x)M,(y)N,z)[\sigma] & \stackrel{\text{def}}{=} & \mathsf{Or}_L((x)\ M[\sigma],(y)\ N[\sigma],z)\\
19. & \mathsf{Imp}_R((x)\langle a\rangle M,b)[\sigma] & \stackrel{\text{def}}{=} & \mathsf{Imp}_R((x)\langle a\rangle\ M[\sigma],b)\\
20. & \mathsf{Imp}_L(\langle a\rangle M,(x)N,y)[\sigma] & \stackrel{\text{def}}{=} & \mathsf{Imp}_L(\langle a\rangle\ M[\sigma],(x)\ N[\sigma],y)\\
21. & \mathsf{Exists}_R(\langle a\rangle M,\mathsf{t},b)[\sigma] & \stackrel{\text{def}}{=} & \mathsf{Exists}_R(\langle a\rangle\ M[\sigma],\mathsf{t},b)\\
22. & \mathsf{Exists}_L((x)[\mathsf{y}]M,y)[\sigma] & \stackrel{\text{def}}{=} & \mathsf{Exists}_L((x)[\mathsf{y}]\ M[\sigma],y)\\
23. & \mathsf{Forall}_R(\langle a\rangle[\mathsf{y}]M,b)[\sigma] & \stackrel{\text{def}}{=} & \mathsf{Forall}_R(\langle a\rangle[\mathsf{y}]\ M[\sigma],b)\\
24. & \mathsf{Forall}_L((x)M,\mathsf{t},y)[\sigma] & \stackrel{\text{def}}{=} & \mathsf{Forall}_L((x)\ M[\sigma],\mathsf{t},y)
\end{array}
$$

Fig. 4 Proof substitution

are given in Fig. 5. The side-conditions attached to each rule ensure that commuting cuts cannot be rewritten using a rule for a logical cut. Again we automatically assume that $\stackrel{c}{\longrightarrow}$ and $\stackrel{l}{\longrightarrow}$ are closed under context formation. We shall write $\stackrel{cut}{\longrightarrow}$ whenever we refer to both $\stackrel{c}{\longrightarrow}$ and $\stackrel{l}{\longrightarrow}$.

Urban [17] and Urban and Bierman [18] present a detailed proof showing that $\stackrel{cut}{\longrightarrow}$ is strongly normalising (even in the classical case). However, $\stackrel{cut}{\longrightarrow}$ is *not* confluent!

Logical Cuts $(i = 1, 2)$

1. $\mathsf{Cut}(\langle b\rangle \mathsf{And}_R(\langle a_1\rangle M_1, \langle a_2\rangle M_2, b), (y)\mathsf{And}^i_L((x)N, y)) \xrightarrow{l} \mathsf{Cut}(\langle a_i\rangle M_i, (x)N)$
   if $\mathsf{And}_R(\langle a_1\rangle M_1, \langle a_2\rangle M_2, b)$ and $\mathsf{And}^i_L((x)N, y)$ freshly introduce b and y
2. $\mathsf{Cut}(\langle b\rangle \mathsf{Or}^i_R(\langle a\rangle M, b), (y)\mathsf{Or}_L((x_1)N_1, (x_2)N_2, y)) \xrightarrow{l} \mathsf{Cut}(\langle a\rangle M, (x_i)N_i)$
   if $\mathsf{Or}^i_R(\langle a\rangle M, b)$ and $\mathsf{Or}_L((x_1)N_1, (x_2)N_2, y)$ freshly introduce b and y
3. $\mathsf{Cut}(\langle b\rangle \mathsf{Imp}_R((x)\langle a\rangle M, b), (z)\mathsf{Imp}_L(\langle c\rangle N, (y)P, z))$
   $\xrightarrow{l} \mathsf{Cut}(\langle a\rangle \mathsf{Cut}(\langle c\rangle N, (x)M), (y)P)$ or
   $\xrightarrow{l} \mathsf{Cut}(\langle c\rangle N, (x)\mathsf{Cut}(\langle a\rangle M, (y)P))$
   if $\mathsf{Imp}_R((x)\langle a\rangle M, b)$ and $\mathsf{Imp}_L(\langle c\rangle N, (y)P, z)$ freshly introduce b and z
4. $\mathsf{Cut}(\langle b\rangle \mathsf{Exists}_R(\langle a\rangle M, \mathsf{t}, b), (y)\mathsf{Exists}_L((x)[\mathsf{y}]N, y)) \xrightarrow{l} \mathsf{Cut}(\langle a\rangle M, (x)N(^{\mathsf{y}}/_{\mathsf{t}}))$
   if $\mathsf{Exists}_R(\langle a\rangle M, \mathsf{t}, b)$ and $\mathsf{Exists}_L((x)[\mathsf{y}]N, y)$ freshly introduce a and x
5. $\mathsf{Cut}(\langle b\rangle \mathsf{Forall}_R(\langle a\rangle [\mathsf{y}]M, b), (y)\mathsf{Forall}_L((x)N, \mathsf{t}, y)) \xrightarrow{l} \mathsf{Cut}(\langle a\rangle M(^{\mathsf{y}}/_{\mathsf{t}}), (x)N)$
   if $\mathsf{Forall}_R(\langle a\rangle [\mathsf{y}]M, b)$ and $\mathsf{Forall}_L((x)N, \mathsf{t}, y)$ freshly introduce a and x
6. $\mathsf{Cut}(\langle a\rangle M, (x)\mathsf{Ax}(x, b)) \xrightarrow{l} M[a \mapsto b]$
   if M freshly introduces a
7. $\mathsf{Cut}(\langle a\rangle \mathsf{Ax}(y, a), (x)M) \xrightarrow{l} M[x \mapsto y]$
   if M freshly introduces x

Commuting Cuts

8. $\mathsf{Cut}(\langle a\rangle M, (x)N)$
   $\xrightarrow{c} M(^{a}/_{(x)N})$ if M does not freshly introduce a, or
   $\xrightarrow{c} N(^{x}/_{\langle a\rangle M})$ if N does not freshly introduce x

Fig. 5 Cut-reductions for logical and commuting cuts

4 The Correspondence Question

Before we can answer the correspondence question, we need to formalise the translation from sequent proofs to natural deduction proofs. Figure 6 shows the standard translation, which appeared first in Prawitz [14]. It translates inductively a sequent proof so that right-rules are mapped to introduction rules on the root of natural deduction proofs, and left-rules to elimination rules at the top (except $\vee_L$ and $\exists_L$ which translate like right-rules). It is not hard to show that this translation is onto, but a proof is omitted. First we show that typed terms translate to typed terms.

Proposition 2 *For all* $M \in \mathcal{S}$, *if the sequent* $\Gamma \triangleright M \triangleright a : B$ *is derivable, then* $\Gamma \triangleright |M| : B$ *is derivable and vice versa.*

Proof Routine induction on the structure of M. □

Next we shall show how the symmetric operation of proof-substitution relates to the usual lambda-term substitution. We shall first consider the $(\wedge, \supset, \forall)$-fragment.

$$
\begin{array}{ll}
|\mathsf{Ax}(x,a)| \stackrel{\text{def}}{=} x & |\mathsf{Cut}(\langle a\rangle M,(x)N)| \stackrel{\text{def}}{=} |N|(^{x}/_{|M|}) \\
|\mathsf{And}_R(\langle a\rangle M,\langle b\rangle N,c)| \stackrel{\text{def}}{=} \langle |M|,|N|\rangle & |\mathsf{And}^i_L((x)M,y)| \stackrel{\text{def}}{=} \begin{cases} |M|(^{x}/_{\mathtt{fst}(y)}) \\ |M|(^{x}/_{\mathtt{snd}(y)}) \end{cases} \\
|\mathsf{Or}^i_R(\langle a\rangle M,b)| \stackrel{\text{def}}{=} \begin{cases} \mathtt{inr}(|M|) \\ \mathtt{inl}(|M|) \end{cases} & |\mathsf{Or}_L((x)M,(y)N,z)| \stackrel{\text{def}}{=} \mathtt{case}(z,\lambda x.|M|,\lambda y.|N|) \\
|\mathsf{Imp}_R((x)\langle a\rangle M,c)| \stackrel{\text{def}}{=} \lambda x.|M| & |\mathsf{Imp}_L(\langle a\rangle M,(x)N,y)| \stackrel{\text{def}}{=} |N|(^{x}/_{(y\ |M|)}) \\
|\mathsf{Exists}_R(\langle a\rangle M,\mathsf{t},b)| \stackrel{\text{def}}{=} \mathtt{inx}(\mathsf{t},|M|) & |\mathsf{Exists}_L((x)[\mathsf{y}]M,y)| \stackrel{\text{def}}{=} \mathtt{casex}(y,\lambda x.[\mathsf{y}]|M|) \\
|\mathsf{Forall}_R(\langle a\rangle[\mathsf{y}]M,b)| \stackrel{\text{def}}{=} [\mathsf{y}]|M| & |\mathsf{Forall}_L((x)M,\mathsf{t},y)| \stackrel{\text{def}}{=} |M|(^{x}/_{y\,\mathsf{t}})
\end{array}
$$

Fig. 6 Translation from sequent proofs to natural deduction proofs

Lemma 3 *For all* $M, N \in \mathcal{S}^{(\wedge,\supset,\forall)}$ *we have*

(i) $|M(x/\langle a\rangle N)| \equiv |M|[x/|N|]$
(ii) $|M(a/(x)N)| \equiv |N|[x/|M|]$ *provided* $a \in FC(M)$.

Proof By induction on the structure of M. □

Note that the side-condition in (ii) is always satisfied in intuitionistic logic for proof-substitutions which arise from reducing commuting cuts. This lemma enables us to show that every cut-elimination reduction maps to zero or more normalisation reductions.

Theorem 4 *For all* $M, N \in \mathcal{S}^{(\wedge,\supset,\forall)}$*, if* $M \xrightarrow{cut} N$*, then* $|M| \xrightarrow{\beta}{}^{*} |N|$.

Proof By induction on the definition of $\xrightarrow{cut}$. □

In particular we have that if $M \xrightarrow{c} N$ then $|M| \equiv |N|$. This fact will be useful for proving the following lemma, which relates normalisation steps to a sequence of cut-elimination steps.

Lemma 5 *Given an* $M \in \mathcal{S}^{(\wedge,\supset,\forall)}$ *and* $|M| \xrightarrow{\beta} L$*. Then there exists an* $N \in \mathcal{S}^{(\wedge,\supset,\forall)}$ *such that* $M \xrightarrow{cut}{}^{+} N$ *and* $L \xrightarrow{\beta}{}^{*} |N|$.

Proof By induction on M, but we can restrict the induction to Ms which have only logical cuts. To see this apply as many $\xrightarrow{c}$-reductions as possible to M, obtaining a $\xrightarrow{c}$-normal form, say M'. We have that $M \xrightarrow{c}{}^{*} M'$ and $|M| \equiv |M'|$. Before we analyse one case, some useful terminology will be introduced. Let us write $P\{x_1, \ldots, x_n\}$ for the lambda-term P having n free occurrences of the variable x, and thus $P\{Q_1, \ldots, Q_n\}$ will stand for $P[x/Q]$. Notice that if the outermost term-constructor of Q does not correspond to an introduction rule, then a reduction in $P\{Q_1, \ldots, Q_n\}$ can occur only in the 'P-part' or in one of the 'Q_i-parts'. With this fact in place, let us consider a $\wedge_R/\wedge_{L_1}$-logical cut having the form $\mathsf{Cut}(\langle c\rangle \mathsf{And}_R(\langle a\rangle S, \langle b\rangle T, c), (y)\mathsf{And}^1_L((x)R, y))$. Then $|M| =$

$|R|[x/\mathtt{fst}(\langle|S|,|T|\rangle)]$ which we can write equivalently as $|R|\{\mathtt{fst}\ (\langle|S|,|T|\rangle)_1, \ldots, \mathtt{fst}(\langle|S|,|T|\rangle)_n\}$. There are three cases to be dealt with for a reduction occurring in $|M|$. First, if the reduction occurs in the $|R|$-part, say $|R|\xrightarrow{\beta}|R'|$, then we can conclude by appealing to the induction hypothesis. In this case N is M except R is replaced by R'. Second, if the reduction occurs in one of the subterms of $\langle|S|,|T|\rangle$, say $|S|\xrightarrow{\beta}|S'|$ (the other subcase being similar), then we have the reduction $|M|\xrightarrow{\beta}|R|\{\ldots,\mathtt{fst}(\langle|S'|,|T|\rangle)_i,\ldots\}\xrightarrow{\beta}{}^*|R|\{\mathtt{fst}(\langle|S'|,|T|\rangle)_1, \ldots, \mathtt{fst}(\langle|S'|,|T|\ \rangle)_n\}$. The last term is the term where all other 'copies' of $|S|$ have been reduced, too. Again we can appeal to the induction hypothesis and conclude with N being M except S is replaced by S'. In the third case one of the $\mathtt{fst}(\langle|S|,|T|\rangle)$ reduces to $|S|$. Hence, we have the reduction $|M|\xrightarrow{\beta}|R|\{\ldots, |S|_i,\ldots\}\xrightarrow{\beta}{}^*|R|\{|S|_1,\ldots,|S|_n\}$. In this case we can conclude with $N\equiv \mathsf{Cut}(\langle a\rangle S,(x)R)$ and $M\xrightarrow{c}{}^*M'\xrightarrow{l}N$ as required. The other logical cuts are treated similarly. □

Unfortunately, this is a slightly weaker correspondence as obtained by Zucker [19, p. 79, Theorem 2]. This is because of the way the translation $|\_\_|$ is set up and also because we are not permuting contraction rules downwards in a sequent proof as Zucker does. In effect, it is *not* true that every normalisation step is matched by a series of cut-elimination steps, rather we have to consider 'clusters' of normalisation steps. So in order to obtain a result concerning the correspondence question we prove the following theorem.

Theorem 6 *Given an $M\in\mathcal{S}^{(\wedge,\supset,\forall)}$ and $|M|\xrightarrow{\beta}{}^*L_{\mathtt{nf}}$ where $L_{\mathtt{nf}}$ is a normal form. Then there exists an $N\in\mathcal{S}^{(\wedge,\supset,\forall)}$ such that $M\xrightarrow{cut}{}^*N$ and $|N|=L_{\mathtt{nf}}$.*

Proof We can use the fact that $\xrightarrow{\beta}$ is confluent and strongly normalising. The induction is on the length of the longest reduction sequence starting from $|M|$. By confluence we can repeatedly apply Lemma 5 until we reach the normal form $L_{\mathtt{nf}}$ of $|M|$. In this way, we construct the reduction sequence $M\xrightarrow{cut}{}^*N$. □

In effect we have given a positive answer to the correspondence question in the $(\wedge,\supset,\forall)$-fragment: every cut-reduction sequence maps to a beta-reduction sequence, and every beta-reduction sequence leading to a normal form can be mapped to a cut-reduction sequence. Let us now reflect on this answer and see whether the result Theorem 6 can be improved and whether the answer can be extended to $\vee$ and $\exists$.

It has been suggested for example by Barendregt and Ghilezan [3], Herbelin [9] and Espírito Santo [5] that lambda-calculi extended with explicit substitution operators provide a better correspondence result, meaning that every beta-reduction can be matched by a series of cut-elimination steps. As can be seen in Lemma 5 a cut, say $\mathsf{Cut}(\langle a\rangle M,(x)N)$, is translated as $|N|\{|M|_1,\ldots,|M|_n\}$. Now if any of the $|M|_i$ does a reduction, then this cannot be 'matched' in the cut. Instead if we translate the cut using an explicit substitution operator, for example $|N|\langle x:=|M|\rangle$, we can avoid this problem. However we are not aware any work that points out that with explicit substitution calculi Theorem 4 fails—already in the ($\supset$)-fragment! We shall demonstrate this phenomenon within the explicit substitution calculus λx; see for example Bloo [4].

Example 7 The terms of λx are given by the grammar

$$x \mid \lambda x.M \mid M\,N \mid M\langle x := N\rangle\;.$$

In contrast to the lambda calculus, in λx beta-reductions are split into more atomic steps: one step is the reduction $(\lambda x.M)N \overset{\beta'}{\longrightarrow} M\langle x := N\rangle$ introducing the explicit substitution operator, and the others involve reductions that permute this operator inside the term M. As a consequence, the translation from sequent proofs to natural deduction proofs can be modified as follows.

$$\begin{aligned} |\mathsf{Cut}(\langle a\rangle M, (x)N)|' &\stackrel{\text{def}}{=} |N|'\langle x := |M|'\rangle \\ |\mathsf{Imp}_R((x)\langle a\rangle M, c)|' &\stackrel{\text{def}}{=} \lambda x.|M|' \\ |\mathsf{Imp}_L(\langle a\rangle M, (x)N, y)|' &\stackrel{\text{def}}{=} |N|'\langle x := (y\ |M|')\rangle \end{aligned}$$

To show that the correspondence with cut-elimination fails, consider the logical cut-reduction for $\supset_R/\supset_L$. In Fig. 3 we defined this cut-reduction as

$$\begin{aligned} &\mathsf{Cut}(\langle b\rangle \mathsf{Imp}_R((x)\langle a\rangle M, b), (z)\mathsf{Imp}_L(\langle c\rangle N, (y)P, z)) \\ &\quad\overset{l}{\longrightarrow} \mathsf{Cut}(\langle a\rangle \mathsf{Cut}(\langle c\rangle N, (x)M), (y)P) \;\;\textbf{or} \\ &\quad\overset{l}{\longrightarrow} \mathsf{Cut}(\langle c\rangle N, (x)\mathsf{Cut}(\langle a\rangle M, (y)P)) \end{aligned} \tag{1}$$

including a choice of how the cuts are arranged in the reduct (there is no reason to prefer one choice over the other). Using the translation $|\_\_|'$ given above only the first reduction is matched by reductions in λx; the other is *not*. The translations of the reducts in (1) are:

$$\begin{aligned} |\mathsf{Cut}(\langle a\rangle \mathsf{Cut}(\langle c\rangle N, (x)M), (y)P)|' &= |P|'\langle y := |M|'\langle x := |N|'\rangle\rangle \\ |\mathsf{Cut}(\langle c\rangle N, (x)\mathsf{Cut}(\langle a\rangle M, (y)P))|' &= |P|'\langle y := |M|'\rangle\langle x := |N|'\rangle \end{aligned}$$

With the (standard) reduction rules of λx, only the first reduct can be reached.

$$\begin{array}{ll} & |\mathsf{Cut}(\langle b\rangle \mathsf{Imp}_R(\langle x\rangle\langle a\rangle M, b), \langle z\rangle \mathsf{Imp}_L(\langle c\rangle N, \langle y\rangle P, z))|' \\ = & \quad |P|'\langle y := z\,|N|'\rangle\langle z := \lambda x.|M|'\rangle \\ \longrightarrow & \quad |P|'\langle y := z\,|N|'\langle z := \lambda x.|M|'\rangle\rangle \\ \longrightarrow^* & \quad |P|'\langle y := (\lambda x.|M|')\,|N|'\rangle \\ \longrightarrow & \quad |P|'\langle y := |M|'\langle x := |N|'\rangle\rangle \end{array}$$

The theorem fails, if we try to reach the second reduct. Here one would need a reduction of the form $(x \notin FV(P))$

$$P\langle y := M\langle x := N\rangle\rangle \longrightarrow P\langle y := M\rangle\langle x := N\rangle$$

which, as far as we know, has not been considered for any explicit substitution calculus and indeed would be rather strange. □

What is shown by this example is that the obvious candidate for a lambda calculus with an explicit substitution operator and a natural translation from sequent proofs to this lambda calculus gives a closer correspondence between cut-elimination and normalisation, but not all cut-reductions are matched.

We shall encounter a similar phenomenon when trying to extend Theorems 4 and 6 to all connectives. In the $(\wedge, \supset, \forall)$-fragment all $\stackrel{c}{\longrightarrow}$-reductions are mapped by $|\_\_|$ onto an identity (see Lemma 2.3). This does not hold for the connectives $\vee$ and $\exists$. In consequence, the correspondence fails. The issues are again best explained with an example.

Example 8 Consider the following sequent proof (where for brevity we have omitted all terms and labels).

$$\dfrac{\dfrac{\dfrac{\dfrac{}{A\vdash A}\quad\dfrac{}{B\vdash B}}{A, B\vdash A\wedge B}\wedge_R \quad \dfrac{\dfrac{}{A\vdash A}\quad\dfrac{}{B\vdash B}}{A, B\vdash A\wedge B}\wedge_R}{A\vee B, B, A\vdash A\wedge B}\vee_L \qquad \dfrac{\dfrac{\dfrac{}{A\vdash A}\quad\dfrac{}{A\vdash A}}{A\vdash A\wedge A}\wedge_R}{A\wedge B\vdash A\wedge A}\wedge_{L_1}}{A\vee B, B, A\vdash A\wedge A}\,\mathrm{Cut} \tag{2}$$

To eliminate the cut, we need to apply Rule 8 from Fig. 5 giving

$$\dfrac{\dfrac{\dfrac{\dfrac{}{A\vdash A}\quad\dfrac{}{B\vdash B}}{A, B\vdash A\wedge B}\wedge_R \quad \dfrac{\dfrac{\dfrac{}{A\vdash A}\quad\dfrac{}{A\vdash A}}{A\vdash A\wedge A}\wedge_R}{A\wedge B\vdash A\wedge A}\wedge_{L_1}}{A, B\vdash A\wedge A}\,\mathrm{Cut} \qquad \dfrac{\dfrac{\dfrac{}{A\vdash A}\quad\dfrac{}{B\vdash B}}{A, B\vdash A\wedge B}\wedge_R \quad \dfrac{\dfrac{\dfrac{}{A\vdash A}\quad\dfrac{}{A\vdash A}}{A\vdash A\wedge A}\wedge_R}{A\wedge B\vdash A\wedge A}\wedge_{L_1}}{A, B\vdash A\wedge A}\,\mathrm{Cut}}{A\vee B, B, A\vdash A\wedge B}\vee_L \tag{3}$$

where two copies of the cut have been created. The corresponding natural deduction proofs are as follows: forming $|(2)|$ we have the natural deduction proof

$$
\dfrac{
\dfrac{
\dfrac{}{A\vee B\vdash A\vee B}\quad
\dfrac{\overline{A\vdash A}\quad\overline{B\vdash B}}{A,B\vdash A\wedge B}\wedge_I\quad
\dfrac{\overline{A\vdash A}\quad\overline{B\vdash B}}{A,B\vdash A\wedge B}\wedge_I
}{
\dfrac{A\vee B,A,B\vdash A\wedge B}{A\vee B,A,B\vdash A}\wedge_{E_1}
}\vee_E
\qquad
\dfrac{
\dfrac{}{A\vee B\vdash A\vee B}\quad
\dfrac{\overline{A\vdash A}\quad\overline{B\vdash B}}{A,B\vdash A\wedge B}\wedge_I\quad
\dfrac{\overline{A\vdash A}\quad\overline{B\vdash B}}{A,B\vdash A\wedge B}\wedge_I
}{
\dfrac{A\vee B,A,B\vdash A\wedge B}{A\vee B,A,B\vdash A}\wedge_{E_1}
}\vee_E
}{A\vee B,A,B\vdash A\wedge A}\wedge_I
$$

in which the $\wedge_{E_1}$-rules can be permuted with the $\vee_E$-rules. This gives

$$
\dfrac{
\dfrac{
\dfrac{}{A\vee B\vdash A\vee B}\quad
\dfrac{\dfrac{\overline{A\vdash A}\quad\overline{B\vdash B}}{A,B\vdash A\wedge B}\wedge_I}{A,B\vdash A}\wedge_{E_1}\quad
\dfrac{\dfrac{\overline{A\vdash A}\quad\overline{B\vdash B}}{A,B\vdash A\wedge B}\wedge_I}{A,B\vdash A}\wedge_{E_1}
}{A\vee B,A,B\vdash A}\vee_E
\qquad
\dfrac{
\dfrac{}{A\vee B\vdash A\vee B}\quad
\dfrac{\dfrac{\overline{A\vdash A}\quad\overline{B\vdash B}}{A,B\vdash A\wedge B}\wedge_I}{A,B\vdash A}\wedge_{E_1}\quad
\dfrac{\dfrac{\overline{A\vdash A}\quad\overline{B\vdash B}}{A,B\vdash A\wedge B}\wedge_I}{A,B\vdash A}\wedge_{E_1}
}{A\vee B,A,B\vdash A}\vee_E
}{A\vee B,A,B\vdash A\wedge A}\wedge_I
$$

Here we are stuck! There is no reduction that would reduce to |(3)|—the natural deduction proof below.

$$
\dfrac{
\dfrac{}{A\vee B\vdash A\vee B}\quad
\dfrac{
\dfrac{\dfrac{\overline{A\vdash A}\quad\overline{B\vdash B}}{A,B\vdash A\wedge B}\wedge_I}{A,B\vdash A}\wedge_{E_1}\quad
\dfrac{\dfrac{\overline{A\vdash A}\quad\overline{B\vdash B}}{A,B\vdash A\wedge B}\wedge_I}{A,B\vdash A}\wedge_{E_1}
}{A,B\vdash A\wedge A}\wedge_I\quad
\dfrac{
\dfrac{\dfrac{\overline{A\vdash A}\quad\overline{B\vdash B}}{A,B\vdash A\wedge B}\wedge_I}{A,B\vdash A}\wedge_{E_1}\quad
\dfrac{\dfrac{\overline{A\vdash A}\quad\overline{B\vdash B}}{A,B\vdash A\wedge B}\wedge_I}{A,B\vdash A}\wedge_{E_1}
}{A,B\vdash A\wedge A}\wedge_I
}{A\vee B,B,A\vdash A\wedge A}\vee_E
$$

What is needed is a reduction rule which permutes $\wedge_I$ with $\vee_E$. On the level of lambda terms the corresponding reduction rule is as follows.

$$
\begin{aligned}
&\langle \mathtt{case}(P,\lambda x.M,\lambda y.N),\mathtt{case}(P,\lambda z.S,\lambda w.T)\rangle\\
&\qquad\qquad\longrightarrow\ \mathtt{case}(P,\lambda x.\langle M,S[z/x]\rangle,\lambda y.\langle N,T[w/y]\rangle)
\end{aligned}
$$

□

Again what is shown by this example is that a cut-reduction, which cannot be dispensed with in our cut-elimination procedure, does not match with any of the standard reductions for natural deduction. It requires a reduction which from the term-rewriting point of view is very strange: first it breaks confluence of the lambda-calculus and second it requires a 'test' that checks that two subterms are identical before applying this rule. Nevertheless, from a semantical point of view the rule does make sense: it appears as an equivalence in the work by Altenkirch et al. [1].

5 Conclusion

In this chapter we studied the correspondence question between cut-elimination and normalisation. This was studied previously by Zucker [19]. He showed that in the $(\wedge,\supset,\forall)$-fragment of intuitionistic logic reductions sequences in the sequent calculus can be mapped to reduction sequences in natural deduction and vice versa.

Because his cut-elimination procedure is not strongly normalising when $\vee$ and $\exists$ are included, he concluded that the correspondence must fail for these connectives. We obtained a similar result, but using the cut-elimination procedure from Urban [17], which *is* strongly normalising for *all* connectives. We gave a detailed example for why certain cut-reductions do not match with the (standard) reductions of natural deduction. Because we annotated proofs with terms our technical details are much simpler than in Zucker's work. We discovered that there is a rather pleasing interplay between the semantical work by Altenkirch et al. [1] and the correspondence questions. However, details still remain to be investigated.

References

1. Altenkirch, T., Dybjer, P., Hofmann, M., & Scott, P. (2001). Normalization by Evaluation for typed lambda calculus with coproducts. *Proceedings of Logic in Computer Science* (pp. 203–210).
2. Barbanera, F. & Berardi, S. (1994). A symmetric lambda calculus for "classical" program extraction. *Theoretical Aspects of Computer Software, volume 789 of LNCS* (pp. 495–515). Springer.
3. Barendregt, H., & Ghilezan, S. (2000). Theoretical pearls: Lambda terms for natural deduction, sequent calculus and cut elimination. *Journal of Functional Programming*, *10*(1), 121–134.
4. Bloo, R. (1997). Preservation of termination for explicit substitution. *Ph.D. thesis*. Eindhoven University of Technology.
5. Espírito Santo, J. C. (2000). Revisiting the correspondence between cut elimination and normalisation. *Proceedings of ICALP 2000, volume 1853 of LNCS* (pp. 600–611). Springer.
6. Gallier, J. (1993). Constructive logics. Part I: A tutorial on proof systems and typed λ-calculi. *Theoretical Computer Science*, *110*(2), 239–249.
7. Gentzen, G. (1935). Untersuchungen über das logische Schließen I and II. *Mathematische Zeitschrift*, *39*(176–210), 405–431.
8. Girard, J.-Y., Lafont, Y., & Taylor, P. (1989). *Proofs and types, volume 7 of cambridge tracts in theoretical computer science*. Cambridge: Cambridge University Press.
9. Herbelin, H. (1994). A λ-calculus structure isomorphic to sequent calculus structure. *Computer Science Logic, volume 933 of LNCS* (pp. 67–75). Springer.
10. Kreisel, G. (1971). A survey of proof theory II. *Proceedings of the 2nd Scandinavian Logic Symposium, volume 63 of Studies in Logic and the Foundations of Mathematics* pp. 109–170. North-Holland.
11. Laird, J. (2001). A deconstruction of non-deterministic cut elimination. *Proceedings of the 5th International Conference on Typed Lambda Calculi and Applications, Volume 2044 of LNCS* pp. 268–282. Springer.
12. Negri, S., & von Plato, J. (2001). *Structural proof theory*. Cambridge: Cambridge University Press.
13. Pottinger, G. (1977). Normalisation as homomorphic image of cut-elimination. *Annals of Mathematical Logic*, *12*, 323–357.
14. Prawitz, D. (1965). *Natural deduction: A proof-theoretical study*. Stockholm: Almquist and Wiksell.
15. Troelstra, A. S., & Schwichtenberg, H. (2000). *Basic proof theory. Cambridge tracts in theoretical computer science* (2nd ed.). Cambridge: Cambridge University Press.
16. Ungar, A. M. (1992). *Normalisation, cut-elimination and the theory of proofs, volume 28 of CLSI Lecture Notes*. Stanford: Center for the Study of Language and Information.
17. Urban, C. (2000). Classical logic and computation. *Ph.D. Thesis*. Cambridge University.

18. Urban, C., & Bierman, G. M. (2001). Strong normalisation of cut-elimination in classical logic. *Fundamenta Informaticae*, *45*(1–2), 123–155.
19. Zucker, J. (1974). The correspondence between cut-elimination and normalisation. *Annals of Mathematical Logic*, *7*, 1–112.

Proofs, Reasoning and the Metamorphosis of Logic

Jean-Baptiste Joinet

Abstract With the "mathematical watershed", Logic had been transformed into a foundational theory for mathematics, a theory of truth and proofs—far away from its philosophical status of theory of the intellectual process of reasoning. With the recent substitution of the traditional *proofs-as-discourses* paradigm by the *proofs-as-programs* one, Logic is now becomming a foundational theory for computing. One could interpret this new watershed as being "yet another technological drift", bringing Logic always closer to practical ingeneering, always further from the human intellectual process of reasoning. This article promotes the dual point of view: enlightened by the contemporary analysis of the dynamic of proofs, which bring us to a new understanding of the semantic counterpart of processes operationality (including the links between semantic dereliction due to inconsistency and computational exuberance), Logic has never appeared so close to being, finally, the theory of reasoning.

Keywords Proof · Computation · Reasoning · Philosophy of logic

1 Introduction

When it comes to giving a short definition of Logic, one frequently hears (sometimes even said by oneself) that Logic is "the theory of reasoning". Yet, all contemporary logicians know very well that Logic, as it is practiced today, is not properly about reasoning, but about proofs. Reasoning indeed is a human intellectual process,

---

Lecture given in 2001 under title *From Natural Deduction to the nature of reasoning* (colloquium *Natural Deduction 2001*, Rio de Janeiro).

J-B. Joinet (✉)
Faculté de philosophie, Université Jean Moulin - Lyon 3, Lyon, France
e-mail: Jean-Baptiste.Joinet@univ-lyon3.fr

L. C. Pereira et al. (eds.), *Advances in Natural Deduction*,
Trends in Logic 39, DOI: 10.1007/978-94-007-7548-0_3,

whereas proofs—as logicians, from Aristotle to Gentzen, have investigated them—are discourses: oral or written, maybe only formalized, but discourses nonetheless.

This shortcut made by logicians when presenting their discipline (or perhaps more accurately said: "the fantasy" of logicians when they are representing to themselves the subject of their work) actually accompanied Logic, under various declensions, all along its history. Moreover, as starts in the nineteenth century the process of gradual absorption of Logic by Mathematics, the view of Logic as being the "Theory of thought" itself continues to prevail[1] and, even after the birth of proper *Proof theory*—actually by the pen of its promotor himself, David Hilbert—it is reiterated: "The fundamental idea of my proof theory is none other than to describe the activity of our understanding, to make a protocol of the rules according to which our thinking actually proceeds" ([1], p. 475).

As those same logicians who present Logic choosing such words, in the meantime explicitly deny any psychological *nature* to logical principles, one sees that, underlying such formulations, stands the somewhat confused idea that some kind of isomorphism or interface would hold between the rules "according to which our thinking proceeds" (or "has to proceed" as well[2]) and the rules linking sentences in our deductive discourses, so would hold between *reasonings* and *proofs*.

To clarify how reasoning and proofs radically differ, it is useful to stress the three following specific points, more or less traditionnally emphasized when comparing the two notions. First, as it is a process, reasoning is of a dynamic nature, whereas proofs, as discourses and especially as texts, are static objects. Second, the intellectual, rational process of reasoning would presuppose the grasping of meaning, the mental representation of objects and structures, which is frequently summed up by speaking of the "semantic" or "contentual" aspects of reasoning, whereas proofs articulate sentences inferentially, according only to their shape, without regard for reference. Third, whereas a wrong proof is not a proof, incorrect reasoning still is reasoning.

Dynamic character of reasoning	versus	Static character of proofs
Referential dimension of reasoning	versus	Inferential nature of proofs
Indeterminacy of reasoning	versus	Correctness of proofs

The reasons why contemporary Logic happens to deal with proofs instead of reasoning are numerous and varied. They are on the whole tied up with the fact that the silent human process of reasoning remains mainly observable through private mental self-introspection (or perhaps through some difficult to exploit but measurable side effects, e.g. those produced by neural activity etc.), so that consciousness intervenes in its study, moreover, both as an object and as a tool for observation.

[1] G. Boole's *Investigation of the Laws of Thought* is here emblematic [2].

[2] The fact that even the early *intuitionistic* tradition conceived explicitly Logic as a *non-psychological* organon, as pure normativity (see for instance Kant, *Logik*, Introduction, first section), does not obliterate its conception of Logic as the science of the *laws of thought: that Logic studies "the rules according to which our thinking must proceed", rather than "how it proceeds" does not make any difference here.*

Even if, disregarding the usual problem of subjectivity's epistemological weakness, one acknowledged as relevant a question such as: "what information could introspection reveal to us concerning reasoning?", the following specific difficulties—which match up with the three points stressed above—would have to be taken into account.

A first obstacle which prevents a direct approach to the intellectual process of reasoning is its introspective opaqueness. As David Hume so nicely says, "it is remarkable concerning the operations of the mind that, though most intimately present to us, yet, whenever they become the object of reflexion, they seem involved in obscurity; nor can the eye readily find those lines and boundaries which discriminate and distinguish them" ([3], Chap. 1). The complex temporality and profound structure of reasoning, with its likely distributed interaction mechanisms, generally escapes our consciousness. Because we can perceive at best only some of the epiphenomenal, conscious effects of this internal process, we are unable to qualify directly its deep technological nature and have a direct apprehension of its dynamic. In strong contrast with this situation, rational discourses, and especially written ones, are by definition communicable and completely observable objects, and so are describable phenomena without any hidden parts, whose stability *a contrario* permits the manifestation and study of their structure.

Another obstacle is just simply that we are not clearly able to even recognize reasoning, unable "from the inside" to separate from the general stream of our thought some sub-process which would constitute its "rational part" and be independent of emotions and, especially when deliberation about practical choices is involved, of desires, fears and social determinations. In any event, when considering human thought in vivo, the traditional splitting of rational and non-rational aspects seems rather arbitrary (this is also why the specific dynamic of wrong or mistaken reasoning is so unclear). In strong contrast with this situation, proofs as opposed to reasoning, or at least formalized proofs, are recognizable (for a text to be a proof is a decidable property). In particular, once a list of "logical rules" has been set down, the borderline between proofs and non-proofs becomes unambiguously determined.

Finally, a relevant theory of reasoning should cover the referential dimension of the rational process, and hence would presuppose a theory of meaning as mental representation. This usually leads to unclear metaphorical formulations such as intimate "grasp" of semantic "content", private, direct mental contact with concepts, and to all manners of ambiguities that torment the philosophy of mind and consciousness. In strong contrast with this situation again, the logical analysis of discursivity opens onto a theory of objective, consciousness-independent meaning.

Among these issues, the last one constitutes the point of departure for analytic philosophy (in the true sense, i.e. restricted to epistemological topics), whose dismissal of psychologism in favour of semantic analysis parallels the methodological dismissal of reasoning as an object of study which de facto prevails in Logic. In a conference given in 1987, in which he extricated the roots of the analytic watershed in philosophy, Michael Dummett deepened the question of this methodological precedence of language over thought [4]. I am retaining from him, notably, the thesis that because there is something objective in meaning (as communicability shows) which is completely involved and kept within discursivity (as communication shows), the only objective knowledge about thought we can hope to grasp, if ever any, shall be proportionate to what is objective in discursivity: its structure, the norms which regulate our practice and use, or whatever …

In the present article, I will remain faithful to this analytic inspiration, which as I understand it, does not disqualify inquiries into reasoning, but merely postpones them as questions which could be posed only afterwards, as "secondary" questions. Instead of entering into the habitual debate over "truth conditions" which often comes down to searching outside of discursivity for an objective foundation for semantics (facts described by sets, worlds of concepts, human rational practices whose only criterion is to be the ones humans happen to learn, etc.), I would like to focus my investigations upon objective features of discursivity and proofs that have been revealed by recent proof-theoretical developments, beginning with Gentzen's seminal work and continuing through later work based on the so-called *Curry-Howard correspondance* between proofs and programs (i.e. the remark that the evaluation of simply typed lambda-calculus and the normalization of minimal Natural Deduction are isomorphic), have revealed.

More specifically, my aim here is to evaluate to what extent these developments give new insights into the relationship between "reasoning" (the human rational mental process) and "proofs" (structured inferential texts). I will tackle this question by successively taking up the three particular points recalled above which differentiate reasoning and proofs, examining in each case how to articulate them in light of recent proof-theoretical investigations (in particular those from the so-called *proofs-as-programs paradigm*).

2 Dynamic Character of Reasoning Versus Static Character of Proofs

Whereas reasoning is a process, and thus inseparable from its dynamic, proofs, as usually taken into account by formal Logic, are traditionally apprehended as texts, and at first glance lack any true dynamic dimension. Indeed, one occasionally finds the absence of dynamic used against the formalist perspective in Logic, which is then typically blamed for being only occupied with cold, lifeless, static things, thus incomparable to human reasoning.[3] Is it so true, however, that contemporary Logic when working on proofs, does not deal with the dynamic?

First, there is still the old idea (dating back to at least the Aristotelian view on syllogisms) that some dynamic is involved when one is accomplishing a deductive step or reading a structured inferential discourse, insofar as one is indeed moving from sentence to sentence following the orientation of the deduction. Nevertheless, even those who consider inferential steps as imitating to a certain extent steps from idea to idea (echoing the famous Cartesian description of reasoning in the *Regulae*, in which the mind travels through an ordered chain of ideas, from distinct representation

[3] See Descartes, Hegel, Brouwer: actually, it is the anti-formalist topos *par excellence*.

to distinct representation), do not assimilate the dynamic involved to the general, complex process by which an individual deliberates, deduces and modifies his or her abstract representations and concrete judgments.

There is, however, a second axis through which contemporary Logic investigates directly what is undoubtedly a part of the dynamic process of reasoning, namely *heuristics*. In a general sense, heuristics covers the search for proofs of a given sentence from a given theory, hence includes the process of determinating which relevant sentences should be proven beforehand in the search for success. Modern formal Logic developed numerous techniques for tackling the heuristics challenge, importing dynamic features into proof theory. To what extent can we conclude that a dimension of human reasoning is here modeled? Of course, the way the heuristic dynamic is realized at the technical level (often using specific *ad hoc* formal systems suited to heuristic strategies, instead of systems such as Natural Deduction whose devising from the start was explicitly guided by the aim to imitate, in some respects, the way mathematicians "naturally" happen to write their proofs) renders difficult any comparison of the human heuristic with an artificial one. Moreover, one could object that contrary to the natural human heuristic dynamic which is part of the rational process, the artificial one stays outside of the world of proofs, since with respect to the search process, the produced proofs simply appear as outputs a priori not involving any tracks of the process by which they have been produced, so as by-products external to the dynamic.

Another kind of dynamic, though, also occupies the logical stage, namely *proofs' normalisation*, that process (first studied by G. Gentzen) by which proofs are converted into analytic ones (but more generally other processes of proofs' conversion such as, for instance, elimination, in relevant cases, of non-constructive components of proofs), and which implements a dynamic completely internal to the proofs' world. Besides the well-known epistemological value of the convertibility of proofs (which legitimates abstract and—when applicable—non-constructive detours as safe proof-theoretic means), the rational flavour of the conversion process itself deserves to be pointed out.

First of all, one has to observe that proofs, as subject to conversion processes, acquire a radically new status. In this light, the basic components of proofs, logical rules, become, literally, programming instructions: they determine the precise way evaluation works and controls the operational destiny of the conversion process. That proofs have been historically first modeled, from a static point of view, as texts, turns out to be but an epiphenomenal feature, beyond which appears their true nature as operators acting on rational resources. (Incidentally, one may remark that in fact, even as static objects—though now enlightened by the dynamic perspective and adapted to it—proofs do no longer really fit within the *proofs-as-discourses* view. Indeed, in recent proof systems like J-Y Girard's proof-nets [5, 6], their correctness depends on global geometrical properties of the proof-structures, and so is not a locally checkable condition. Such an account of proofs is very far from what logicians have been accustomed to, namely textual structures linking rules of inference picked out from a catalogue determined in advance, and of which Hilbert's style systems or sequents' derivation systems are paradigmatic instances).

Moreover, like ideas, proofs also happen to live very well together. Their "sociability" is effective not only at the static level through their modularity (proofs can be associated, combined, reused), but also dynamically, as normalization precisely shows by ensuring, in case their combination increases the level of abtraction, that analytical avatars of the abstract proof still remain "kept in sight".

Thus far from being absent from proof-theory, the dynamic today has become the central object of mainstream logical investigations, and a conceptual unification of both dynamics—that of heuristics and that of proofs' conversion—is even at hand.[4] For all that, is the newfound centrality of the dynamic enough to conclude that bridges between theory of proof and theory of reasoning are now in sight? A crucial indication could come from assessing to what extent the intrusion of the dynamic sheds light on the semantic aspects that reasoning reputedly involves.

3 Referential Aspects of Reasoning Versus Inferential Nature of Proofs

The uses of sentences fall into two apparently discrete modes. In the *inferential mode*, which is completely internal to discourse, sentences are used and linked together according only to their form, no matter the interpretations they otherwise could receive. In the *referential mode*, discourses are "discourses about"—they refer to—and sentences then are used according to interpretations of their components by referents external to discourse. The use of sentences in proofs[5] provides the paradigmatic instance of the inferential mode. At the opposite extreme, because reasoning involves the mental handling of representations (of individuals, structures, concepts, or whatever), it usually is taken to be an instance of the referential mode. This traditionally accepted dichotomy, whose unifying principle is given by 'completeness theorems', actually has been deeply disturbed by (relatively) recent proof-theoretical advances (which happen to confirm in a striking way the relevance of Michael Dummett's remarks about the replacement of the notion of truth by the notion of proof as the central notion of the theory of meaning [7]).

As for the prevalence of the inferential mode in proofs, one first has to notice that proofs, at least analytic (cut-free) ones, being, so to speak, extensional descriptions of objects or structures, clearly involve 'internal' denotation in the sense that they are but explicit constructions. One may first illustrate this by considering the particular shape of the analytic proofs of the type of natural integers (or other data types) in second-order intuitionistic Natural Deduction, namely "Church's integers". In this first example, the reference is so to speak inferentially built: such a proof is but the standard construction of a given integer (in the Leibniz-Peano sense). Another example of such an inferential explicitation of reference is given by the fact that whenever

---

[4] See J-Y Girard's *Ludics* [8].

[5] i.e. the various ways sentences are concretely handled in proofs (for instance, in *Natural Deduction* systems, by such or such "introduction" and "elimination" rules).

they are in a relevant form, analytic proofs literally describe models: think of the model (the counter-model) that one literally 'reads' in the analytic (pseudo-)proof of an (unprovable) sentence generated by the (here failed) proof-search strategy used to prove the completeness theorem for first-order Logic.[6]

But, of course, the main point is elsewhere. Beyond this first kind of 'inferential reference', subtler accounts of proofs' denotation have been elaborated by contemporary proof theory, gradually realizing, at various levels and with always more relevance, Heyting's beautiful original idea of considering proofs as functions and not simply as texts.

All these semantics are *semantics of proofs* (as opposed to *semantics of sentences*), which means that they approximate the nature of proofs as operators, catching their "being" as a "doing"[7] and revealing the nature of argumentative dealings as computations over rational resources (one recovers a universal operational meaning for sentences themselves, in case one is able to identify an operational behavior common to all proofs of a given sentence[8]).

At one pole, there is *denotational semantics*, where proofs are directly interpreted by functions belonging to suited set-theoretically defined functions spaces (in such a way that the interpretation of proofs is invariable during their normalization). However, semantics of this kind keep only a shadow of proofs' dynamic: interpretation remains too extensional and equalizes computations that should not be considered as equal and so, semantics fail to spot many intentional dynamic features.

At the opposite pole, one finds *operational semantics*, the "behaviourist" approach of the dynamic of proofs. A function is simply identified with the paraphrastic description of the set of all computations steps (of all the sequences of states describing possible evaluations). The defect in this case is converse to the former one: too much intentionality also kills off intentionality.

Between these two extreme poles, there is the more recent *interaction semantics* ("game semantics" and "ludics"), where dynamic (evaluation) is interpreted as a structured interaction, described, in the language of Game-Theory, as plays between processes whose rules structure their inter-communication (and where "conversion dynamic" and "proof-search dynamic" tend to coincide).

On the other hand, as regards the prevalence of the referential mode concerning reasoning and its "semantic contents", etc., one must conversely observe that, while one does not assume something like a "mental discourse" that would be used when we are reasoning, the words "referential mode", which, as they have been introduced above, do only apply to discourses (not to "psychological entities" like ideas), happen to be used here improperly and mistakenly. Yet, to assume the existence of such a mental discourse somehow amounts to reduplicating the speaker "inside" the speaker, thus adopting the old vain prosopopeia of intellectual faculties (whose aporetic nature, being rather commonly acknowledged, do not require here deeper

6 See for instance: [9].

7 "Church's integers" for instance, to continue with our example, dynamically appear as *iterators*.

8 Particular case of Krivine's "specification problem" [10].

presentation[9]). Only one way remains thus open: as soon as reasoning is concerned, we must completely reverse the traditional view about meaning, now looking at it as an issue addressed, no longer to sentences, but to processes themselves.

Of course, at first sight, such a radical shift seems to present more new problems than answers. First of all because as we have seen, the notion of "process" seems very tricky compared to that of "discourse" (at least insofar as reasoning is concerned), but also because the account of meaning one henceforth must give in order to escape the above "reduplication" difficulty could no longer be in terms of something external (as referents did with respect to discourses), but of something completely internal to the world of processes. Thus how could meaning generation be a property of processes? As vague as the notion of process may be, it certainly is in their nature not only to evolve (amongst the effects produced by a process appear its own transformations), but also to produce effects on other processes, mutual effects. In short, processes act and interact. Whatever its technological matter, whatever its implementation, the essence of a process is completely involved in its (potential) dynamic behavior: not only its own possible destinies under evaluation, but also the full set of possible operational effects it will occasion in all possible processes' interaction contexts. With respect to semantics, the answer brought by processes is thus of a radically new kind, which implements the performative way of meaning[10]: doing, is the way processes speak.

Enlightened by the theory of proofs' dynamic, the inferential mode is thus in the end far from being non-referential. Moreover, the kind of semantics onto which it opens, do not incur the two major reproaches that the usual modellings of reference do, namely to award no proper status to abstraction (now caught by computational complexity and non-analyticity of proofs) and to offer an evaluation-independent approach to sense. In that connection, it is not the least of the virtues of the operational viewpoint in Logic that it sheds light not only on the analogy between proofs conversion and rational dynamic, not only on the inseparability of evaluation and sense, but even finally, as we are going to see now, on nonsense itself.

4 Correctness of Proofs Versus Indeterminacy of Reasoning

Reasoning in its general sense includes "wrong reasoning": mistakes, inconsistencies, errances, partiality, revision, interruption of reflection (because of abstruseness, loops to be broken, breaks to be taken, emergencies or whatever). By contrast, when proof theory was born at the beginning of the last century, it was certainly not conceived of as including a theory of "non-proofs". Admittedly, the definition of formal

---

[9] For a synthetic presentation of the sophism of "homoncules", see [11]. About the "mental language" thema, see the beautiful book of Claude Panaccio: "Le discours intérieur" [12].

[10] I am borrowing from [13], himself inspired by Austin terminology, the idea of "performativity", there used to qualify formal systems of proofs (like Krivine's AF_2) where proofs indeed do (as programs) what the sentences they are proving "say" they should. For a more complete presentation and a defense of the concept of "performativity" applied to the semantics of proofs, see [14, 15].

proof systems that was provided by Logic supplied decision procedures separating texts accepted as proofs from other texts (provided the catalogue of rules is set). Although such a definition thus gave a de facto account of what "wrong proofs" are, their status remained that of garbage, not worth studying.

On these issues, the recent transformations in proof theory, now focused not so much on proofs themselves as on their dynamic, have broadened the outlook. The first though not the main point: because once the dynamic is concretely implemented, occasions for new kinds of mistakes may happen (in the real world, "un pas suffit pour un faux-pas"[11]; the evaluation could be interrupted, etc). But overall, because of the status of proofs having changed, that of non-proofs, as a result, also has changed. From the proofs-as-programs viewpoint, a proof is no longer just a normed discourse, but a process whose operationality, tamed by Logic, is "under control". This becomes clear once one tackles proofs in computational terms: then a proof is a typed program, whose typability ensures that certain computational properties are satisfied (typically: termination, complexity bounds,[12] etc). Nevertheless, untypable programs— those processes that do not correspond to proofs and that Logic fails to civilize—share with proofs their life as dynamic entities, their computational nature.

The new status that paradoxes acquire when scrutinized from the *belvedere* of the dynamic illustrates this radical change of viewpoint. The traditional reasons given to reject inconsistency, from Aristotle (dialogue become impossible) to modern Logic (every sentence becomes provable: all of them becoming equivalent, there could be no meaning), may now be unifyingly reformulated: inconsistency permits the typing of infinite—and even infinitely silent—computations.[13] The point with paradoxical theories is thus not so much that they violently equalize sentences, but that they produce computational exuberance, and leads to the degeneration of the evaluation of meaning.

5 Conclusion

The "naturalness" of Natural Deduction, akin to the way mathematicians historically have come to write their proofs, is the fruit of the imitative intention which governed its settlement by Gerhard Gentzen. Any extension of such a "figurative" attempt to reasoning itself stumbles over our inability to describe and even recognize what we are suppose to imitate. Because no criterion for the success of the imitation enterprise is offered when the reasoning is on the line of sight, simulation, imitation, figurativity have properly no methodological meaning. The best one can try then is, on the one hand, to recover at most some of the features of this globally uncatchable phenomenon (e.g. dynamic features), and on the other hand, to reach some more abstract form of naturalness appearing in "aesthetic" properties like simplicity, universality,

[11] Carlo Goldoni (original italian version not recovered).

[12] See [16] and [17].

[13] Adding $X = \neg X$ (X an atomic formula), non-normalizable λ-terms like $\Delta\Delta$ become typable.

non-diffuseness (cf. proof nets' sobriety), harmony (input / output rules [18]) or in "architectonic" properties like modularity, transparent cutting-out into subsystems, representational strength (complexity)...

Of course the need for such an "artificial naturalness", could be interpreted as the sign that, ever further from human thought, the drift of Logic—from its philosophical origins to mathematics, and now from mathematics to theoretical computer science—henceforth is, and with no return, consummated. Logic would just have become a foundational theory of computing, a technical device for designing and studying computation and programming languages.[14] In this connection, the current extension of the Curry–Howard style approach to new programming devices oriented towards communication [10] instead of usual recursive computation of data (so ever closer to technological issues, ever further from what was, century after century, the central concern of Logic: rational thought) seems somehow to confirm this metamorphosis of Logic towards technology.

However, the *Logical Foundations of Computing* could just as well be dually seen "from the other side of the isomorphism" as *Computational Foundations for Logic.*[15] From this complementary viewpoint, where studied objects are dynamic ones, with an operationally, interactively built-in evaluation semantics, and among which even paradoxes receive a relevant dynamic status, Logic has probably never appeared so close to being the theory of reasoning.

Acknowledgments I would like to thank Gilles Dowek (french National Institute for Research in Computer Science and Automatics, I.N.R.I.A.) for the friendly and serious moments we spent, walking and talking in the tropical garden of the PUC university campus in Rio, John Tain for his help concerning English and François Rivenc for his remarks and advice about an earlier version of this article. My participation to the conference *Natural Deduction, Rio 2002* was supported by the French National Center for Scientific Research (Proofs-Programs-Systems team) and the Brazilian Ministry for Research.

References

1. Hilbert, D. (1927). The foundations of mathematics. In J. van Heijenoort (Ed.), *From Frege to Gödel: a sourcebook in mathematical logic* (pp. 1971–1931). Cambridge: Harvard University Press.
2. Boole, G. (1930). *Investigation of the laws of thought.* Mineola: Dover Publications.
3. Hume, D. (1902). An enquiry concerning human understanding. Oxford: Oxford University Press. (First edition published in 1748).
4. Dummett, Michael. (1993). *Origins of analytic philosophy*. Cambridge, London: Duckworth, Harvard University Press.
5. Girard, J.-Y. (1996). Proof-nets: the parallel syntax for proof-theory. In Ursini & Agliano (Eds.), Logic and algebra. New York: Marcel Dekker.

[14] Indeed, it has for instance given birth to programming languages like Prolog, Caml, etc.

[15] At first restricted to propositional minimal Logic, the Curry–Howard style approach has been extended now to almost all parts of Logic, notably to first and higher order classical Logic and ZF set theory [19–23].

6. Girard, J. Y. (1995). Linear logic: its syntax and semantics. In J.-Y. Girard, Y. Lafont, & L. Regnier (Eds.), *Advances in linear logic: mathematical society lecture notes series* (Vol. 222, pp. 1–42). London: Cambridge University Press.
7. Dummett, M. (1978). *The philosophical basis of intuitionistic logic. Truth and other enigmas*. London: Duckworth.
8. Girard, J.-Y. (2001). Locus solum. *Mathematical Structures in Computer Science*, *11*(03), 301–506.
9. Wainer, S. S., & Wallen, L. A. (1992). Basic proof theory. In S. S. Wainer, P. Aczel, & H. Simmons (Eds.), *Proof theory: a selection of papers from the Leeds proof theory programme 1990* (pp. 1–26). New York: Cambridge University Press.
10. Danos, Vincent, & Krivine, Jean-Louis. (2000). Disjunctive tautologies as synchronisation schemes. Proceedings of CSL'2000. LCS, 1862, (pp. 202–301).
11. Pinkas, D. (1995). *La matérialité de l'esprit*. Paris: La découverte.
12. Panaccio, C. (1999). *Le discours intérieur: de Platon à Guillaume d'Ockham*. Paris: Editions du Seuil.
13. Danos, V. (1990). La logique linéaire appliquée à l'étude de divers processus de normalisation (principalement du lambda-calcul), (Université Paris 7, Paris). Ph.D. Thesis.
14. Joinet, J.B. (2007). Sur le temps logique. Logique, dynamique et cognition, (pp. 31–49). Paris: Publications de la Sorbonne. ISBN 978-2-85944-584-3.
15. Joinet, T.-A. (2009). *Ouvrir la logique au monde*. Hermann: Philosophie et mathématique de l'interaction. Paris. ISBN 978-2-85944-584-3.
16. Danos, Vincent, & Joinet, Jean-Baptiste. (2003). Linear logic and elementary time. *Information and Computation*, *183*(1), 123–137.
17. Girard, J. Y. (1998). Light linear logic. *Information and Computation*, *143*(2), 175–204.
18. Prawitz, D. (1977). Meaning and proofs: on the conflict between classical and intuitionistic logic. *Theoria*, *43*(1), 1–40.
19. Danos, V., Joinet, J. B., & Schellinx, H. (1997). A new deconstructive logic: linear logic. *The Journal of Symbolic Logic*, *62*(3), 755–807.
20. Girard, J.-Y. (1991). A new constructive logic: classical logic. *Mathematical Structures in Computer Science*, *1*(3), 255–296.
21. Krivine, J.-L. (1994). Classical logic, storage operators and second order lambda-calculus. *In Annals of Pure and Applied Logic*, *68*, 63–78.
22. Krivine, J.-L. (2001). Typed lambda-calculus in classical Zermelo-Fraenkel set theory. *Archive for Mathematical Logic*, *40*(3), 189–205.
23. Parigot, M. (1992). Lambda-mu calculus: An algorithmic interpretation of classical natural deduction. Proceedings of the International Conference on Logic Programming and Automated Reasoning (pp. 190–201).

Natural Deduction for Equality: The Missing Entity

Ruy J. G. B. de Queiroz and Anjolina G. de Oliveira

Abstract The conception of the very first decision procedures for first-order sentences brought about the need for giving citizenship to function symbols (e.g. Skolem functions). We argue that a closer look at proof procedures for first-order sentences with equality brings about the need for introducing (function) symbols for rewrites. This is appropriately done via the framework of labelled natural deduction which allows to formulate a proof theory for the "logical connective" of propositional equality. The basic idea is that when analysing an equality sentence into (i) proof conditions (*introduction*) and (ii) immediate consequences (*elimination*), it becomes clear that we need to bring in identifiers (i.e. function symbols) for sequences of rewrites, and this is what we claim is the missing entity in P. Martin-Löf's equality types (both intensional and extensional). What we end up with is a formulation of what appears to be a middle ground solution to the 'intensional' versus 'extensional' dichotomy which permeates most of the work on characterizing propositional equality in natural deduction style. (Part of this material was presented at the *Logical Methods in the Humanities Seminar*, Stanford University, and the authors

Ruy J. G. B. de Queiroz—Research partially supported by a Senior Research Fellow (*Bolsa de Produtividade em Pesquisa*, "Pesquisador 1-C") grant n. 304365/2003-3 from CNPq (http://www.cnpq.br). The final version was prepared while in the chair of Edward Larocque Tinker Visiting Professor, Dept Philosophy, Stanford Univ, USA. Thanks are due to Profs. Solomon Feferman and Grigori Mints for their kind nomination to the Tinker Visiting Chair and for their great hospitality. Thanks are also due to Prof. Herbert Klein, Molly Vitorte, Megan Gorman and Omar Ochoa at the Center for Latin American Studies, Stanford Univ, for making the Visiting Professorship possible, and run so smoothly. A postdoctoral grant n. 2857-05-7 from CAPES (http://www.capes.gov.br) is also gratefully acknowledged.

Anjolina G. de Oliveira—Visiting Scholar at the Center for the Study of Language and Information (CSLI), Stanford, while working on the final version. Thanks are due to Keith Devlin, Executive Director of CSLI, for the great hospitality. Research partially funded by a grant from CAPES, *Estágio Pós-Doutoral* Proc. 2746/05-0..

R. J. G. B. de Queiroz (✉) · A. G. de Oliveira
Centro de Informática, Universidade Federal de Pernambuco, Recife, Brazil
e-mail: ruy@cin.ufpe.br

L. C. Pereira et al. (eds.), *Advances in Natural Deduction*,
Trends in Logic 39, DOI: 10.1007/978-94-007-7548-0_4,

would like to thank Solomon Feferman and Grigori Mints for their comments and suggestions.)

Keywords Natural deduction · Equality · Labelled deduction · Equality type

1 Introduction

The clarification of the notion of normal form for equality reasoning took an important step with the work of Statman in the late 1970s [44, 45]. The concept of *direct computation* was instrumental in the development of Statman's approach. By way of motivation, let us take a simple example from the λ-calculus.

$$\begin{array}{lllllll} (\lambda x.(\lambda y.yx)(\lambda w.zw))v & \rhd_\eta & (\lambda x.(\lambda y.yx)z)v & \rhd_\beta & (\lambda y.yv)z & \rhd_\beta & zv \\ (\lambda x.(\lambda y.yx)(\lambda w.zw))v & \rhd_\beta & (\lambda x.(\lambda w.zw)x)v & \rhd_\eta & (\lambda x.zx)v & \rhd_\beta & zv \end{array}$$

There is at least a sequence of conversions from the initial term to the final term. (In this case we have given two!) Thus, in the formal theory of λ-calculus, the term $(\lambda x.(\lambda y.yx)(\lambda w.zw))v$ is declared to be **equal** to zv.

Now, some natural questions arise. Are the sequences themselves *normal*? Are there non-normal sequences? If yes, how are the latter to be identified and (possibly) normalized?

When it comes to equality, similar questions arise as to when a sequence of application of equality principles should be taken to be 'non-redundant'. Clearly, successive applications of *symmetry* should be considered unnecessary. But there are other combinations of applications of equality principles which will ultimately prove to be useless as far as the actual proof of equality is concerned.

As rightly pointed out by Le Chenadec in [5], the notion of normal proof has been somewhat neglected by the systems of equational logic: "In proof-theory, since the original work of on sequent calculus, much work has been devoted to the normalization process of various logics. Such an analysis was lacking in equational logic (the only exceptions we are aware of are Statman [44], Kreisel and Tait [19])". The works of Statman [44, 45] and Le Chenadec [5] represent significant attempts to fill this gap. Statman studies proof transformations for the equational calculus E of Kreisel–Tait [19]. Le Chenadec defines an equational proof system (the *LE system*) and gives a normalization procedure.

The intention here is to show how the framework of labelled natural deduction can help us formulate a proof theory for the "logical connective" of propositional equality.[1] The connective is meant to be used in reasoning about equality between referents (i.e. the objects of the functional calculus), as well as with a general notion

[1] An old question is in order here: what is a logical connective? We shall take it that from the point of view of proof theory (natural deduction style) a logical connective is whatever logical symbol which is analysable into rules of *introduction* and *elimination*.

of substitution which is needed for the characterization of the so-called *term declaration logics* [2]. The characterization of propositional equality may be useful for the establishment of a proof theory for 'descriptions'.

In order to account for the distinction between the equalities that are:

definitional, i.e. those equalities that are given as rewrite rules (equations), or else originate from general functional principles (e.g. β, η, etc.),

and those that are:

propositional, i.e. the equalities that are supported (or otherwise) by an evidence (a composition of rewrites),

we need to provide for an equality sign as a symbol for *rewrite* (i.e. as part of the functional calculus on the labels), and an equality sign as a symbol for a *relation* between referents (i.e. as part of the logical calculus on the formulas).

Definitional equalities. Let us recall from the theory of λ-calculus, that:

Definition 1 ([16]) The formal theory of $\lambda\beta\eta$ equality has the following axioms:

$$(\alpha)\ \lambda x.M = \lambda y.[y/x]M \qquad (y \notin FV(M))$$

$$(\beta)\ (\lambda x.M)N = [N/x]M$$

$$(\eta)\ (\lambda x.Mx) = M \qquad (x \notin FV(M))$$

$$(\xi)\ \frac{M = M'}{\lambda x.M = \lambda x.M'}$$

$$(\mu)\ \frac{M = M'}{NM = NM'}$$

$$(\nu)\ \frac{M = M'}{MN = M'N}$$

$$(\rho)\ M = M$$

$$(\sigma)\ \frac{M = N}{N = M}$$

$$(\tau)\ \frac{M = N \quad N = P}{M = P}$$

Propositional equality. Again, let us recall from the theory of λ-calculus, that:

Definition 2 ([16], Definition 1.32) P is β-equal or β-convertible to Q (notation $P =_\beta Q$) iff Q is obtained from P by a finite (perhaps empty) series of β-contractions and reversed β-contractions and changes of bound variables. That is, $P =_\beta Q$ iff there exist $P_0, \ldots, P_n$ ($n \geq 0$) such that

$$P_0 \equiv P,\ P_n \equiv Q,$$
$$(\forall i \geq n-1)(P_i \rhd_{1\beta} P_{i+1} \text{ or } P_{i+1} \rhd_{1\beta} P_i \text{or} P_i \equiv_\alpha P_{i+1}).$$

NB: equality with an **existential** force.

Remark 3 In setting up a set of Gentzen's ND-style rules for equality we need to account for:

1. definitional versus propositional equality;
2. there may be more than one *normal* proof of a certain equality statement;
3. given a (possibly non-normal) proof, the process of bringing it to a normal form should be finite and confluent.

The missing entity. Within the framework of the functional interpretation (*à la* Curry–Howard [17]), the definitional equality is often considered by reference to a judgement of the form:

$$a = b : D$$

which says that a and b are equal elements from domain D. Notice that the 'reason' why they are equal does not play any part in the judgement. This aspect of 'forgetting contextual information' is, one might say, the first step towards 'extensionality' of equality, for whenever one wants to introduce intensionality into a logical system one invariably needs to introduce information of a 'contextual' nature, such as, where the identification of two terms (i.e. equation) comes from.

We feel that a first step towards finding an alternative formulation of the proof theory for propositional equality which takes care of the intensional aspect is to allow the 'reason' for the equality to play a more significant part in the form of judgement. We also believe that from the point of view of the logical calculus, if there is a 'reason' for two expressions to be considered equal, the proposition asserting their equality will be true, regardless of what particular composition of rewrites (definitional equalities) amounts to the evidence in support of the proposition concerned. Given these general guidelines, we shall provide what may be seen as a middle ground solution between the intensional [21, 22] and the extensional [23] accounts of Martin-Löf's propositional equality. The intensionality is taken care by the functional calculus on the labels, while the extensionality is catered by the logical calculus on the formulas. In order to account for the intensionality in the labels, we shall make the composition of rewrites (definitional equalities) appear as indexes of the equality sign in the judgement with a variable denoting a sequence of equality identifiers (we have seen that in the Curry–Howard functional interpretation there are at least four 'natural' equality identifiers: β, η, ξ and μ). So, instead of the form above, we shall have the following pattern for the equality judgement:

$$a =_s b : D$$

where 's' is meant to be a sequence of equality identifiers.

In the sequel we shall be discussing in some detail the need to identify the kind of definitional equality, as well as the need to have a logical connective of 'propositional equality' in order to be able to reason about the functional objects (those on the left-hand side of the ':' sign).

Term rewriting. Deductive systems based on the Curry–Howard isomorphism [17] have an interesting feature: normalization and strong normalization (Church–Rosser property) theorems can be proved by reductions on the terms of the functional calculus. Exploring this important characteristic, we have proved these theorems for the *Labelled Natural Deduction—LND* [41, 42] via a term rewriting system constructed from the *LND*-terms of the functional calculus [29]. Applying this same technique to the *LND* equational fragment, we obtain the normalization theorems for the equational logic of the *Labelled Natural Deduction* System [28, 30, 31].

This technique is used given the possibility of defining two measures of redundancy for the *LND* system that can be dealt with in the object level: the terms on the functional calculus and the *rewrite reason* (composition of rewrites), the latter being indexes of the equations in the *LND* equational fragment.

In the *LND* equational logic [38], the equations have the following pattern:

$$a =_s b : D$$

where one is to read: a is equal to b *because* of 's' ('s' being the *rewrite reason*); 's' is a term denoting a sequence of equality identifiers (β, η, α, etc.), i.e. a composition of rewrites. In other words, 's' is the *computational path* from a to b.

In this way, the *rewrite reason* (reason, for short) represents an **orthogonal measure of redundancy** for the *LND*, which makes the *LND* equational fragment an "enriched" system of equational logic. Unlike the traditional equational logic systems, in *LND* equational fragment there is a gain in local control by the use of *reason*. All the proof steps are recorded in the composition of rewrites (reasons). Thus, consulting the reasons, one should be able to see whether the proof has the normal form. We have then used this powerful mechanism of controlling proofs to present a precise normalization procedure for the *LND* equational fragment. Since the reasons can be dealt with in the object level, we can employ a computational method to prove the normalization theorems: we built a term rewriting system based on an algebraic calculus on the "*rewrite reasons*", which computes normal proofs. With this we believe we are making a step towards filling a gap in the literature on equational logic and on proof theory (natural deduction).

Kreisel–Tait's system. In [19] define the system E for equality reasoning as consisting of axioms of the form $t = t$, and the following rules of inference:

$$(E1) \quad \frac{E[t/x] \quad t = u}{E[u/x]}$$

$$(E2) \quad \frac{s(t) = s(u)}{t = u}$$

$$(E3) \quad \frac{0 = s(t)}{A} \quad \text{for any formula } A$$

$$(E4_n) \quad \frac{t = s^n(t)}{A} \quad \text{for any formula } A$$

where t and u are terms, '0' is the first natural number (zero), 's(-)' is the successor function.

Statman's normal form theorem. In order to prove the normalization results for the calculus E Statman defines two subsets of E: (i) a natural deduction-based calculus for equality reasoning NE; (ii) a sequent style calculus SE.

The NE calculus is defined as having axioms of the form $a = a$, and the rule of substituting equals for equals:

$$(=) \quad \frac{E[a/u] \quad a \approx b}{E[b/u]}$$

where E is any set of equations, and $a \approx b$ is ambiguously $a = b$ and $b = a$.

Statman arrives at various important results on normal forms and bounds for proof search in NE. In this case, however, a rather different notion of normal form is being used: the 'cuts' do not arise out of an 'inversion principle', as it is the case for the logical connectives, but rather from a certain form of sequence of equations which Statman calls 'computation'. With the formulation of a proof theory for the 'logical connective' of propositional equality we wish to analyse equality reasoning into its basic components: rewrites, on the one hand, and statements about the existence of rewrites, on the other hand. This type of analysis came to the surface in the context of constructive type theory and the Curry–Howard functional interpretation.

Martin-Löf's Equality type. There has been essentially two approaches to the problem of characterizing a proof theory for propositional equality, both of which originate in P. Martin-Löf's work on *Intuitionistic Type Theory*: the intensional [22] and the extensional [23] formulations.

The extensional version. In [23] and [24] presentations of *Intuitionistic Type Theory* P. Martin-Löf defines the type of *extensiona* propositional equality 'I' (here called 'I_{ext}') as:

I_{ext}-formation

$$\frac{D\ type \qquad a : D \qquad b : D}{I_{ext}(D, a, b)\ type}$$

I_{ext}-introduction

$$\frac{a = b : D}{\mathtt{r} : I_{ext}(D, a, b)}$$

I_{ext}*-elimination*[2]

$$\frac{c : I_{ext}(D, a, b)}{a = b : D}$$

I_{ext}*-equality*

$$\frac{c : I_{ext}(D, a, b)}{c = \mathtt{r} : I_{ext}(D, a, b)}$$

Note that the above account of propositional equality does not 'keep track of all proof steps': both in the I_{ext}*-introduction* and in the I_{ext}*-elimination* rules there is a considerable loss of information concerning the deduction steps. While in the I_{ext}*-introduction* rule the 'a' and the 'b' do not appear in the 'trace' (the label/term alongside the logical formula), the latter containing only the canonical element '$\mathtt{r}$', in the rule of I_{ext}*-elimination* all the trace that might be recorded in the label 'c' simply disappears from label of the conclusion. If by 'intensionality' we understand a feature of a logical system which identifies as paramount the concern with issues of *context* and *provability*, then it is quite clear that any logical system containing I_{ext}-type can hardly be said to be 'intensional': as we have said above, neither its *introduction* rule nor its *elimination* rule carry the necessary *contextual* information from the premise to the conclusion.

And, indeed, the well-known statement of the extensionality of functions can be proved as a theorem of a logical system containing the I_{ext}-type such as Martin-Löf's *Intuitionistic Type Theory* [24]. The statement says that if two functions return the same value in their common codomain when applied to each argument of their common domain (i.e. if they are equal pointwise), then they are said to be (extensionally) equal. Now, we can construct a derivation of the statement written in the formal language as:

$$\forall f, g^{A \to B}.(\forall x^A.I_{ext}(B, \mathtt{APP}(f, x), \mathtt{APP}(g, x)) \to I_{ext}(A \to B, f, g))$$

by using the rules of proof given for the I_{ext}, assuming we have the rules of proof given for the implication and the universal quantifier.

The intensional version. Another version of the propositional equality, which has its origins in Martin-Löf's early accounts of *Intuitionistic Type Theory* [21, 22], and is apparently in the most recent, as yet unpublished, versions of type theory, is defined in [46] and [27]. In a section dedicated to the *intensional versus extensional* debate, ([46], p. 633) says that:

---

[2] The set of rules given in [23] contained the additional *elimination* rule:

$$\frac{c : I(D, a, b) \qquad d : C(\mathtt{r}/z)}{\mathtt{J}(c, d) : C(c/z)}$$

which may be seen as reminiscent of the previous *intensional* account of propositional equality [22].

Martin-Löf has returned to an intensional point of view, as in [21], that is to say, $t = t' \in A$ is understood as "t and t' are definitionally equal". As a consequence the rules for identity types have to be adapted.

If we try to combine the existing accounts of the *intensional* equality type 'I' [22, 27, 46], here denoted 'I_{int}', the rules will look like:

I_{int}*-formation*

$$\frac{D\ type \quad a : D \quad b : D}{I_{int}(D, a, b)\ type}$$

I_{int}*-introduction*

$$\frac{a : D}{\mathrm{e}(a) : I_{int}(D, a, a)} \qquad \frac{a = b : D}{\mathrm{e}(a) : I_{int}(D, a, b)}$$

I_{int}*-elimination*

$$\frac{a : D \quad b : D \quad c : I_{int}(D, a, b) \quad \begin{array}{c}[x:D]\\ d(x):C(x,x,\mathrm{e}(x))\end{array} \quad \begin{array}{c}[x:D, y:D, z:I_{int}(D,x,y)]\\ C(x,y,z)\ type\end{array}}{\mathrm{J}(c,d):C(a,b,c)}$$

I_{int}*-equality*

$$\frac{a : D \quad \begin{array}{c}[x : D]\\ d(x) : C(x, x, \mathrm{e}(x))\end{array} \quad \begin{array}{c}[x : D, y : D, z : I_{int}(D, x, y)]\\ C(x, y, z)\ type\end{array}}{\mathrm{J}(\mathrm{e}(a), d(x)) = d(a/x) : C(a, a, \mathrm{e}(a))}$$

With slight differences in notation, the 'adapted' rules for identity type given in [46] and [27] resembles the one given in [22]. It is called *intensional* equality because there remains no direct connection between judgements like '$a = b : D$' and '$s : I_{int}(D, a, b)$'.

A labelled proof theory for propositional equality. Now, it seems that an alternative formulation of propositional equality within the functional interpretation, which will be a little more elaborate than the extensional I_{ext}-type, and simpler than the intensional I_{int}-type, could prove more convenient from the point of view of the 'logical interpretation'. It seems that whereas in the former we have a considerable loss of information in the I_{ext}*-elimination*, in the latter we have an I_{int}*-elimination* too heavily loaded with (perhaps unnecessary) information. If, on the one hand, there is an *over*explicitation of information in I_{int}, on the other hand, in I_{ext} we have a case of *under*explicitation. With the formulation of a proof theory for equality via labelled natural deduction we wish to find a middle ground solution between those two extremes.

2 Labelled Deduction

The functional interpretation of logical connectives via deductive systems which use some sort of labelling mechanism [11, 24] can be seen as the basis for a general framework characterizing logics via a clear separation between a functional calculus on the *labels*, i.e. the referents (names of individuals, expressions denoting the record of proof steps used to arrive at a certain formula, names of 'worlds', etc.) and a logical calculus on the formulas. The key idea is to make these two dimensions as harmonious as possible, i.e. that the functional calculus on the labels matches the logical calculus on the formulas at least in the sense that to every abstraction on the variables of the functional calculus there corresponds a discharge of an assumption-formula of the logical calculus. One aspect of such interpretation which stirred much discussion in the literature of the past 10 years or so, especially in connection with *Intuitionistic Type Theory* [24], was that of whether the logical connective of propositional equality ought to be dealt with 'extensionally' or 'intensionally'. Here we attempt to formulate what appears to be a middle ground solution, in the sense that the intensional aspect is dealt with in the functional calculus on the labels, whereas the extensionality is kept to the logical calculus. We also intend to demonstrate that the connective of propositional equality (cf. [1] '$\doteq$') needs to be dealt with in a similar manner to 'Skolem-type' connectives (such as disjunction and existential quantification), where notions like *hiding*, *choice* and *-dependent variables* play crucial rôles.

Our motivation: where did it all start? The characterization of a proof theory for *Labelled Deductive Systems* has been the concern of some authors for some time now [12, 39, 40]. Here we address two topics of special interest to logic and computation, namely *substitution* and *unification*. As a starting point, we posed ourselves two interrelated questions: how could we incorporate the handling of rewrites and function symbols into the proof theory, and how could we 'give logical content', so to speak, to the procedures coming from unification algorithms?

For those not familiar with the *LDS* perspective, it suffices at this stage to say that the declarative unit of logical systems is seen as made up of two components: a formula and a label. The label is meant to carry information which may be of a less declarative nature than that carried by the formulas. The introduction of such an 'extra' dimension was motivated by the need to cope with the demands of computer science applications.

Indeed, with the diversification of computer science applications to problems involving *reasoning*, there has been a proliferation of logics originated mainly from the need to tailor the logical system to the demands of the particular application area. If there were a number of 'logics' already developed and well established in the mathematical and philosophical logic literature (relevant, intuitionistic, minimal, etc.), the diversification was significantly increased with the contribution from computer science.

Gabbay observed that many of the distinctive features of most logics being studied by logicians and computer scientists alike, stemmed from 'meta-level' considerations: in order to consider a step to be a valid one, it was invariably the case that one had to take into account questions like: 'whether the assumptions have actually been used'; 'whether they have been used in a certain order'; 'whether the number of times an assumption was used has been in keeping with the need to take care of resources'; etc.

There are a number of inconveniences in having to cope with increasingly diverse logical systems, and Gabbay set out a research programme with at least the following desiderata:

- to find a unifying framework (sequent calculus by itself would not do, and we shall see why later on) *factoring out* meta- from object- level features;
- to keep the logic (and logical steps, for that matter) simple, handling meta-level features via a *separate*, yet *harmonious* calculus;
- to have means of structuring and combining logics;
- to make sure the relevant assumptions in a deduction are uncovered, paying more attention to the explicitation and use of resources.

The idea of *labelled deduction* seemed to be a natural evolution from the traditional logical systems. The development of a novel approach to logic, namely *Labelled Deductive Systems*, where the meta-level features would be incorporated into the deductive calculus in an orderly manner, looked general enough to be an appropriate candidate for such a unifying framework.

In summary, it seems fair to say that *Labelled Deductive Systems* offer a new perspective on the discipline of *logic and computation*. Arising from computer science applications, it provides the essential ingredients for a framework whereby one can study:

- meta-level features of logical systems, by 'knocking down' some of the elements of the meta-level reasoning to the object-level, and allowing each logical step to take care of *what has been done so far*;
- the 'logic' of Skolem functions and substitution (dependencies, term declaration).

Why sequent calculus by itself will not do. Boole did manage to formalize the algebra of logical connectives, with the aspect of *duality* coming out very neatly. The sequent calculus follows on this quest for duality:

$$\begin{array}{lcl} \text{negative} & \vdash & \text{positive} \\ \text{conjunctive} & \vdash & \text{disjunctive} \end{array}$$

Nevertheless, since Frege, logic is *also* about quantifiers, predicates, functions, equality among referents, etc. In a few words: beyond *duality*, first-order logic also deals with *quantification* in a *direct* fashion, instead of via, say, Venn diagrams. Thus, a proof theory for first-order logic ought to account for the manipulation of function symbols, terms, dependencies and substitutions, as Herbrand already perceived.

We shall see a little more about this later on when we come to a brief discussion of the so-called 'sharpened Hauptsatz'. But it seems appropriate to add here that we are looking for strengthening the connections between Gentzen's and Herbrand's methods in proof theory. We believe that the two-dimensional approach of *LDS* is the right framework for the enterprise. This is because, if, on the one hand:

(+) Gentzen's methods come with a well-defined mathematical theory of proofs,

and

(+) Herbrand's method show how to handle function symbols and terms in a direct fashion,

on the other hand:

(−) in Gentzen's calculi (plain natural deduction, sequent calculus) function symbols are not *citizens*,

and

(−) Herbrand's methods hardly give us means of looking at proofs (deductions) as the main objects of study.

By combining a functional calculus on the labels (which carry along referents, function symbols) with a logical calculus on the formulas, the *LDS* perspective can have the (+)'s without the (−)'s.

The generality of Herbrand base. Let us take the example which Leisenring uses to demonstrate the application of Herbrand's decision procedure to check the validity of the formula [20]:

$$\exists x.\forall y.(P(x) \to P(y))$$

Herbrand's 'original' procedure. The first step is to find the Herbrand resolution ($\exists$-prenex normal form), which can be done by introducing a new function symbol g, and obtaining:

$$\exists x.(P(x) \to P(g(x)))$$

As this would be equivalent to a disjunction of substitution instances like:

$$P(a) \to P(g(a)) \quad \vee \quad P(a') \to P(g(a')) \quad \vee \quad P(a'') \to P(g(a'')) \quad \vee \quad \cdots$$

the second step is to find a p-substitution instance (p finite) which is a tautology. For that, we take the 'Herbrand base' to be $\{a, g\}$, where a is an arbitrary individual from the domain, and g is an arbitrary function symbol which can construct, out of a, further elements of the domain. Thus, the 1-substitution instance is:

$$P(a) \to P(g(a))$$

which is clearly not a tautology. Now, we can iterate the process, and find the 2-reduction as a disjunction of the 1-reduction and the formula made up with a 2-substitution (taking $a' = g(a)$), that is:

$$P(a) \to P(g(a)) \quad \vee \quad P(g(a)) \to P(g(g(a)))$$

which is a tautology.

In summary:

1. $\exists x.\forall y.(P(x) \to P(y))$
2. take g as a unary function
3. $\exists x.(P(x) \to P(g(x)))$
4. $P(a) \to P(g(a)) \quad \vee \quad P(a') \to P(g(a')) \quad \vee \quad P(a'') \to P(g(a'')) \quad \vee \quad \cdots$
5. 1st substitution: $P(a) \to P(g(a))$
6. take $a' = g(a)$
7. 2nd substitution: $P(a) \to P(g(a)) \quad \vee \quad P(g(a)) \to P(g(g(a)))$ (tautology)

In checking the validity of $\exists x.\forall y.(P(x) \to P(y))$ we needed the following extra assumptions:

1. the domain is non-empty (step 4).
2. there is a way of identifying an arbitrary term with another one (step 6).

As we shall see below, the labelled deduction method will have helped us 'to bring up to the surface' those two (hidden) assumptions.

Now, how can we justify the *generality* of the 'base' $\{a, g\}$? Why is it that it does not matter which a and g we choose, the procedure always works? In other words, why is it that *for any* element a of the domain and *for any* 'function symbol' g, the procedure always works?

In a previous opportunity [39] we have already demonstrated the *universal force* which is given to Skolem functions by the device of *abstraction* in the elimination of the existential quantifier. The point was that although there was no quantification over function symbols being made in the logic (the logical calculus on the formulas, that is), an abstraction on the name for the Skolem function was performed in the functional calculus on the labels. The observation suggested that, as in the statement of Skolem's theorem, *for any* (new) function symbol f we choose when Skolemizing $\forall x.\exists y.P(x, y)$ to $\forall x.P(x, f(x))$, if an arbitrary statement can be deduced from the latter then it can also be deduced from the former, regardless of the choice of f. We shall come back to this point later on when we will then demonstrate that in labelled natural deduction the Herbrand function gets abstracted away thus getting universal force.

Gentzen–Herbrand connections: the sharpened Hauptsatz. The connections between the proof theory as developed by Gentzen and the work on the proof theory of first-order logic by Herbrand are usually seen through the so-called 'sharpened Hauptsatz'.

Theorem (Gentzen's sharpened Hauptsatz)
Given $\Gamma \vdash \Delta$ (prenex formulae), if $\Gamma \vdash \Delta$ is provable then there is a cut-free, pure-variable proof which contains a sequent $\Gamma' \vdash \Delta'$ (the midsequent) with:

1. *Every formula in $\Gamma' \vdash \Delta'$ is quantifier free.*

2. *No quantifier rule above* $\Gamma' \vdash \Delta'$.
3. *Every rule below* $\Gamma' \vdash \Delta'$ *is either a quantifier rule, or a contraction or exchange structural rule (not a weakening).*

The theorem relies on the so-called *permutability* lemma which says that quantifier rules can always be 'pushed down' in the deduction tree.

Lemma (Permutability [18])
Let π *be a cut-free proof of* $\Gamma \vdash \Delta$ *(with only prenex formulas); then it is possible to construct another proof* π' *where:*

> every quantifier rule can be permuted with a logical or structural rule applied below it (with some provisos).

(The interested reader will find detailed expositions of the sharpened Hauptsatz in [18] and [13].)

Example For the sake of illustration, let us construct a sequent-calculus deduction of the formula used in our original example, i.e. let us build proof of the sequent

$$\vdash \exists x.\forall y.(P(x) \rightarrow P(y))$$

and see what the *midsequent* means in this case:

$$\begin{array}{c}
P(b), P(a) \vdash P(b), P(c) \\
\hline
P(b) \vdash P(a) \rightarrow P(b), P(c) \\
\hline
P(b) \vdash P(c), P(a) \rightarrow P(b) \\
\hline
\boxed{\vdash P(b) \rightarrow P(c), P(a) \rightarrow P(b)} \\
\hline
\vdash \forall y.(P(b) \rightarrow P(y)), P(a) \rightarrow P(b) \\
\hline
\vdash \exists x.\forall y.(P(x) \rightarrow P(y)), P(a) \rightarrow P(b) \\
\hline
\vdash P(a) \rightarrow P(b), \exists x.\forall y.(P(x) \rightarrow P(y)) \\
\hline
\vdash \forall y.(P(a) \rightarrow P(y)), \exists x.\forall y.(P(x) \rightarrow P(y)) \\
\hline
\vdash \exists x.\forall y.(P(x) \rightarrow P(y)), \exists x.\forall y.(P(x) \rightarrow P(y)) \\
\hline
\vdash \exists x.\forall y.(P(x) \rightarrow P(y))
\end{array}$$

Note that every rule below the boxed sequent is either a quantifier rule, or a contraction or exchange rule (no weakenings). Due to the *eigenvariable* restrictions to the quantifiers rules ($\exists$ on the left, and $\forall$ on the right), we had to use contraction and choose the same b in different instantiations. (As we will see later on, when we use labelled natural deduction the 'assumption' that the first used b is the same as the other one is introduced as a logical formula using propositional equality.)

With such an example we intend to draw the attention to the fact that although the sharpened Hauptsatz brings Gentzen's methods closer to Herbrand's methods, it shows how less informative the former is with respect to the latter. The midsequent does not mention any function symbol, nor does it point to the inductive nature of the generation of the so-called 'Herbrand universe': the fact that the proof obtains is related to the 'meta-level' choice of the same instantiation constant b.

Back to Frege. The device of variable-binding, and the idea of having terms representing incomplete 'objects' whenever they contain free variables, were both introduced in a systematic way by Frege in his *Grundgesetze*. As early as [9] developed in his *Grundgesetze I* what can be seen as the early origins of the notions of *abstraction* and *application*, when showing techniques for transforming functions (expressions with free variables) into value-range terms (expressions with no free variables) by means of an 'introductory' operator of abstraction producing the *Werthverlauf* expression,[3] e.g., '$\grave{\varepsilon} f(\varepsilon)$', and the effect of its corresponding 'eliminatory' operator '$\cap$' on a *value-range* expression. For example, in ([9], Sect. 34, pp. 52ff), translated in ([10], p. 92):

> (...) it is a matter only of designating the value of the function $\Phi(\xi)$ for the argument Δ, i.e. $\Phi(\Delta)$, by means of "Δ" and "$\grave{\varepsilon}\Phi(\varepsilon)$". I do so in this way:
>
> "$\Delta \cap \grave{\varepsilon}\Phi(\varepsilon)$"
>
> which is to mean the same as "$\Phi(\Delta)$".

Note the similarity to the rule of functional *application*, where 'Δ' is the argument, '$\grave{\varepsilon}\Phi(\varepsilon)$' is the function, and '$\cap$' is the application operator 'APP'.

Expressing how important he considered the introduction of a variable-binding device for the functional calculus (recall that the variable-binding device for the logical calculus had been introduced earlier in *Begriffsschrift*), Frege says:

> The introduction of a notation for courses-of-values seems to me to be one of the most important supplementations that I have made of my *Begriffsschrift* since my first publication on this subject. (*Grundgesetze I*, Sect. 9, p. 15f.)

The idea of forming *value-range* function-terms by abstracting from the corresponding free variable is in fact very useful in representing the handling of assumptions within a natural deduction style calculus. In particular, when the natural deduction presentation system is based on a 'labelling' mechanism the binding of free variables in the labels corresponds to the discharge of respective assumptions. In the sequel we shall be using 'abstractors' (such as 'λ' in '$\lambda x.f(x)$') to bind free-variables and discharge the assumption labelled by the corresponding variable.

Dividing the tasks: a functional calculus on the labels, a logical calculus on the formulas. We have seen that the origins of variable binding mechanisms, both on the formulas of logic (the propositions) and on the expressions of the functional calculus (the terms), go back at least as far as Frege's early investigations on a 'language of concept writing'. Although the investigations concerned essentially the establishment of the basic laws of logic, for Frege the functional calculus would have

[3] Frege ([9], Sect. 3, p. 7), translated as *course-of-values* in ([10], p. 36), and *value-range* in most other translations of Frege's writings, including the translation of [8] where the term first appeared.

the important rôle of demonstrating that arithmetic could be formalised simply by defining its most basic laws in terms of rules of the 'calculus of concept writing'. Obviously, the calculus defined in *Begriffsschrift*, in spite of its functional style, was primarily concerned with the 'logical' side, so to speak. The novel device of binding free variables, namely the universal quantifier, was applicable to propositional functions. Thus, *Grundgesetze* was written with the intention of fulfilling the ambitious project of designing a language of concept-writing which could be useful to formalise mathematics. Additional mechanisms to handle the functional aspects of arithmetic (e.g. equality between number-expressions, functions over number-expressions, etc.) had to be incorporated. The outcome of Frege's second stage of investigations also brought pioneering techniques of formal logic, this time with respect to the handling of functions, singular terms, definite descriptions, etc. An additional mechanism of variable binding was introduced, this time to bind variables of functional expressions, i.e. expressions denoting individuals, not truth-values.

Summarising, we can see the pioneering work of Frege in its full significance if we look at the two sides of formal logic he managed to formulate a calculus for:

1. the 'logical' calculus on formulas (*Begriffsschrift*)
2. the 'functional' calculus on terms (*Grundgesetze*).

As a pioneer in any scientific activity one is prone to leave gaps and loopholes to be later filled by others. It happened with Frege that a big loophole was discovered earlier than he would himself have expected: Russell's discovery of the antinomies of his logical notion of set was a serious challenge. There may be several ways of explaining why the resulting calculus was so much susceptible to that sort of challenge. We feel particularly inclined to think that the use of devices which were designed to handle the so-called 'objects', i.e. expressions of the functional calculus, ought to have been kept apart from, and yet harmonised with, the logical calculus on the formulas. Thus, here we may start wondering what might have been the outcome had Frege kept the two sides separate and yet harmonious.

Let us for a moment think of a connection to another system of language analysis which would seem to have some similarity in the underlying ontological assumption, with respect to the idea of dividing the logical calculus into two dimensions, i.e. functional versus logical. The semantical framework defined in Montague's [26] intensional logic makes use of a distinction among the semantic *types* of the objects handled by the framework, namely e, t and s, in words: *entities*, *truth-values*, and *senses*. The idea was that logic (language) was supposed to deal with objects of three kinds: names of entities, formulas denoting truth-values, and possible-worlds/contexts of use. Now, here when we say that we wish to have the bi-dimensional calculus, we are saying that the entities which are *namable* (i.e. individuals, possible-worlds, etc.) will be dealt with separately from (yet harmoniously with) the logical calculus on the formulas, by a calculus of functional expressions. Whereas the variables for individuals are handled 'naturally' in the interpretation of first-order logic with labelled natural deduction, the introduction of variables to denote *contexts*, or *possible-worlds* (structured collection of labelled formulas), as in [40], is meant to account for Montague's *senses*.

Reassessing Frege's two-dimensional calculus. In an attempt to reassess the benefits of having those two sides working together, we would like to insist on the two sides being treated separately. Thus, instead of binding variables on the formulas with the device of forming 'value-range' expressions as Frege does, we shall have a clear separation of functional versus logical devices. Cf. the following opening lines of *Grundgesetze I*, Sect. 10:

> Although we laid it down that the combination of signs "$\dot{\varepsilon}\Phi(\varepsilon) = \dot{\alpha}\Psi(\alpha)$" has the same denotation as "$\overset{a}{\smile}$ $\Phi(a) = \Psi(a)$", this by no means fixes completely the denotation of a name like "$\dot{\varepsilon}\Phi(\varepsilon)$".

Note that both the abstractor '' and the universal quantifier '$\smile$' are used for binding free variables of formulas of the logical calculus such as 'Φ' and 'Ψ'. In labelled natural deduction we shall take the separation 'functional versus logical' more strictly than Frege himself did. While the abstractors will be used to bind variables in the functional calculus, the quantifiers will be used to bind variables in the logical calculus. Obviously, variables may be occurring in both 'sides', but in each side the appropriate mechanism will be used accordingly.

We still want to have the device of forming propositional functions, so we still need to have the names of variables of the functional calculus being carried over to take part in the formulas of the logical side. That will be dealt with accordingly when we describe what it means to have predicate formulas in a labelled system. Nevertheless, abstractors shall only bind variables occurring in expressions of the functional calculus, and quantifiers shall bind variables occurring in formulas of the logical calculus. For example, in:

$$\begin{array}{cc} \forall\text{-}introduction & \exists\text{-}introduction \\ \dfrac{\begin{array}{c}[x : D] \\ f(x) : P(x)\end{array}}{\Lambda x.f(x) : \forall x^D.P(x)} & \dfrac{t : D \quad f(t) : P(t)}{\varepsilon x.(f(x), t) : \exists x^D.P(x)} \end{array}$$

whilst the abstractors 'Λ' and 'ε' bind variables of the functional calculus, the quantifiers '$\forall$' and '$\exists$' bind variables of the logical calculus, even if the same variable name happens to be occurring in the functional expression as well as in the logical formula.

Notice that although we are dealing with the two sides independently, the harmony seems to be maintained: to each discharge of assumption in the logical calculus there will correspond an abstraction in the functional calculus. In the case of quantifier rules, we observe that the *introduction* of the universal quantifier is made with the arbitrary name x being bound on both sides (functional and logical) at the same time. In the existential case the 'witness' a is kept unbound in the functional calculus, whilst in the formula the binding is performed.

Here is not really the place to discuss the paradoxes of Frege's formalised set theory, but it might be helpful to single out one particularly relevant facet of his 'mistake'. First, let us recall that the development of mechanisms for handling both sides of a calculus of concept writing, namely the logical and the functional, would perhaps recommend special care in the harmonising of these two sides. We all know today (thanks to the intervention of the likes of Furth [10], Aczel [1], Dummett [6, 7], Sluga [43], and others) that one of the fundamental flaws of Frege's attempt to put the two sides together was the so-called 'Law V' of *Grundgesetze*, which did exactly what we shall avoid here in such 'functional interpretation' of logics: using functions where one should be using propositions, and vice versa. The 'Law V' was stated as follows:

$$\vdash (\dot{\varepsilon} f(\varepsilon) = \dot{\alpha} g(\alpha)) = (\overset{a}{\smile}\!\!- f(a) = g(a))$$

Here we have equality between terms—i.e. $\dot{\varepsilon} f(\varepsilon) = \dot{\alpha} g(\alpha)$ and $f(a) = g(a)$—in a par with equality between truth-values—i.e. the middle equality sign.

In his thorough analysis of Frege's system, Aczel [1] makes the necessary distinction by introducing the sign for propositional equality:

$$(\lambda x.f(x) \doteq \lambda x.g(x)) \leftrightarrow \forall x.(f(x) \doteq g(x)) \text{ is true}$$

where '$\doteq$' stands for propositional equality, and '$\leftrightarrow$' is to mean logical equivalence (i.e. 'if and only if').

Despite the challenges to his theories of formal logic, Frege's tradition has remained very strong in mathematical logic. Indeed, there is a tendency among the formalisms of mathematical logic to take the same step of 'blurring' the distinction between the functional and the logical side of formal logic. As we have already mentioned, Frege introduced in the *Grundgesetze* the device of binding variables in the functional calculus,[4] in addition to the variable-binding device presented in *Begriffsschrift*, but allowed variables occurring in the formulas to be bound not only by the quantifier(s), but also by a device of the functional calculus, namely the 'abstractors'. One testimony to the strength of Frege's legacy which is particularly relevant to our purposes here is the formalism described in Hilbert and Bernays' [14, 15] book where various calculi of singular terms are established. One of these calculi was called the ε-calculus, and consisted of an extension of first-order logic by adding the following axiom schema:

$$\begin{array}{lll} (\varepsilon_1) & A(a) \to A(\varepsilon_x A(x)) & \text{(for any term '}a\text{')} \\ (\varepsilon_2) & \forall x.(A(x) \leftrightarrow B(x)) \to (\varepsilon_x A(x) = \varepsilon_x B(x)) & \end{array}$$

where any term of the form '$\varepsilon_x A(x)$' is supposed to denote a term 't' with the property that '$A(t)$' is true, if there is one such term.

Now, observe that the addition of these new axioms has to be proven 'harmless' to the previous calculus, namely the first-order calculus with bound variables, in the sense that no formulas involving only the symbols of the language of the old

---

[4] In fact, Frege had already introduced the device which he called *Werthverlauf* in his article on 'Function and Concept' [8], which, in its turn, may have been inspired by Peano's functional notation.

calculus which was not previously a theorem, is a theorem of the new calculus. For that one has to prove the fundamental theorems stating that the new calculus is only a 'conservative extension' the old calculus (First and Second ε-Theorems).

The picture becomes slightly different when we follow somewhat more strictly the idea of dividing, as sharply as we can, the two tasks: let all that has to do with entities to be handled by the functional calculus on the labels, and leave only what is 'strictly logical' to the logical calculus on the formulas. So, in the case of ε-terms, we shall not simply replace an existentially quantified variable in a formula (e.g. 'x' in '$\exists x.A(x)$') by an ε-term involving a formula (e.g. '$A(\varepsilon_x A(x))$'). Instead, we shall use 'ε' as an abstractor binding variables of the functional calculus, as we have seen from the rule of $\exists$-*introduction* shown previously. In other words, we do not have the axioms (or rules of inference) for the existential quantifier plus other axiom(s) for the ε-symbol. We present the existential quantifier with its usual logical calculus on the formulas, alongside the ε-terms taking care of the 'functional' side. And so be it:

$$\exists\text{-}introduction \qquad \frac{t : D \quad f(t) : P(t)}{\varepsilon x.(f(x), t) : \exists x^D.P(x)} \qquad\qquad \exists\text{-}elimination \qquad \frac{e : \exists x^D.P(x) \quad \begin{array}{c}[u : D, g(u) : P(u)] \\ d(g, u) : C\end{array}}{\texttt{INST}(e, \acute{g}\acute{u}d(g, u)) : C}$$

Notice that here the concern with the 'conservative extension' shall be significantly different from the one Hilbert and Bernays [14, 15] had. We have the ε-symbol appearing on the label (the functional side, so to speak), and it is only introduced alongside the corresponding existential formula [35]. (More details of a treatment of the peculiarities of the existential quantifier are given in [39]. As for the reassessment of Frege's two-dimensional logical system and how it fits into the framework of the Dummett–Prawitz analysis of Gentzen's idea that the introduction rules give the meaning of logical connectives, cf. [32–34, 36, 37].)

Identifiers for (Compositions of) Equalities

In the functional interpretation, where a functional calculus on the labels go hand in hand with a logical calculus on the formulas, we have a classification of equalities, whose identifications are carried along as part of the deduction: either β-, η-, ξ-, μ- or α- equality will have been part of an expression labelling a formula containing '$\doteq$'. There one finds the key to the idea of 'hiding' in the *introduction* rule, and opening local (Skolem-type) assumptions in the *elimination* rule. (Recall that in the case of disjunction we also have alternatives: either into the left disjunct, or into the right disjunct.) So, we believe that it is not unreasonable to start off the formalization of propositional equality with the parallel to the disjunction and existential cases in mind. Only, the witness of the type of propositional equality are not the 'a's and 'b's of '$a = b : D$', but the actual (sequence of) equalities (β-, η-, ξ-, α-) that might have been used to arrive at the judgement '$a =_s b : D$' (meaning '$a = b$' *because* of 's'), 's' being a sequence made up of β-, η-, ξ- and/or α-equalities, perhaps with some of the general equality rules of reflexivity, symmetry and transitivity. So, in the *introduction* rule of the type we need to form the canonical proof as if we were *hiding* the actual sequence. Also, in the rule of *elimination* we need to open a new local assumption introducing a new variable denoting a possible sequence as a (Skolem-type) new constant. That is, in order to eliminate the connective '$\doteq$' (i.e. to deduce something from a proposition like '$\doteq_D (a, b)$'), we start by choosing a new variable to denote the reason why the two terms are equal: 'let t be an expression (sequence of equalities) justifying the equality between the terms'. If we then arrive at an arbitrary formula 'C' labelled with an expression where the t still occurs free, then we can conclude that the same C can be obtained from the $\doteq$-formula regardless of the identity of the chosen t, meaning that the label alongside C in the conclusion will have been abstracted from the free occurrences of t.

Observe that now we are still able to 'keep track' of all proof steps (which does not happen with Martin-Löf's I_{ext}-type) [23, 24], and we have an easier formulation (as compared with Martin-Löf's I_{int}-type) [22] of how to perform the *elimination* step.

The Proof Rules

In formulating the propositional equality connective, which we shall identify by '$\doteq$', we shall keep the pattern of inference rules essentially the same as the one used for the other logical connectives (as in, e.g. [39]), and we shall provide an alternative presentation of propositional equality as follows:

$\doteq$-introduction

$$\frac{a =_s b : D}{s(a, b) : \doteq_D (a, b)}$$

$\doteq$-reduction

$$\frac{\dfrac{a =_s b : D}{s(a, b) : \doteq_D (a, b)} \doteq\text{-}intr \qquad \begin{matrix}[a =_t b : D] \\ d(t) : C\end{matrix}}{\texttt{REWR}(s(a, b), \acute{t}d(t)) : C} \doteq\text{-}elim \quad \rhd_\beta \quad \begin{matrix}[a =_s b : D] \\ d(s/t) : C\end{matrix}$$

$\doteq$-induction

$$\frac{e : \doteq_D (a, b) \qquad \dfrac{[a =_t b : D]}{t(a, b) : \doteq_D (a, b)} \doteq\text{-}intr}{\texttt{REWR}(e, \acute{t}t(a, b)) : \doteq_D (a, b)} \doteq\text{-}elim \quad \rhd_\eta \quad e : \doteq_D (a, b)$$

where '´' is an abstractor which binds the occurrences of the (new) variable 't' introduced with the local assumption '$[a =_t b : D]$' as a kind of 'Skolem'-type constant denoting the (presumed) 'reason' why 'a' was assumed to be equal to 'b'.

(Recall the Skolem-type procedures of introducing new local assumptions in order to allow for the elimination of logical connectives where the notion of 'hiding' is crucial, e.g. disjunction and existential quantifier—in [39]).

Now, having been defined as a 'Skolem'-type connective, '$\doteq$' needs to have a conversion stating the non-interference of the newly opened branch (the local assumption in the $\doteq$-*elimination* rule) with the main branch. Thus, we have:

$\doteq$-*(permutative) reduction*

$$\dfrac{\dfrac{e : \doteq_D (a,b) \qquad \begin{matrix}[a =_t b : D] \\ d(t) : C\end{matrix}}{\mathtt{REWR}(e, \acute{t}d(t)) : C}}{w(\mathtt{REWR}(e, \acute{t}d(t))) : W}\,\mathsf{r} \quad \rhd_\zeta \quad \dfrac{e : \doteq_D (a,b) \qquad \dfrac{\begin{matrix}[a =_t b : D] \\ d(t) : C\end{matrix}}{w(d(t)) : W}\,\mathsf{r}}{\mathtt{REWR}(e, \acute{t}w(d(t))) : W}$$

provided w does not disturb the existing dependencies in the term e (the main branch), i.e. provided that rule 'r' does not discharge any assumption on which '$\doteq_D (a,b)$' depends. The corresponding ζ-equality is:

$$w(\mathtt{REWR}(e, \acute{t}d(t))) =_\zeta \mathtt{REWR}(e, \acute{t}w(d(t)))$$

The equality indicates that the operation w can be pushed inside the '-abstraction term, provided that it does not affect the dependencies of the term e.

Since we are defining the logical connective '$\doteq$' as a connective which deals with singular terms, where the 'witness' is supposed to be hidden, we shall not be using direct *elimination* like Martin-Löf's I_{ext}-*elimination*. Instead, we shall be using the following $\doteq$-*elimination*:

$$\dfrac{e : \doteq_D (a,b) \qquad \begin{matrix}[a =_t b : D] \\ d(t) : C\end{matrix}}{\mathtt{REWR}(e, \acute{t}d(t)) : C}$$

The *elimination* rule involves the introduction of a new local assumption (and corresponding variable in the functional calculus), namely '$[a =_t b : D]$' (where 't' is the new variable) which is only discharged (and 't' bound) in the conclusion of the rule. The intuitive explanation would be given in the following lines. In order to eliminate the equality $\doteq$-connective, where one does not have access to the 'reason' (i.e. a sequence of 'β', 'η', 'ξ' or 'ζ' equalities) why the equality holds because '$\doteq$' is supposed to be a connective dealing with singular terms (as are '$\vee$' and '$\exists$'), in the first step one has to open a new local assumption supposing the equality holds because of, say 't' (a new variable). The new assumption then stands for 'let t be the unknown equality'. If a third (arbitrary) statement can be obtained from this new local assumption via an unspecified number of steps which does not involve any binding of the new variable 't', then one discharges the newly introduced local assumption binding the free occurrences of the new variable in the label alongside the statement obtained, and concludes that that statement is to be labelled by the term '$\mathtt{REWR}(e, \grave{t}d(t))$' where the new variable (i.e. t) is bound by the '`'-abstractor.

Another feature of the $\doteq$-connective which is worth noticing at this stage is the equality under 'ξ' of all its elements (see second *introduction* rule). This does not mean that the labels serving as evidences for the $\doteq$-statement are all identical to a constant (cf. constant '$\mathtt{r}$' in Martin-Löf's I_{ext}-type), but simply that if two (sequences of) equality are obtained as witnesses of the equality between, say 'a' and 'b' of domain D, then they are taken to be equal under ξ-equality. It would not seem unreasonable to think of the $\doteq$-connective of propositional equality as expressing the proposition which, whenever true, indicates that the two elements of the domain concerned are equal under some (unspecified, *hidden*) composition of definitional equalities. It is as if the proposition points to the existence of a term (witness) which depends on both elements and on the kind of equality judgements used to arrive at its proof. So, on the logical side, one forgets about what was the actual witness. Cf. the existential generalization:

$$\frac{F(t)}{\exists x.F(x)}$$

where the actual witness is in fact 'abandoned'. Obviously, as we are interested in keeping track of relevant information introduced by each proof step, in labelled natural deduction system the witness is not abandoned, but is carried over as an unbounded name in the label of the corresponding conclusion formula.

$$\frac{t : D \qquad f(t) : F(t)}{\varepsilon x.(f(x), t) : \exists x^D.F(x)}$$

Note, however, that it is carried along *only* on the functional side, the logical side not keeping any trace of it at all.

Now, notice that if the functional calculus on the labels is to match the logical calculus on the formulas, then we must have the resulting label on the left of the '$\rhd_\beta$' as β-convertible to the concluding label on the right. So, we must have the convertibility equality:

$$\mathtt{REWR}(s(a, b), \grave{t}d(t)) =_\beta d(s/t) : C$$

The same holds for the η-equality:

$$\mathtt{REWR}(e, \grave{t}t(a, b)) =_\eta e :\doteq_D (a, b)$$

Parallel to the case of disjunction, where two different `constructors` distinguish the two alternatives, namely '`inl`' and '`inr`', we here have any (sequence of) equality identifiers ('β', 'η', 'μ', 'ξ', etc.) as `constructors` of proofs for the $\doteq$-connective. They are meant to denote the alternatives available.

General rules of equality. Apart from the already mentioned 'constants' (identifiers) which compose the reasons for equality (i.e. the indexes to the equality on the functional calculus), it is reasonable to expect that the following rules are taken for granted: *reflexivity*, *symmetry* and *transitivity*.

Substitution without involving quantifiers. We know from logic programming, i.e. from the theory of unification, that substitution can take place even when no quantifier is involved. This is justified when, for some reason a certain referent can replace another under some condition for identifying the one with the other.

Now, what would be counterpart to such a 'quantifier-less' notion of substitution in a labelled natural deduction system. Without the appropriate means of handling equality (definitional and propositional) we would hardly be capable of finding such a counterpart. Having said all that, let us think of what we ought to do at a certain stage in a proof (deduction) where the following two premises would be at hand:

$$a =_g y : D \qquad \text{and} \qquad f(a) : P(a)$$

We have that a and y are equal ('identifiable') under some arbitrary sequence of equalities (rewrites) which we name g. We also have that the predicate formula $P(a)$ is labelled by a certain functional expression f which depends on a. Clearly, if a and y are 'identifiable', we would like to infer that P, being true of a, will also be true of y. So, we shall be happy in inferring (on the logical calculus) the formula $P(y)$. Now, given that we ought to compose the label of the conclusion out of a composition of the labels of the premises, what label should we insert alongside $P(y)$? Perhaps various good answers could be given here, but we shall choose one which is in line with our 'keeping record of what (relevant) data was used in a deduction'. We have already stated how much importance we attach to names of individuals, names of formula instances, and of course, what kind of deduction was performed (i.e. what kind of connective was introduced or eliminated). In this section we have also insisted on the importance of, not only 'classifying' the equalities, but also having variables for the kinds of equalities that may be used in a deduction. Let us then formulate our rule of 'quantifier-less' substitution as:

$$\frac{a =_g y : D \qquad f(a) : P(a)}{g(a, y) \cdot f(a) : P(y)}$$

which could be explained in words as follows: if a and y are 'identifiable' due to a certain g, and $f(a)$ is the evidence for $P(a)$, then let the composition of $g(a, y)$ (the label for the propositional equality between a and y) with $f(a)$ (the evidence for $P(a)$) be the evidence for $P(y)$.

By having this extra rule of substitution added to the system of rules of inference, we are able to validate one half of the so-called 'Leibniz's law', namely:

$$\forall x^D.\forall y^D.(\dot{=}_D (x, y) \rightarrow (P(x) \rightarrow P(y)))$$

The *LND*equational fragment. As we already mentioned, in the *LND* equational logic, the equations have an index (the *reason*) which keeps all proof steps. The *reasons* is defined by the kind of rule used in the proof and the equational axioms (*definitional equalities*) of the system. The rules are divided into the following classes: (i) general rules; (ii) subterm substitution rule; (iii) ξ- and μ-rules.

Since the *LND* system is based on the Curry–Howard isomorphism [17], terms represent proof constructions, thus proof transformations correspond to equalities between terms. In this way, the *LND* equational logic can deal with equalities between *LND* proofs. The proofs in the *LND* equational fragment which deals with equalities between deductions are built from the basic proof transformations for the *LND* system, given in [39, 41, 42]. These basic proof transformations form an equational system, composed of *definitional equalities* (β, η and ζ).
General rules.

Definition (equation)
An equation in LND_{EQ} is of the form:

$$s =_r t : D$$

where s and t are terms, r is the identifier for the *rewrite reason*, and D is the type (formula).

Definition (system of equations)
A system of equations S is a set of equations:

$$\{s_1 =_{r_1} t_1 : D_1, \ldots, s_n =_{r_n} t_n : D_n\}$$

where r_i is the *rewrite reason* identifier for the ith equation in S.

Definition (rewrite reason)
Given a system of equations S and an equation $s =_r t : D$, if $S \vdash s =_r t : D$, i.e. there is a deduction/computation of the equation starting from the equations in S, then the *rewrite reason* r is built up from:

(i) the constants for *rewrite reasons*: $\{ \rho, \beta, \eta, \zeta \}$;
(ii) the r_i's;

using the substitution operations:

(iii) $\mathtt{sub}_{\mathtt{L}}$;
(iv) $\mathtt{sub}_{\mathtt{R}}$;

and the operations for building new *rewrite reasons*:

(v) σ, τ, ξ, μ.

Definition (general rules of equality)
The general rules for equality (reflexivity, symmetry and transitivity) are defined as follows:

reflexivity *symmetry* *transitivity*

$$\frac{x : D}{x =_\rho x : D} \qquad \frac{x =_t y : D}{y =_{\sigma(t)} x : D} \qquad \frac{x =_t y : D \quad y =_u z : D}{x =_{\tau(t,u)} z : D}$$

The "subterm substitution" rule. Equational logic as usually presented has the following inference rule of substitution:

$$\frac{s = t}{s\theta = t\theta}$$

where θ is a substitution.

Note that the substitution θ "appeared" in the conclusion of the rule. As rightly pointed out by Le Chenadec in [5], from the view point of the subformula property (objects in the conclusion of some inference should be subobjects of the premises), this rule is unsatisfactory. He then defines two rules:

$$IL\frac{M = N \quad C[N] = O}{C[M] = O} \qquad IR\frac{M = C[N] \quad N = O}{M = C[O]}$$

where M, N and O are terms and the context $C[\_]$ is adopted in order to distinguish subterms.

In [29] we have formulated an inference rule called "subterm substitution" which deals in an *explicit way*[5] with substitutions. In fact, the *LND* [41, 42] can be seen as an enriched system which brings to the object language terms, and now substitutions.

Definition (subterm substitution)
The rule of "subterm substitution" is framed as follows:

$$\frac{x =_r C \mid y \mid : D \quad y =_s u : D'}{x =_{\mathrm{sub_L}(r,s)} C \mid u \mid : D} \qquad \frac{x =_r w : D' \quad C \mid w \mid =_s u : D}{C \mid x \mid =_{\mathrm{sub_R}(r,s)} u : D}$$

where C is the context in which the subterm detached by '| |' appears and D' could be a subdomain of D, equal to D or disjoint to D.

The symbols sub_L and sub_R denote on which side (L—*left* or R— *right*) is the premiss that contains the subterm to be substituted.

Note that the transitivity rule previously defined can be seen as a special case for this rule when $D' = D$ and the context C is empty.

[5] In [41] de Queiroz and Gabbay recall Girard, who describes the intimate connections between constructivity and explicitation, and claim that "...one of the aims of inserting a label alongside formulas (accounting for the steps made to arrive at each particular point in the deduction) is exactly that of making *explicit* the use of formulas (and instances of formulas and individuals) throughout a deduction ..."

The ξ- and μ-rules. In the Curry–Howard "formulae-as-types" interpretation [17], the ξ-rule[6] states when two *canonical* elements are equal, and the μ-rule[7] states when two *noncanonical* elements are equal. So, each introduction rule for the *LND* system has associated to it a ξ-rule and each elimination rule has a related μ-rule. For instance, the ξ-rule and μ-rule for the connective $\wedge$ are defined as follows:

$$\frac{x =_u y : A \qquad s =_v t : B}{\langle x, s \rangle =_{\xi(u,v)} \langle y, t \rangle : A \wedge B}$$

$$\frac{x =_r y : A \wedge B}{\mathtt{FST}(x) =_{\mu(r)} \mathtt{FST}(y) : A} \qquad \frac{x =_r y : A \wedge B}{\mathtt{SND}(x) =_{\mu(r)} \mathtt{SND}(y) : B}$$

Term rewriting system for*LND* with equality. In [30] we have proved termination and confluence for the rewriting system arising out of the proof rules given for the proposed natural deduction system for equality.

Back to our example. Having defined the proof rules for our formulation of propositional equality, the proof of the logically valid formula used in our example related to the connections between Gentzen's and Herbrand's method would be as follows:

$$\cfrac{[a : D] \quad \cfrac{\cfrac{\cfrac{[a =_g y : D] \qquad [f(a) : P(a)]}{g(a, y) \cdot f(a) : P(y)}}{\lambda f.g(a, y) \cdot f(a) : P(a) \to P(y)}}{\Lambda y.\lambda f.g(a, y) \cdot f(a) : \forall y^D.(P(a) \to P(y))}}{\varepsilon x.((\Lambda y.\lambda f.g(x, y) \cdot f(x)), a) : \exists x^D.\forall y^D.(P(x) \to P(y))}$$

[6] The ξ-rule is the formal counterpart to Bishop's constructive principle of definition of a set ([4], p. 2) which says: "To define a set we prescribe, at least implicitly, what we have (the constructing intelligence) to do in order to construct an element of the set, and what we must do to show that two elements of the set are equal." Cf. also ([4], p. 12) Bishop defines a product of set as "The *cartesian product*, or simply *product*, $X \equiv X_1 \times \cdots \times X_n$ of sets $X_1, X_2, \ldots, X_n$ is defined to be the set of all ordered n-tuples $(x_1, \ldots, x_n)$ and $(y_1, \ldots, y_n)$ of X are equal if the coordinates x_i and y_i are equal for each i." See also ([24], p. 8): "... a set A is defined by prescribing how a canonical element of A is formed as well as how two equal canonical elements of A are formed". We also know from the theory of Lambda Calculus the definition of ξ-rule, see e.g. ([3], pp. 23 and 78): "$\xi : M = N \Rightarrow \lambda x.M = \lambda x.N$".

[7] The μ-rule is also defined in the theory of Lambda Calculus, see e.g. [25]: "The equational axioms and inference rules are as follows, where $[N/x]M$ denotes substitution of N for x in M. . . .

$$(\mu) \qquad \frac{\Gamma \rhd M_1 = M_2 : \sigma \Rightarrow \tau \quad \Gamma \rhd N_1 = N_2 : \sigma_{\prime\prime}}{\Gamma \rhd M_1 N_1 = M_2 N_2 : \tau}$$

and is divided into two equalities μ and ν in ([16], p. 66):

$$(\mu)\frac{M = M'}{NM = NM'} \qquad (\nu)\frac{M = M'}{MN = M'N}$$

Notice, however, that the proof term '$\varepsilon x.((\Lambda y.\lambda f.g(x, y) \cdot f(x)), a)$' contains 2 free variables, namely a and g, associated to, respectively, two assumptions:

1. the domain is nonempty: $[a : D]$
2. there is a way of identifying a term with another one: $[a =_g y : D]$

The variable g, which denotes a sequence of rewrites taking from a to y, has to be dealt with before the quantification of y. And, from our proof rules defined above, the assumption '$[a =_g y : D]$' appears in the application of an elimination rule to a statement like '$\doteq_D (a, y)$'. Thus, our formal derivation takes the following form:

$$\dfrac{[a:D] \qquad \dfrac{\dfrac{\dfrac{[r:\doteq_D(a,y)] \qquad \dfrac{\dfrac{[a=_g y:D] \quad [f(a):P(a)]}{g(a,y)\cdot f(a):P(y)}}{\lambda f.g(a,y)\cdot f(a):P(a)\to P(y)}}{\mathtt{REWR}(r,\overset{,}{g}\lambda f.g(a,y)\cdot f(a)):P(a)\to P(y)}}{\lambda r.\mathtt{REWR}(r,\overset{,}{g}\lambda f.g(a,y)\cdot f(a)):\doteq_D(a,y)\to(P(a)\to P(y))}}{\Lambda y.\lambda r.\mathtt{REWR}(r,\overset{,}{g}\lambda f.g(a,y)\cdot f(a)):\forall y^D.(\doteq_D(a,y)\to(P(a)\to P(y)))}}{\varepsilon x.((\Lambda y.\lambda r.\mathtt{REWR}(r,\overset{,}{g}\lambda f.g(x,y)\cdot f(x))),a):\exists x^D.\forall y^D.(\doteq_D(x,y)\to(P(x)\to P(y)))}$$

where the assumption '$a : D$' is still undischarged (the proof term still carries a free occurrence of a), thus the proof still rests on the assumption that the domain is nonempty.

3 Finale

The conception of the very first decision procedures for first-order sentences in the 1920s brought about the need for giving 'logical' citizenship to function symbols (e.g., Skolem functions). We have taken the view that a closer look at proof procedures for first-order sentences with equality brings about the need for introducing what we have called the "missing entity": (function) symbols for rewrites. This, we have argued, is appropriately done via the framework of labelled natural deduction which allows to formulate a proof theory for the "logical connective" of propositional equality. The basic idea is that when analysing an equality sentence into (i) proof conditions (*introduction*) and (ii) immediate consequences (*elimination*), it becomes clear that we need to bring in identifiers (i.e. function symbols) for sequences of rewrites, and this is what we have claimed should be the missing entity in P. Martin-Löf's equality types, both intensional and extensional. What we end up with is a formulation of what appears to be a middle ground solution to the 'intensional' versus 'extensional' dichotomy which permeates most of the work on characterising propositional equality in natural deduction style.

References

1. Aczel, P. H. G. (1980). Frege structures and the notions of proposition, truth and set. in J. Barwise, H.-J. Keisler, & K. Kunen (Eds.), *The Kleene Symposium, Vol. 101 of Studies in logic and the foundations of mathematics* (pp. 31–59). Amsterdam: North-Holland Publishing Co. (xx+425pp, 1980.Proceedings of the Symposium held in June 18–24).
2. Aczel, P. H. G. (1991). *Term declaration logic and generalised composita.* In Sixth Annual IEEE Symposium on Logic in Computer Science (LICS'91) (pp. 22–30). Amsterdam: IEEE Press. Proceedings of the Symposium held July 15–18, in Amsterdam, The Netherlands.
3. Barendregt, H.P. (1984). *The lambda calculus, its syntax and semantics, volume 103 of studies in logic and the foundation of mathematics.* Amsterdam: North-Holland (revised edition).
4. Bishop, E. (1967). *Foundations of constructive analysis.* McGraw-Hill series in Higher Mathematics (xiv+371pp). New York: McGraw-Hill Book Company.
5. Chenadec, Ph. Le. (1989). On the logic of unification. *Journal of Symbolic Computation, 8*(1 and 2), 141–199.
6. Dummett, M. (1973). *Frege: Philosophy of language* (xliii+708pp, 2nd (1981) ed.). London: Duckworth.
7. Dummett, M. (1991). *Frege: Philosophy of mathematics* (xiii+331pp). London: Duckworth.
8. Frege, G. (1891).*Funktion und Begriff.* Proceedings of the Jena Medical and Scientific Society.
9. Frege G. (1893). *Grundgesetze der Arithmetik. Begriffsschriftlich abgeleitet.* I. Jena: Verlag von Hermann Pohle. Reprinted in volume 32 of Olms Paperbacks, Georg Olms Verlagsbuchhandlung, Hildesheim, XXXII+254pp. Partial English translation in [11].
10. Furth M. (1964). *The basic laws of arithmetic. Exposition of the system* (lxiv+143pp). Berkeley: University of California Press. Partial English translation of Gottlob Frege's Grundgesetze der Arithmetik.
11. Gabbay, D. M. (1996). *Labelled deductive systems.* Volume I: Foundations. Oxford: Oxford University Press.
12. Gabbay, D. M., & de Queiroz, R. J. G. B. (1992). Extending the Curry-Howard interpretation to linear, relevant and other resource logics. *The Journal of Simbolic Logic, 57*(4), 1319–1365.
13. Girard, J.-Y. (1987). *Proof theory and logical complexity. Volume I of studies in proof theory* (503 pp). Napoli: Bibliopolis.
14. Hilbert, D., & Bernays, P. (1934). *Grundlagen der Mathematik I,* volume XL of Die Grundlehren der mathematischen Wissenschaften (XII+471pp). Berlin: Verlag von Julius Springer.
15. Hilbert, D., & Bernays, P. (1939). *Grundlagen der Mathematik II,* volume L of Die Grundlehren der mathematischen Wissenschaften (XII+498pp). Berlin: Verlag von Julius Springer.
16. Hindley, J. R., & Seldin, J. P. (1986). *Introduction to combinators and λ-Calculus.* Cambridge: London Mathematical Society Student Texts: Cambridge University Press.
17. Howard, W. A. (1980). The formulae-as-types notion of construction. In J. R. Seldin & J. R. Hindley (Eds.), *To H. B. Curry: Essays on combinatory logic lambda calculus and formalism* (xxv+606pp). London: Academic Press.
18. Kleene, S. C. (1967). *Mathematical logic.* New York: Wiley Interscience.
19. Kreisel, G., & Tait, W. (1961). Finite definability of number theoretic functions and parametric completeness of equational calculi. *Zeitschr. f Math. Logik und Grundlagen d. Math, 7,* 28–38.
20. Leisenring, A. C. (1969). *Mathematical logic and Hilbert's ε-Symbol,* A volume of University Mathematical Series (ix+142pp). London: MacDonald Technical and Scientific.
21. Martin-Löf, P. (1975). About models for intuitionistic type theories and the notion of definitional equality. In S. Kanger (Ed.), *Proceedings of the third Scandinavian logic symposium, Series studies in logic and the foundations of mathematics* (pp. 81–109). Amsterdam: North-Holland.
22. Martin-Löf, P. (1975a). An intuitionistic theory of types: Predicative part. in H. E. Rose & J. C. Shepherdson (Eds.), *Logic Colloquium '73, Vol. 80 of studies in logic and the foundations of mathematics* (pp. 73–118, viii+513pp). Amsterdam: North-Holland.

23. Martin-Löf, P. (1982). Constructive mathematics and computer programming. In L. J. Cohen, J. Łos, H. Pfeiffer, & K.-P. Podewski (Eds.), *Logic, methodology and philosophy of science VI, Series studies in logic and the foundations of mathematics* (pp. 153–175, xiii+738pp). Amsterdam: North-Holland.
24. Martin-Löf, P. (1984). *Intuitionistic type theory*. Series studies in proof theory (iv+91pp). Bibliopolis Naples. Notes by Giovanni Sambi of a series of lectures given in Padova, June.
25. Mitchell, J. C., & Scedrov, A. (1993). Notes on sconing and relators. In E. Boerger, et al. (Eds.), *Computer science logic '92*, Selected Papers (pp. 352–378). Springer LNCS 702.
26. Montague, R. (1970). Universal grammar. *Theoria*, *36*, 373–398.
27. Nordström, B., Petersson, K., & Smith, J. M. (1990). *Programming in Martin-Löf's type theory. An introduction*, Vol. 7 of The International Series of Monographs on Computer Science (x+221pp). Oxford: Clarendon Press.
28. de Oliveira, A. G. (1995). *Proof transformations for labelled natural deduction via term rewriting* (In Portuguese). Master's thesis, Depto. de Informática, Universidade Federal de Pernambuco, C.P. 7851, Recife, PE 50732–970, Brasil.
29. de Oliveira, A. G., & de Queiroz, R. J. G. B. (1994). *Term rewriting systems with labelled deductive systems*. In Proceedings of Brazilian symposium on artificial intelligence (SBIA'94), Lecture Notes in Artificial Intelligence, Springer, pp. 59–72.
30. de Oliveira, A. G., & de Queiroz, R. J. G. B. (1999). A normalization procedure for the equational fragment of labelled natural deduction. *Logic Journal of the Interest Group in Pure and Applied Logics*, *7*(2), 173–215.
31. de Oliveira, A. G., & de Queiroz, R. J. G. B. (2005). A new basic set of proof transformations. In S. Artemov, H. Barringer, A. Garcez, L. Lamb, & J. Woods (Eds.), *We will show them! Essays in Honour of Dov Gabbay* (Vol. 2, pp. 499-528). London: College Publications. ISBN 1904987125.
32. de Queiroz, R. J. G. B. (1988). A proof-theoretic account of programming and the rôle of reduction rules. *Dialectica*, *42*(4), 265–282.
33. de Queiroz, R. J. G. B. (1989). The mathematical language and its semantics: To show the consequences of a proposition is to give its meaning. In P. Weingartner & G. Schurz (Eds.), *Reports of the thirteenth international Wittgenstein symposium, volume 18 of Schriftenreihe der Wittgenstein-Gesellschaft* (pp. 259–266). Austria: Hölder-Pichler-Tempsky. Symposium held in Kirchberg/Wechsel.
34. de Queiroz, R. J. G. B. (1991). Meaning as grammar plus consequences. *Dialectica*, *45*(1), 83–86.
35. de Queiroz, R. J. G. B. (1992). *Grundgesetze alongside Begriffsschrift*. In Abstracts of Fifteenth International Wittgenstein, Symposium. pp. 15–16.
36. de Queiroz, R. J. G. B. (1994). Normalisation and language-games. *Dialectica*, *48*(2), 83–125.
37. de Queiroz, R. J. G. B. (2001). Meaning, function, purpose, usefulness, consequences—interconnected concepts. *Logic Journal of the Interest Group in Pure and Applied Logics*, *9*(5), 693–734.
38. de Queiroz, R. J. G. B., & Gabbay, D. M. (1994). Equality in labelled deductive systems and the functional interpretation of propositional equality. In P. Dekker & M. Stokhof (Eds.), *Proceedings of the 9th Amsterdam Colloquium* (pp. 547–546)
39. de Queiroz, R. J. G. B., & Gabbay, D. M. (1993). The functional interpretation of the existential quantifier. *Bulletin of the Interest Group in Pure and Applied Logics*, *3*(2 and 3), 243–290.
40. de Queiroz, R. J. G. B., & Gabbay, D. M. (1997). The functional interpretation of modal necessity. In de Rijke, Maarten (Ed.), *Advances in intensional logic (Applied logic series)* (pp. 61–91). Dordrecht: Kluwer Academic Publishers.
41. de Queiroz, R. J. G. B., & Gabbay, D. M. (1999). Labelled natural deduction. In H. J. Ohlbach & U. Reyle (Eds.), *Logic, language and reasoning*. Essays in Honor of Dov Gabbay's 50th Anniversary (pp. 173–250). Dordrecht: Kluwer Academic Publishers.
42. de Queiroz, R. J. G. B., Gabbay, D. M., & de Oliveira, A. G. (2013). *The functional interpretation of non-classical logics*. River Edge, NJ: Imperial College Press, World Scientific.

43. Sluga, H. (1980). *Gottlob Frege. Series The Arguments of the Philosophers* (xi+203pp). London: Routledge & Kegan Paul.
44. Statman, R. (1977). Herbrand's Theorem and Gentzen's notion of a direct proof. In J. Barwise (Ed.), *Handbook of mathematical logic*. Amsterdam: North-Holland.
45. Statman, R. (1978). Bounds for proof-search and speed-up in the predicate calculus. *Annals of Mathematical Logic*, *15*, 225–287.
46. Troelstra, A. S., & van Dalen, D. (1988). *Constructivism in mathematics: An introduction*. Vol. II, Vol. 123 of Studies in logic and the foundations of mathematics (xvii+535pp). Amsterdam: North-Holland.

Paul Hertz's Systems of Propositions As a Proof-Theoretical Conception of Logic

Javier Legris

Abstract Paul Hertz was an outstanding German physicist, who also devoted himself to mathematical logic and wrote a series of papers that remained rather unnoticed, even if they influenced the development of proof theory and particularly Gentzen's work. This chapter aims to examine Hertz's logical theory placing it in its historical context and remarking its influence on Gentzen's sequent calculus. The analysis of the formal structure of proofs was one of Hertz's most important achievements and it can be regarded as an anticipation of a "theory of proofs" in the current sense. But also, it can be asserted that Hertz's systems played the role of a bridge between traditional formal logic and Gentzen's logical work. Hertz's philosophical ideas concerning the nature of logic and its place in scientific knowledge will be also analysed in this chapter.

Introduction

During the 1920s of the twentieth century, Paul Hertz (1881–1940) published in some German mathematical journals, such as Mathematische Annalen, a series of papers that influenced the development of proof theory and, particularly, Gentzen's work. The very idea of sequent calculi and of structural rules was first conceived in them. This fact was explicitly acknowledged by Gentzen himself who devoted his first published chapter to Hertz's systems. As a matter of fact, the logical work of Hertz and its influence on Gentzen remained rather unnoticed by logicians. However, Paul Bernays had a highly positive appraisal of Hertz's contributions. Alonzo Church listed in his bibliography almost all of Hertz's logical writings (see [6], p. 187),

J. Legris (✉)
IIEP-BAIRES, CONICET and University of Buenos Aires, Av. Córdoba 2122, Buenos Aires, Argentina
e-mail: legris@econ.uba.ar; jlegris@retina.ar

L. C. Pereira et al. (eds.), *Advances in Natural Deduction*,
Trends in Logic 39, DOI: 10.1007/978-94-007-7548-0_5,

and also Haskell B. Curry pointed out the importance of Hertz's contributions in a historical note in his Foundations of Mathematical Logic ([7], p. 246 ff.).[1]

Hertz studied Physics and Mathematics in Göttingen, where he earned his PhD in 1904. He taught in Heidelberg until 1912. In this period, he contributed essentially to statistical mechanics and thermodynamics. According to Bernays, typical of his research was its conceptual and methodological clarity (see Bernays 1969). In 1913 he moved to Göttingen again, where he—under the influence of David Hilbert's program—became interested in Epistemology and Methodology of Science. He edited, together with Moritz Schlick, the work of Hermann von Helmholtz on Philosophy of Science,[2] and received then an appointment to teach on "Methoden der exakten Naturwissenschaften" (Methodology of exact natural sciences) in Göttingen. In 1933, his venia legendi in Göttingen was withdrawn and he received a research grant from the American Rescue Committee for the University of Geneva and—from 1936 on—for the German University of Prague. In 1939 Hertz emigrated to the USA, where he died in 1940 (for further references see Bernays 1969).[3]

Hertz's main contributions to logic could be summarized as follows:

1. An Analysis of the structure of proofs.
2. The idea of a system consisting exclusively of structural deductions.
3. The notion of sequent, anticipating GentzenÃŕs sequent calculus.
4. The development of methods for constructing minimal axiom systems.

Hertz's original interest in logic was largely focused on the formal properties of axiom systems and his goal consisted in developing reduction methods for axiomatic systems from which some sort of "minimal" and independent system could be obtained, that is, a system where proofs should be as elemental as possible. Taking into account the situation of diferent axiomatic systems for the same theory, he explicitly thought about the possibility of developing reduction procedures to achieve

---

[1] In his biographical article on Hertz, Bernays wrote "Diese Untersuchungen [of Hertz] sind Vorlüfer verschiedener neuerer Forschungen zur mathematischen Logik und Axiomatik, insbesondere hat G. Gentzens Sequenzenkalkul von den H.schen Betrachtungen über Satzsysteme seinen Ausgang genommen" ([5], p. 712), and in a chapter on sequent calculi, Bernays asserted: "... in der Hertz'schen Theorie der Satzsysteme ein gewiss bei weitem noch nicht hinsichtlich der möglichen Fragestellungen und Erkenntnisse ausgeschöpftes Forschungs-gebiet der Axiomatik und Logik vorliegt." ([4], p. 5, footnote). Haskell Curry in a historical note of his textbook on mathematical logic stated: "For the present context it is worthwhile to point out that Gentzen was apparently influenced by Hertz [...] This throws some light on the role of "Schnitt" in the Gentzensystem". ([7], p. 246 f.). Vittorio Michele Abrusci wrote an introductory chapter on Hertz's logical work [1].

[2] Together with Moritz Schlick edited Hertz in 1921 the epistemological writings von Helmholtz. In some of these writings, problems concerning the philosophy and methodology of mathematics were discussed, especially in relation to the evaluation of non-Euclidean geometry. In these editions the following papers of von Helmholtz were included "über den Ursprung und die Bedeutung der geometrischen Axiome", "über die Tatsachen, die der Geometrie zugrunde liegen", "Zahlen und Messen" and "Die Tatsachen in der Wahrnehmung".

[3] Further information about Hertz can be found in the Nachlaβ of Paul Bernays (ETH Zürich), the Nachlaβ of David Hilbert (Göttingen), biographical notes (not published) by Adriaan Rezus (Nijmegen), and above all in Hertz's Nachlaβ located at Archives for Scientific Philosophy, University of Pittsburgh.

some sort of irreductible ("normal") axiomatic system (see [10], p. 246). In doing so, he initiated an investigation of the general properties of axiomatic systems. In this respect, he referred to "the notion of axiom in general" ([13], p. 427). His research led him to the idea of Satzsystem (system of propositions), undoubtedly his main contribution to mathematical logic.

1 Hertz's Conception of Logic

In Hertz's sense, a "proposition", related to a "basic domain" of elements, is an expression of the forms:

1. $a \to b$,
2. $a_1, a_2, \ldots, a_n \to b$.

where $a_1, a_2, \ldots, a_n$ are called the antecedent of the proposition and b its succedent ([12], p. 273). Propositions of the form (1) are called lineal (lineare) propositions, i.e., they have only one antecedent. Apart from this, he distinguished between propositions with free individual variables (generally understood as universally quantified) and propositions with constants which he called, respectively, 'macropropositions' (Makrosätze) and 'micropropositions' (Mikrosätze).

Hertz proposed different interpretations for these propositions, depending on the nature of the basic domain (events or predicates). Basically, we can find the following interpretations:

(i) elements as events (Ereignisse, see [13], p. 459);
(ii) elements as predicates: If the predicates $a_1, a_2, \ldots, a_n$ satisfy an object, then b satisfies it too (see [12], p. 273);
(iii) propositions as formal implications "in the sense of Russell" (see [12], p. 247), that is, they are to be understood as general valid implications between sentences.

According to this, the very symbol $\to$ is alternately interpreted either as a logical relation (between events, predicates or sentences) or as a logical constant, and it remains unclear which of these interpretations was privileged by Hertz. However, he conceived the idea of system of propositions in a very abstract way, that is, as a "complex" of elements.[4]

It should be noticed that in that time Hertz's systems of propositions were mainly seen as a contribution to positive logic (i.e. propositional logic restricted only to conjunction, disjunction and material implication), which was analysed by Bernays and others in Göttingen and for which its decidability had already been proved (see [2], p. 11). In the *Grundlagen der Mathematik* by Hilbert and Bernays, Hertz is mentioned related to this field of research (see [17] p. 68 n. 1).

---

[4] "Unter Satz verstehen wir einen Inbegriff eines Komplexes, der auch nur aus einem ein einzigen Element bestehen kann und antecedens heiβt, und eines Elementes, das succedens heiβt." ([10], p. 81)

With systems of this kind, Hertz aimed to solve the problem of the deductive closure of formal theories, which was discussed at that time, and probably it was the main subject of Hertz's investigations. This problem can be stated as the search for a general way to determine that every theorem of a deductive system can be proved. In other words, it consisted in assuring that every theorem is contained in the system. It was also handled explicitly by Fritz London in his doctoral thesis *Über die Bedingungen der Möglichkeit einer deduktiven Theorie*, published in 1923, and later by Alfred Tarski (see [20], p. 70), but on a completely different background than Hertz.

He established a minimal set of rules for closed systems of propositions (*geschlossene Satzsysteme*), which constituted an antecedent of the structural rules in Gentzen's sequent calculus. He considered the case of an "inference system" (*Schlusssystem*) consisting in proofs for a system of proposition through these rules.

Following the characterization of closed systems of propositions, they must contain what Hertz called "tautological propositions", of the form $a \to a$, which are "essentially logical" propositions.

Appart from this, Hertz realized that the closure of the system depends on both the rules of the system and the proof procedures in it. In all the papers he wrote on the subject in the 1920s he presented two basic inference rules, which not only constituted the minimal set of rules for building closed systems but also played the role of basic principles defining logical inference. The first rule was a generalization of transitivity of the arrow and was called by Hertz with the traditional name of *syllogism (Syllogismus)*. It had the following structure:

$$\frac{\begin{array}{c} a_{11}, a_{12}, \ldots, a_{1n} \to b_1 \\ a_{21}, a_{22}, \ldots, a_{2n} \to b_2 \\ \ldots \\ a_{m1}, a_{m2}, \ldots, a_{mn} \to b_m \\ a_{11}, \ldots, a_{1n}, a_{21}, \ldots, a_{2n}, a_{m1}, \ldots, a_{mn}, b_1 \ldots b_m \to c \end{array}}{a_{11}, \ldots, a_{1n}, a_{21}, \ldots, a_{2n}, a_{m1}, \ldots, a_{mn} \to c}$$

Hertz conceived this rule as a generalized form of the *modus Barbara* of the Aristotelian syllogistic (if we think the rule as applied to *Makrosätze*, that is, those propositions having free variables, which should be understood as universally quantified), and for him it played a decisive role in the characterization of logic. It constituted the basis of the whole deductive logic (see his later chapter of 1935). Moreover, Hertz himself considered his systems of propositions as "the old theory about the chained inferences according to the *Modus Barbara*" (1929b, p. 178). He understood 'chained-inferences' (*Kettenschlüsse*) as the succesive application of the syllogism rule in a system of propositions as some kind of *sorites*.

The second rule is the rule of 'immediate inference' (*ummitelbarer Schluß*) having the following form (see [13], p. 463):

$$\frac{a_1, a_2, \ldots, a_n \to b}{a^1, a^2, \ldots, a^m, a_1, a_2, \ldots, a_n \to b}$$

This rule permits the introduction of every antecedent in a proposition stating also a kind of monotonicity principle for deductive inferences, another distinctive feature of logic. When this rule is applied to tautological propositions, the results are logical propositions, having the form $a_1, a_2, \ldots, a_n, a \to a$ and called *trivial propositions* by Hertz.

2 Hertz and Gentzen's Sequent Calculus

If we restrict ourselves to sequents with only one succedent, the similarities between Hertz's rules of syllogism and immediate inference, on the one hand, and the cut rule and thinning of Gentzen's sequent calculus, on the other hand are obvious. Moreover, Hertz's tautological propositions correspond to the logical axioms in Gentzen's sequent calculus. In his chapter of 1933 on Hertz, Gentzen fomulated a version of *Satzsysteme*, which constituted an intermediate system between Hertz's system and the sequent calculus (see [8]). This system contained tautological propositions, a version of immediate inference -called *Verdünnung* (thinning) by Gentzen- and the following simplified version of syllogism:

$$\frac{L \to u \qquad Mu \to v}{LM \to v}$$

which Gentzen called 'cut' (*Schnitt*).

To show that his rules provided closed deductive systems, Hertz had to analyse the structure of the proofs generated by them. Thus, he devised a whole proof theory for systems of propositions. He developed a specific terminology for the description of proofs (see the second part of Hertz 1923). With the expression 'chained-inference' (*Kettenschluß*) he referred to orderings in the applications of rules, and he called 'inference systems' (*Schlußsystem*, see [13], p. 467) the derivations resulting from tautologies by applying the rules of syllogism and immediate inference. These inference systems have always a tree-structure.

In investigating the structure of proofs, Hertz identified two "proof-methods" for systems of propositions with lineal propositions, which constitute two different normal forms for proofs. Following the tradition in formal logic and the usage in logic textbooks of that time, Hertz called these two forms 'Aristotelian' and 'Goclenian'—the latter after the German logician Rudolph Göckel (Rodolphus Goclenius, 1547–1628), who had considered categorical syllogisms of an analogous form. An example of an Aristotelian normal form is the following:

$$\frac{\dfrac{a\to b \quad b\to m}{a\to m} \qquad m \to c}{a \to c}$$

This normal form is characterized by the fact that the right premiss of each application of the syllogism rule is an axiom (or "upper proposition", see e.g. [13], p. 473 ff.).

On the contrary, in the Goclenian normal form the left premiss of each application is an axiom, that is,

$$\frac{a \to b \quad \frac{b \to m \quad m \to c}{b \to c}}{a \to c}$$

That is, the last upper proposition in an Aristotelian normal form is first in a Goclenian one, and vice versa: the last upper proposition in a Goclenian normal form is an Aristotelian one.

Hertz devised a procedure for transforming Goclenian normal form into an Aristotelian one (see [13], p. 473). This procedure should be a decision method related to these normal forms (see [13] Sect. 3). A reconstruction of these procedures can be found in [19].

The notion of logical consequence is currently defined among logicians in a model-theoretical way, following Tarski's ideas who defined it as an operation on sets of formulas (see [20]). Now, if we take into account that syllogism and immediate inference, together with tautological propositions, determine what must be counted as closed deductive systems, we must admit the decisive role they play in characterizing logical consequence from a proof-theoretical point of view. So, Hertz initiated another direction in symbolic logic. By means of his sequent calculus, Gerhard Gentzen provided a definition of logical consequence. The structural rules for sequent calculi determine general conditions for deductive inferences and constitute a framework for defining logical constants. In particular, the rules of cut and thinning, together with logical axioms of sequent calculi give a characterization of logical consequence (see [9]).

To some extent, the Tarskian usual definition of logical consequence by means of the three basic conditions of reflexivity, monotonicity and transitivity was anticipated by Hertz's conception. These similarities were acknowledged by Tarski (see [20], p. 62 fn.).[5] Notwithstanding, Hertz has mainly and decisively influenced Gentzen's proof-theoretical view on logical consequence.

Hertz considered his rules as the "essence of logic", as he stated in his contribution to a conference on mathematical logic that took place later in Geneva in 1934. According to Hertz, the rule of syllogismus specially disclosed a process that could be summarized as the *search* for a middle-term, designated as *interpositum* by him, in order to apply this rule (see [15], pp. 249 ff.). The assertion of an implication $S \to A$ implies the existence of an *interpositum* C, so that $S \to C$ and $C \to A$ can be asserted, and this process goes on until the subject finds basic propositions of some kind. In this process lies the origin of logic. Here, a new formulation of Aristotle's main ideas on the nature of syllogism, and its methodological function, can be found. These ideas were expressed by the *inventio medii*, the search for a middle term to construct a syllogism in modus barbara (see [18] A26 43 a 16–24). However, Hertz emphasized the epistemological aspects. Originally, Hertz regarded

[5] Tarski wrote: "The discussion in Hertz, P. (27) has some points of contact with the present exposition." ([20], p. 62 fn 1).

deduction as a kind of mental process. His epistemology book of 1923, *Über das Denken*, made reference to logical processes in knowledge (see [11], p. 121).

It is interesting to notice that this process is not itself a deductive procedure. It can be identified with what is generally called a *regressive method*, a method for finding the grounds or justification of a proposition (which is then deduced from them). This *regressive method* is opposed to the *progressive method*, which proceeds from the grounds to the proposition in a synthetic way. Deduction should be a kind of such method. The notion of *regressive method* was dicussed in Göttingen at that time within Hilbert's research group, and it should be an essential part of the axiomatic method, its analytical part, consisting in the differentiation between what counts as axiom and what as theorem in an axiomatic system, that is, it consists in establishing the axioms of the system, according to statements of Hilbert (see [16]). The other part of the axiomatic method would be the progressive part consisting in deducing theorems from the axioms.

3 Conclusion

In his philosophical papers and in some letters to Paul Bernays, Hertz made reference in several opportunities to *synthetic* moments or aspects in logic and it should be understood in the sense of a combinatorial or progressive process, such as Bernays stated as well in his philosophical chapter on Hilbert's proof-theory [3].

Another point connected with this philosophical characterization of logic can be found in the original motivations of Hertz. In his first paper on Satzsysteme, Hertz introduced the notion of *ideal elements* ("ideale Elemente") in an axiomatic system to the effect that the number of axioms in the system can be reduced. The expression "ideal element" was in vogue at that time and it came to be an important notion in Hilbert's programme, where a distinction between a real and an ideal part in mathematics is made. In the case of Hertz, it can be said that he intended to grasp only some of the purely *formal* aspects of this notion, excluding semantical properties. This formal aspects consisted in its connecting real elements and in its reductive power. He argued that these ideal elements (terms or formulas) serve as a way to connect real elements. Moreover, these ideal elements provide "an accurate representation of propositions with real elements", "propositions with ideal elements have no meaning in themselves (*Bedeutung an sich*). Only propositions including exclusively real elements have meaning" ([10], p. 249).

Hertz gave the following example of a system with the elements a, b, c, d and e constituting its domain and having the following axioms (on the left), and producing the following simplified system through the ideal element i (on the right):

$$\begin{array}{l} a \to d \\ a \to e \\ b \to d \\ b \to e \\ c \to d \\ c \to e \end{array} \Longrightarrow \begin{array}{l} a \to i \\ b \to i \\ c \to i \\ i \to d \\ i \to e \end{array}$$

(see [10], p. 248).

Now, Hertz suggested that *molecular* propositions could also be seen as ideal propositions:

> Through introduction of ideal propositions, we avoid the use of the words 'or' in the antecedents, and the use of the word 'and' in succedents. ([10], p. 262)

This idea can be easily illustrated through the following system:

$$\begin{array}{l} A_1 \to B_1 \\ A_1 \to B_2 \\ A_2 \to B_1 \\ A_2 \to B_2 \end{array}$$

which can be reduced alternatively to the systems

$$\begin{array}{l} A_1 \to B_1 \wedge B_2 \\ A_2 \to B_1 \wedge B_2 \end{array}$$

or

$$\begin{array}{l} A_1 \vee A_2 \to B_1 \\ A_1 \vee A_2 \to B_2 \end{array}$$

and finally to the system with the only proposition

$$\beta \to \alpha$$

where $B_1 \wedge B_2$ is the ideal element α and $A_1 \vee A_2$ is the ideal element β. So, the introduction of molecular propositions in systems of propositions serves the purpose of reducing the number of axioms.

In this respect, it must be noticed that Hertz did not develop a theory about logical constants, although it is in some way implicit in his reflexion on the nature of logic and in the formulation of his systems of propositions. The case of implication is already mentioned above. In different passages, he considered the case of the disjunction. As he stated, "The disjuntive judgement is interesting only as a premiss in an inference" ([15], p. 250).[6]

---

[6] As a whole, Hertz's conception on logical constants has certain similarities with the idea of logical constants as "puncuation marks".

It seems reasonable to connect these diferents aspects of finding ideal elements, searching an interpositum, the regressive method and, finally, the existence of synthetic processes in deduction, considering them as the key notions in epistemological foundations of logic as it was pursued by Hertz.

References

1. Abrusci, V. M. (1983). Paul Hertz's Logical Works Contents and Relevance. Atti del Convegno Internazionale di Storia della Logica, San Gimignano, 4–8 dicembre 1982. In: V. M. Abrusci, E. Casari & M. Mugnai (eds.), (pp. 369–374) Bologna: Clueb.
2. Bernays, P. (1927). Probleme der theoretischen Logik. *Unterrichtsblätter für Mathematik und Naturwisschenschaften, 33*, 369–377.
3. Bernays, P. (1930). Die Philosophie der Mathematik und die Hilbertsche Beweistheorie. *Blätter für Deutsche Philosophie, 4*(1930–1931), 326–367.
4. Bernays, P. (1965). Betrachtungen zum sequenzen-Kalkül. In: J. M. Bochenski (ed.), *Contributions to Logic and Methodology in Honor* (pp. 1–4). Amsterdam: Anna-Teresa Tymieniecka.
5. Bernays, P. (1969). *Paul Hertz. Neue deutsche Biographie. Herausgegeben von der historischen Kommission bei der bayerischen Akademie der Wissenschaften* (Vol. 8, p. 711 f). Berlin: Ducker and Humblot
6. Church, A. (1936). A bibliography of symbolic logic. *The Journal of Symbolic Logic, 1*, 121–218.
7. Curry, H. B. (1977). *Foundations of mathematical logic*. New York: Dover.
8. Gentzen, G. (1933). Über die Existenz unabhängiger Axiomensystem zu unendlichen Satzsystemen. *Mathematische Annalen, 107*, 329–350.
9. Gentzen, G. (1935). Untersuchungen über das logische Schlie*ß*en. *Mathematische Zeitschrift, 39*, 176–210 (and 405–431).
10. Hertz, P. (1923a). Über Axiomensysteme für beliebige Satzsysteme. I. Teil. Sätze ersten grades. *Mathematische Annalen, 87*, 246–269.
11. Hertz, P. (1923b). *Über das Denken*. Berlin: Springer.
12. Hertz, P. (1928). Reichen die üblichen syllogistischen Regeln fúr das Schlie*ß*en in der positiven Logik elementarer Sätze aus? *Annalen der Philosophie, 7*, 272–277.
13. Hertz, P. (1929a). Über Axiomensysteme für beliebige Satzsysteme. *Mathematische Annalen, 101*, 457–514.
14. Hertz, P. (1929b). Über Axiomensysteme beliebiger Satzsysteme. *Annalen der Philosophie, 8*, 178–204.
15. Hertz, P. (1935). Über das Wesen der Logik und der logischen Urteilsformen. *Abhandlungen der Fries'schen Schule. Neue Folge, 6*, 227–272.
16. Hilbert, D. (1992). Natur und mathematisches Erkennen. Vorlesungen, gehalten 1919–1920 in Göttingen. In: David E. Rowe (ed.). *Nach einer Ausarbeitung von Paul Bernays*. Basel: Birkhauser.
17. Hilbert, D., & Bernays, P. (1934). *Grundlagen der Mathematik* (Vol. I). Berlin: Springer.
18. Prior Analytics trans. by R. Smith (1989) Aristotle, Hackett: Indianapolis.
19. Schroeder-Heister, P. (2002). Resolution and the origins of structural reasoning: Early proof-theoretic ideas of Hertz and Gentzen. Bulletin of Symbolic Logic, *8*(2), 246–265
20. Tarski, A. (1930). Fundamentale Begriffe der Methodologie der deduktiven Wissenschaften I. Monatshefte für Mathematik und Physik 37, pp. 361–404. (A. Tarski & J. H. Woodger, Trans. Logic, Semantics, Metamathematics) (2nd ed.) Hackett: Indianapolis. (1983).

On the Structure of Natural Deduction Derivations for "Generally"

Leonardo B. Vana, Paulo A. S. Veloso and Sheila R. M. Veloso

Abstract Logics for 'generally' were introduced as extensions of First-Order Logic (FOL) for handling assertions with vague notions (e.g., 'generally,' 'most,' and 'several') expressed by formulas with generalized quantifiers. Deductive systems have been developed for such logics. Here, we characterize the structure of derivations in natural deduction style for filter logic. This characterization extends the familiar one for FOL.

Keywords Logics for 'generally' · Generalized quantifiers · Vague notions · Natural deduction · Derivation structure · Normalization · Minimum formula · Minimum segment

1 Introduction

Logics for 'generally' (LG's) were introduced as extensions of First-Order Logic (FOL) for handling assertions with vague notions (e.g., 'generally,' 'most,' and 'several') expressed by generalized formulas. Deductive systems have been developed for LGs [1–10].

Research partly sponsored by the Brazilian agencies CNPq and FAPERJ.

L. B. Vana (✉)
Institute of Mathematics, Universidade Federal Fluminense (UFF), Niterói, Brazil
e-mail: leobvana@gmail.com

P. A. S. Veloso
Systems and Computer Engineering Program, COPPE, Universidade Federal do Rio de Janeiro (UFRJ), Rio de Janeiro, Brazil
e-mail: pasveloso@gmail.com

S. R. M. Veloso
Department of Systems and Computer Engineering, Universidade Estadual do Rio de Janeiro (UERJ), Rio de Janeiro, Brazil
e-mail: Sheila@cos.ufrj.br

L. C. Pereira et al. (eds.), *Advances in Natural Deduction*,
Trends in Logic 39, DOI: 10.1007/978-94-007-7548-0_6,

With derivation rules that reflect the diverse properties of 'generally,' we can construct natural deduction systems for LGs. These systems, which are constructed modularly (in that each one can be built from the basic one (BL) by adding appropriate rules) are correct, complete, and normalizable [11–14].

The equivalence rule, which characterizes BL, reflects extensionality (equal sets have the same properties) and can be seen both as an elimination and an introduction rule. Its interaction with the other ones affects the structure of the derivations, with consequent loss of some usual properties, e.g., the subformula property.

In this chapter, we present a result analogous to the one of minimum formulas for normal derivations [15]. Our result is obtained by controlling and minimizing the number of applications of the equivalence rule, having as a consequence the characterization of the structure of normal proofs.

The familiar structure of natural deduction derivations, an "hour-glass" form consisting of elimination region and introduction region separated by minimum formulas, is a particular case of our characterization.

In FOL, a path from hypothesis A to conclusion C uses elimination rules ($\searrow$) to reach a minimum formula B and then introduction rules ($\nearrow$) to obtain the conclusion; its structure can be represented as follows:

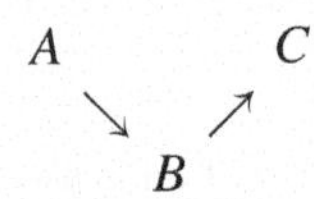

Now, LG's also handle generalized information and the path structure becomes as follows:

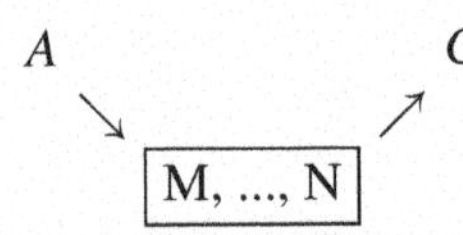

Here, we now have a frontier region manipulating 'generally' (represented by a quantifier), possibly with several applications of the equivalence rule ($\Updownarrow$): $M(\Updownarrow)\cdots(\Updownarrow)N$. In this chapter, we will characterize the inner structure of this frontier region, roughly as generalized introductions ($\nearrow$) and eliminations ($\searrow$), separated by a single application of the equivalence rule ($\Updownarrow$), namely:

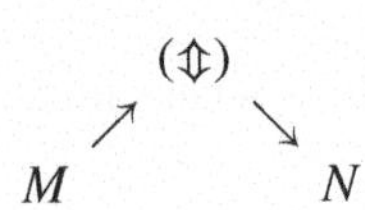

The structure of this chapter is as follows. In Sect. 2, we briefly review some ideas about logics of 'generally' (motivation, syntax, and semantics) and introduce the idea of marked formulas. In Sect. 3, we present natural deduction systems for some logics of 'generally.' In Sect. 4, we examine the normalization result for the natural deduction systems for 'generally'. In Sect. 5, we analyze the structure of the normal derivations. Finally, we comment on the main features of our approach and on on-going work in Sect. 6. The Appendix presents some details of our development.

2 Logics of 'Generally'

In this section, we briefly review some ideas about logics of 'generally' (LGs) [1, 2, 6]. Their goal is providing a framework for reasoning with distinct notions of 'generally' by means of (nonstandard) generalized quantifiers [16, 17].[1]

subsection*Motivation

Vague notions such as 'generally,' 'rarely,' 'most,' 'several,' etc., appear often in assertions and arguments in ordinary language and in some branches of science [6, 20]. Logics of 'generally' (LG's) are designed to capture distinct intuitive notions of 'generally.' We will now illustrate these ideas.

First, consider the universe of Brazilians and imagine that one accepts the two assertions:

(α) "Brazilians generally shave their legs" and
(β) "Brazilians generally have their faces shaved."

In this case, one is likely to accept also the assertion:

($\sqcup$) "Brazilians generally have their faces shaved or sport a moustache";

but one is not likely to accept the following assertion:

($\sqcap$) "Brazilians generally shave their legs and their faces."

Next, consider the universe of natural numbers and imagine that one accepts the two assertions:

(γ) "naturals generally are larger than 15", and
(δ) "naturals generally do not divide 12".

Then, one would probably accept also the following two assertions:

($\vee$) "naturals generally are larger than 15 or even" and
($\wedge$) "naturals generally are larger than 15 and do not divide 12."

LG's provide a framework for capturing distinct notions of 'generally.' One builds a specific logic depending on the particular notion one has in mind. Thus, we will have a system that allows to conclude ($\sqcup$) from $\{(\alpha), (\beta)\}$, without obtaining ($\sqcap$), whereas a different one that enables one to conclude ($\vee$) and ($\wedge$) from $\{(\gamma), (\delta)\}$, although ($\cap$) and ($\wedge$) have similar syntactic structures.

Expressions involving 'generally,' or similar vague notions, occur often in assertions and arguments, as in the above examples. We wish to express such assertions in a precise manner. To express "objects generally have a given property," we add to First Order Logic (FOL) a new quantifier ∇ to represent 'generally.'

---

[1] Actually, the starting point was ultrafilter logic [18] with the aim of providing an alternative treatment for defaults [19].

Syntax and Semantics

We now briefly examine syntax and semantics for 'generally.'

The language for 'generally' is obtained by adding the new quantifier ∇ to the usual FOL syntax.

Given a first-order language L, we will use L^{∇} for the extension of L by the new quantifier ∇. The formulas of language L^{∇} are built by the usual formation rules [21], together with the new (variable binding) rule giving *generalized formulas*: if v is variable and A is a formula of L^{∇}, then so $\nabla v A$ is a formula of L^{∇}. [2]

The notions of variable occurring (free) in a formula are as usual. We shall use the notations $occ(A)$ for the set of variables occurring in formula A and $fr(\Gamma)$ for the set of variables with free occurrences in some formula in the set Γ of formulas. We shall also use familiar notions concerning substitution of variables.

An assertion such as "Objects generally have property P" may be understood as "the set of objects that have the property P is important (among the subsets of the universe of discourse)." So, one gives the semantics for 'generally' by adding families of sets (those that are considered important) to a usual first-order structure and extending the definition of satisfaction to ∇.

A *modulated structure* $\mathfrak{M}^{\mathcal{K}} = \langle \mathfrak{M}, \mathcal{K} \rangle$ consists of a usual structure $\mathfrak{M}$ together with a *complex* $\mathcal{K}$: a family of subsets of the universe M of $\mathfrak{M}$. We extend the definition of *satisfaction* of a formula in a structure under an assignment $s : V \to M$ to the variables as follows: for a generalized formula $\nabla v F$ we define

$$\mathfrak{M}^{\mathcal{K}} \models \nabla v A[s] \text{ iff } \{b \in M : \mathfrak{M}^{\mathcal{K}} \models A[s(v \mapsto b)]\}$$

belongs to the complex $\mathcal{K}$. [3]

Other concepts, such as *model* ($\mathfrak{M}^{\mathcal{K}} \models A$ and $\mathfrak{M}^{\mathcal{K}} \models \Gamma$), are as usual.

On the other hand, the concept of consequence depends on the specific notion of 'generally' involved. For instance, the assertions "Sport lovers watch SporTV" and "Boys generally love sports" appear to lead to "Boys generally watch SporTV." This will be correct if the complexes are closed under supersets, which seems reasonable in the case of 'many.' More precisely, we say that a formula A is an *up-closed consequence* of a set Γ of sentences iff $\mathfrak{M}^{\mathcal{K}} \models A$ whenever $\mathfrak{M}^{\mathcal{K}} \models \Gamma$, for every model $\mathfrak{M}^{\mathcal{K}}$ whose complex $\mathcal{K}$ is closed under supersets (of its universe). We use the notation $\Gamma \models_{\mathcal{S}} A$, where $\mathcal{S}$ is the class of up-closed complexes.

In this manner, each notion of 'generally' gives rise to a corresponding consequence relation: $\Gamma \models_{\mathcal{C}} A$, where $\mathcal{C}$ is a given *module*, i.e., a class of complexes (sharing some properties). [4]

---

[2] For instance, the above assertions (α) and ($\sqcap$) become $\nabla v legs(v)$ and $\nabla v(legs(v) \wedge faces(v))$.

[3] The updated assignment $s(v \mapsto b)$ is as usual: it agrees with s on every variable, except v.

[4] These consequence relations are extensional: $\Gamma \models_{\mathcal{C}} \forall v(A \leftrightarrow B) \to (\nabla v A \leftrightarrow \nabla v B)$.

Besides the *basic module* $\mathcal{B}$ (complexes without restriction), we one can also consider some specific modules, given by their characteristic properties. The following table shows some properties of complexes.[5]

Name	Property
($\forall$) Universe	$M \in \mathcal{K}$
($\emptyset$) Nonvoid	$\emptyset \notin \mathcal{K}$
($\cap$) Intersection	$S \in \mathcal{K}$ and $T \in \mathcal{K} \Rightarrow S \cap T \in \mathcal{K}$
($\cup$) Union	$S \in \mathcal{K}$ and $T \in \mathcal{K} \Rightarrow S \cup T \in \mathcal{K}$
($\supseteq$) Superset	$S \cap T \in \mathcal{K} \Rightarrow S \in \mathcal{K}$ and $T \in \mathcal{K}$

In principle, each combination Ω of properties from the above table can be used to define a notion of 'generally,' giving rise to a consequence relation. Some of these modules are familiar, among these we can mention the following ones.

($\mathcal{P}$) *Proper*: Universe ($\forall$) and Nonvoid ($\emptyset$).
($\mathcal{S}$) *Proper Up-closed*: Universe ($\forall$), Nonvoid ($\emptyset$) and Superset ($\supseteq$).
($\mathcal{L}$) *Proper Lattices*: Universe ($\forall$), Nonvoid ($\emptyset$), Intersection ($\cap$) and Union ($\cup$).
($\mathcal{F}$) *Proper Filters*: Universe ($\forall$), Nonvoid ($\emptyset$), Intersection ($\cap$) and Superset ($\supseteq$).

Other modules (such as ultrafilters) can be introduced by means of similar properties. For proper modules, ∇ is between $\forall$ and $\exists$ (the formulas $\forall v A \to \nabla v A$ and $\nabla v A \to \exists v A$ hold), but ∇'s do not commute (the formula $\nabla x \nabla y A \to \nabla y \nabla x A$ does not hold).[6] In this chapter, we will focus mainly on the case of filters. Our results hold also for the cases of proper, up-closed and lattices.[7]

Marked Formulas

In the next section, we will present deductive systems, in natural deduction style, for proper filters ('many'), emphasizing the treatment of generalized formulas. For this purpose, we will use marked formulas, which have the same intended meaning as generalized formulas. The specific rules of the logics for 'generally' will discipline the use of marked formulas [12, 22].

---

[5] Note that $S \subseteq T$ iff $S = S \cap T$.

[6] Formula $\nabla x \nabla y x < y \to \nabla y \nabla x x < y$ is not satisfied in the naturals with the Fréchet filter (of the cofinite sets) ([5], p. 158)

[7] Note that Union ($\cup$) can be obtained from Superset ($\supseteq$). So we have the following hierarchy

$$
\begin{array}{ccccccc}
 & & & & \mathcal{S} & & \\
 & & & \nearrow & & \searrow & \\
\mathcal{B} & \to & \mathcal{P} & & & & \mathcal{F} \\
 & & & \searrow & & \nearrow & \\
 & & & & \mathcal{L} & &
\end{array}
$$

The reason for using marked formulas is as follows. On the one hand, it is convenient to view a generalized formula $\nabla x A(x)$ as indecomposable, because instantiation of generalized variables does not hold in LG's. On the other hand, the interplay between the generalized quantifier ∇ and other logical constants, such as the propositional connectives and the usual universal and existential quantifiers, depends on the specific LG at hand. Marked formulas turn out to be useful for expressing such interplay.

Also, in the analysis of the structure of derivations, marked formulas play an important role, highlighting the regions where generalized information is manipulated.

A marked formula has the form $\langle A(\_)\rangle$ and is intended to represent a generalized formula $\nabla x A(x)$. The idea is that '_' represents a generic object and '⟨' and '⟩' emphazise that $A(x)$ is the scope of a generalized quantifier. More formally, we consider a new symbol '_' (not in L^∇). Given a formula A and a variable v of L^∇, a *generic instance* of formula A with respect to variable v is the result (noted $A[v/\_]$) of replacing every occurrence of v in A, if any, by the new symbol '_'. The *generic fragment* associated to L^∇ consists of all such generic instances $A[v/\_]$, for a formula A of L^∇ and a variable v. In the generic fragment associated to L^∇, we allow substitution for the new symbol '_': note that $A[v/\_][\_/w] = A[v/w]$. Now, a *marked formula* has the form $\langle M\rangle$, where M is a formula of the generic fragment associated to L^∇. Let L- be the set of all marked formulas associated to L^∇. We add L- to L^∇ to obtain L^*, i.e. $L^* = L^\nabla \cup L$-.[8]

We will use Latin letters and capital Greek letters to denote, respectively, formulas and sets of formulas of L^∇.[9] We extend the familiar definition of *rank of a formula* A (notation $r(A)$)[10] to formulas in L^*. We add to it the following two clauses:

(∇) for a generalized formula $\nabla v A$: $r(\nabla v A) := r(A) + 1$;
($\langle\rangle$) for a marked formula $\langle A[v/\_]\rangle$: $r(\langle A[v/\_]\rangle) := r(A) + 0.5$.

[8] Generic instances may have ∇ (e.g., $\nabla x \nabla y A(x, y)$ gives $\nabla y A(\_, y)$), but they have at most one '_' ($A(\_, \_) \notin L$-).

[9] In general, we will denote formulas in L^∇ by the first capital letters of the Latin alphabet (A, B, C, ...); formulas in L- will be denoted by M, N, ..., while formulas in L^* will be denoted by F, G, H, ...

[10] Recall that the recursive definition of $r(A)$ is as follows:

(*Basis*) if A is a atomic formula, $\bot$ or $\top$, then $r(A) := 0$;
($\neg$) if A is $\neg B$, then $r(A) := r(B) + 1$;
(b) if A is $(B_1 b B_2)$, where b is a binary connective, then $r(A) := r(B_1) + r(B_2) + 1$;
(Q) if A is QvB, where Q is $\forall$ or $\exists$, then $r(A) := r(B) + 1$.

3 Natural Deduction Systems for 'Generally'

We will now review natural deduction systems for some logics of 'generally' [11, 12, 22].

We will start with a natural deduction system for the underlying FOL, and we will extend it so as to cover new formulas: those with the generalized quantifier ∇ and marked ones. The expected structure of a derivation is a FOL derivation with local manipulations of ∇.

These systems will have two parameters. We will consider:

- a natural deduction system ND for the underlying FOL (concerning the connectives and the quantifiers, other than ∇),[11]
- a particular logic of 'generally' (concerning the generalized quantifier ∇ and its behavior with respect to the first-order connectives and the quantifiers).

We will then construct a natural deduction system $ND(\mathcal{G})$, where rules with marked formulas will serve to manipulate ∇.

The concept of a derivation is as usual and discharge of hypotheses is denoted as in [23].[12]

For definiteness, we will consider Classical, Intuitionistic, and Minimal Logic as underlying FOLs [15]. We will use the following versions of the natural deduction system ND for the underlying FOL:

(ML) for Minimal Logic: system ML with the usual rules;
(IL) for Intuitionistic Logic: $IL = ML \cup \{(Abs)\}$;
(CL) for Classical Logic: $CL := IL \cup \{(RaA)\}$;

Rules (Abs) and (RaA) (for absurdity $\perp$) are as follows:

$$(Abs) : \frac{\perp}{A} \quad \text{and} \quad (RaA) : \frac{\begin{matrix}\Gamma, [\neg A]^i \\ \Sigma \\ \perp\end{matrix}}{A}\, i$$

Natural deduction systems for specific logics of 'generally' are obtained by extending a basic system with appropriate rules.

Natural Deduction for Basic Logic of 'Generally'

We will now present natural deduction rules for the basic logic of 'generally' (complexes without restriction).

---

[11] The rules for $\neg$ and $\leftrightarrow$ correspond to viewing $\neg A$ as $A \to \perp$ and $A \leftrightarrow B$ as $(A \to B) \wedge (B \to A)$.

[12] The notation $\begin{matrix}\Phi \\ \Sigma \\ F\end{matrix}$ represents a derivation Σ of the formula F from the set Φ of formulas.

We start with the rules of a natural deduction system ND for Classical, Intuitionistic, or Minimal Logic, and we will extend it to cover generalized and marked formulas, thereby obtaining a system $ND(\mathcal{B})$.

Since we have marked formulas, we extend some elimination rules of ND to cover them. More specifically, we extend the elimination rules for disjunction and existential quantifier to allow a marked conclusion:

$$(\vee E): \frac{A \vee B \quad \begin{matrix}\Gamma, [A]^i \\ \Sigma_1 \\ F\end{matrix} \quad \begin{matrix}\Delta, [B]^i \\ \Sigma_2 \\ F\end{matrix}}{F}\, i \qquad (\exists E): \frac{\exists x A \quad \begin{matrix}\Gamma, [A]^i \\ \Sigma \\ F\end{matrix}}{F}\, i$$

We thus obtain the *extended system* $ND^* = ND \cup \{(\vee E), (\exists E)\}$.[13]

We have *elimination* and *introduction* rules for ∇ (to enter and leave an environment for 'generally'):

$$(\nabla E): \frac{\nabla v A}{\langle A[v/\_]\rangle} \qquad (\nabla I): \frac{\langle A[v/\_]\rangle}{\nabla z A[v/z]} (\text{with } z \notin occ(A[v/\_]))$$

We also have a transformation rule for marked formulas, which corresponds to the extensionality property. The *equivalence rule* ($\Updownarrow$) is as follows (with $v \notin fr(\Gamma \cup \Delta)$):

$$(\Updownarrow): \frac{\begin{matrix}\Gamma, [A]^i \\ \Sigma_1 \\ B\end{matrix} \quad \langle A[v/\_]\rangle \quad \begin{matrix}\Delta, [B]^i \\ \Sigma_2 \\ A\end{matrix}}{\langle B[v/\_]\rangle}\, i$$

In this rule ($\Updownarrow$), the marked formula $\langle A[v/\_]\rangle$ is called its *major premise* and the discharged unmarked formulas A and B are called its *minor premises*; derivations Σ_1 and Σ_2 are called *lateral*[14]

We can now describe our construction for a basic system by

$$ND(\mathcal{B}) := \mathcal{ND}^* \cup (\mathcal{B})(\text{where } (\mathcal{B}) := \{(\nabla\mathcal{I}), (\nabla\mathcal{E}), (\Updownarrow)\}).$$

We can thus obtain the three basic systems:

$$ML(\mathcal{B}) = \mathcal{ML}^* \cup (\mathcal{B}), \mathcal{IL}(\mathcal{B}) = \mathcal{IL}^* \cup (\mathcal{B})] \text{ and } \mathcal{CL}(\mathcal{B}) = \mathcal{CL}^* \cup (\mathcal{B}).$$

In these three basic systems, one can derive

- alphabetic variant: $\nabla v A \to \nabla u A[v/u]$, for $u \notin occ[A]$ (by (∇E) and (∇I));

---

[13] Note that in ND^*, the only rules that allow marked formulas in the conclusion are $(\vee E)$ and $(\exists E)$.

[14] One may regard this rule as eliminating the marked premise and introducing the marked conclusion.

- extensionality: $\forall v(A \leftrightarrow B) \rightarrow (\nabla v A \leftrightarrow \nabla v B)$.

We also have a derivation Π showing that $\forall x \forall y(A(x, y) \leftrightarrow B(x, y))$, $\nabla x \nabla y A(x, y) \vdash_{ND(\mathcal{B})} \nabla x \nabla y B(x, y)$. Using C for the formula $\forall x \forall y(A(x, y) \leftrightarrow B(x, y))$, we can construct such a derivation as follows:

1. We have FOL derivations Δ', from C and $A(x, y)$ to $B(x, y)$, and Δ'', from C and $B(x, y)$ to $A(x, y)$ (using $(\forall E)$ and connective rules).
2. With these derivations Δ' and Δ'', we construct a derivation Σ' of $\nabla y B(x, y)$ from C and $\nabla y A(x, y)$ to $\nabla y B(x, y)$ (using (∇E), $(\Updownarrow)$ and (∇I)).[15]
3. Similarly, we construct a derivation Σ'' of $\nabla y A(x, y)$ from C and $\nabla y B(x, y)$.
4. With these derivations Σ' and Σ'', we construct a derivation Π of $\nabla x \nabla y B(x, y)$ from C and $\nabla x \nabla y A(x, y)$ (using (∇E), $(\Updownarrow)$ and (∇I)).[16]

Natural Deduction for Specific Logics of 'Generally'

We will now present natural deduction systems for specific logics of 'generally'.

Natural deduction systems for specific logics of 'generally' are obtained by extending the basic system $ND(\mathcal{B})$ with appropriate rules. Such rules correspond to the properties of complexes in 2.0 (in Sect. 2).

We wish to create new rules corresponding to the specific properties. The construction hinges on adding new rules for handling marked formulas.

To introduce the basic idea for the construction of a specific logic of generally, consider the intersection property $(\cap)$, expressed by $A \in \mathcal{K}$ and $B \in \mathcal{K} \Rightarrow A \cap B \in \mathcal{K}$. This property can be formulated as a rule:

$$(\nabla\wedge)\dfrac{\nabla v A \quad \nabla v B}{\nabla v(A \wedge B)}.$$

---

[15] Derivation Σ' is as follows:

$$(\nabla I)\dfrac{(\Updownarrow)\dfrac{\begin{matrix}C,[A(x,y)]^1\\ \Delta'\\ B(x,y)\end{matrix} \qquad (\nabla E)\dfrac{\nabla y A(x,y)}{\langle A(x,\_)\rangle} \qquad \begin{matrix}C,[B(x,y)]^1\\ \Delta''\\ A(x,y)\end{matrix}}{\langle B(x,\_)\rangle}\,1}{\nabla y B(x,y)}$$

[16] This derivation Π is as follows:

$$(\nabla I)\dfrac{(\Updownarrow)\dfrac{\begin{matrix}C,[\nabla y A(x,y)]^3\\ \Sigma'\\ \nabla y B(x,y)\end{matrix} \qquad (\nabla E)\dfrac{\nabla x \nabla y A(x,y)}{\langle \nabla y A(\_,y)\rangle} \qquad \begin{matrix}C,[\nabla y B(x,y)]^3\\ \Sigma''\\ \nabla y A(x,y))\end{matrix}}{\langle \nabla y B(\_,y)\rangle}\,3}{\nabla x \nabla y B(x,y)}$$

This rule $(\nabla\wedge)$ can be reformulated as the following rule $(\wedge^* I)$ for introducing $\wedge$ in a marked environment:

$$(\wedge^* I) : \frac{\langle A\,[v/\_]\rangle\ \langle B\,[v/\_]\rangle}{\langle (A \wedge B)[v/\_]\rangle}.$$

In a similar manner, one can obtain rules corresponding to the properties in 2.0. For instance, corresponding to the nonvoid property $(\emptyset)$, one obtains the following elimination rule, eliminating a marked absurdity:

$$(\bot^* E) : \frac{\langle \bot \rangle}{\bot}.$$

We will consider the following operational rules corresponding to the properties of complexes in 2.0:

$$(\top^* I) : \frac{}{\langle A \to A\rangle} \qquad (\bot^* E) : \frac{\langle \bot \rangle}{\bot}$$

$$(\wedge^* I) : \frac{\langle M\rangle\ \langle N\rangle}{\langle M \wedge N\rangle} \qquad (\wedge^* E) : \frac{\langle M \wedge N\rangle}{\langle M\rangle}\ \frac{\langle M \wedge N\rangle}{\langle N\rangle}$$

$$(\vee^* I) : \frac{\langle M\rangle\ \langle N\rangle}{\langle M \vee N\rangle}$$

The correspondence between properties of complexes and rules is as follows:

Properties	$(\forall)$	$(\emptyset)$	$(\cap)$	$(\cup)$	$(\supseteq)$
Rules	$(\top^* I)$	$(\bot^* E)$	$(\wedge^* I)$	$(\vee^* I)$	$(\wedge^* E)$

Note that rules $(\top^* I)$, $(\wedge^* I)$ and $(\vee^* I)$ serve to introduce a symbol in a marked formula, while $(\wedge^* E)$ and$(\bot^* E)$ serve to remove a symbol from a marked formula.[17]

We extend naturally the concepts of *major* and *minor* premises [15] to these operational rules.

Now, a specific logic for 'generally's consists of

- an underlying logic L,
- a particular logic for 'generally', with characteristic property set

$$\Omega \subseteq \{(\forall), (\emptyset), (\cap), (\cup), (\supseteq)\}(\text{cf.}2.0).$$

We can construct a natural deduction system for this logic by considering

[17] Also, rules such as $(\wedge^* I)$ and $(\wedge^* E)$ may be regarded as marked versions of the familiar ones.

- a natural deduction system ND for L,
- the extension of $ND(\mathcal{B})$ by the rules Ω^* corresponding to the properties in Ω.

We thus obtain the system $ND(\Omega) = ND(\mathcal{B}) \cup \Omega^*$. For instance, we can add to the basic system $ND(\mathcal{B}) := \mathcal{ND}^* \cup \{(\nabla\mathcal{I}), (\nabla\mathcal{E}), (\Updownarrow)\}$ the specific rules

($\mathcal{P}$) $(\top^* I)$ and $(\bot^* E)$ to obtain $ND(\mathcal{P})$ for proper logic;
($\mathcal{S}$) $(\top^* I)$, $(\bot^* E)$ and $(\wedge^* E)$ to obtain $ND(\mathcal{S})$ for proper up-closed logic;
($\mathcal{L}$) $(\top^* I)$, $(\bot^* E)$, $(\wedge^* I)$ and $(\vee^* I)$ to obtain $ND(\mathcal{L})$ for proper lattice logic;
($\mathcal{F}$) $(\top^* I)$, $(\bot^* E)$, $(\wedge^* I)$ and $(\wedge^* E)$ to obtain $ND(\mathcal{F})$ for proper filter logic.

Example 1 In these specific systems, we have derivations establishing some expected properties, such as $\forall v(A \to B), \nabla v A \vdash_{ND(\mathcal{S})} \nabla v B$ and $\nabla v \neg A \vdash_{ND(\mathcal{F})} \neg\nabla v A$. To obtain a derivation for the former, consider the following derivation Π (where derivation Σ is as expected):

$$(\nabla I)\cfrac{(\wedge^* E)\cfrac{(\Updownarrow)\cfrac{\begin{matrix}[A]^1, \forall v(A \to B)\\ \Sigma \\ A \wedge B\end{matrix} \qquad \langle A[v/\_]\rangle \quad (\wedge E)\cfrac{[A \wedge B]^1}{A}}{\langle (A \wedge B)[v/\_]\rangle}1}{\langle B[v/\_]\rangle}}{\nabla v B}$$

Derivation Π shows that $\forall v(A \to B), \langle A[v/\_]\rangle \vdash_{ND(\mathcal{S})} \nabla v B$. We can now use (∇E):

$$\begin{matrix}\forall v(A \to B) \quad (\nabla E)\cfrac{\nabla v A}{\langle A[v/\_]\rangle} \\ \Pi \\ \nabla v B\end{matrix}$$

In the sequel, we will restrict our attention to a logic $ND(\Omega) = ND(\mathcal{B}) \cup \Omega^*$, where

- the underlying logic ND is either Minimal Logic ML or Intuitionistic Logic IL, and
- $\Omega^* \subseteq \{(\top^* I), (\bot^* E), (\wedge^* I), (\wedge^* E), (\vee^* I)\}$.

4 Normalization in Logics for 'Generally'

In this section, we will summarize the normalization process for our logics $ML(\Omega)$ and $IL(\Omega)$, with operational rules $\Omega^* \subseteq \{(\top^* I), (\bot^* E), (\wedge^* I), (\wedge^* E), (\vee^* I)\}$ [11, 22].

For the systems $ND(\Omega)$ based on Minimal Logic ML or Intuitionistic Logic IL one can obtain normalized derivations using the same strategies as in [15, 24] to normalize derivations in ML or in IL.

The notions of path and (maximal) segment in a derivation in a system $ND(\Omega)$ are naturally adapted from those notions for the system ND [15].

The redundancies or deviations in a derivation are called maximal segments. A segment $S = (F_1, \ldots, F_n)$ in a derivation Π in $ND(\mathcal{F})$ is a *maximal segment* iff

- either S is a maximal segment in the underlying system ND^*;
- or F_1 is the conclusion of an application of one of the rules (Abs) or $(\Updownarrow)$ and F_n is the major premise of an application of an elimination rule or of the rule $(\Updownarrow)$.

Examples of maximal segments consisting of a single formula F are as follows:

1. Formula F is both the conclusion of an application of an introduction rule of ND^* and the major premise of an application of an elimination rule of ND^*.
2. Formula F is both the conclusion and the major premise of an application of the $(\Updownarrow)$ rule.

Derivations in $ND(\Omega)$ may have other kinds of deviations, due to superfluous applications of $\vee$-elimination or of $\exists$-elimination (involving marked or unmarked formulas). An application (α) of the $\vee$-elimination or of the $\exists$-elimination rules is *superfluous* iff there exists a subderivation of a minor premise of (α) from the hypotheses Γ of the derivation, such that in Γ there is no discharged hypothesis of the application (α). One can show that superfluous applications can be eliminated [11, 22].

We will now examine more closely the maximal segments, considering the underlying logic of $ND(\Omega)$.

First, we will consider the maximal segments in $ML(\mathcal{B})$.

When the underlying logic is Minimal Logic, besides the occurrences of the maximal segments of ML, one can also have occurrences of maximal segments involving generalized and marked formulas.

The maximal segments of length 1 (with only one formula) in the system $ML(\mathcal{B})$ which contain generalized or marked formulas are as follows.

$(\Updownarrow)$; $(\Updownarrow)$ A $(\Updownarrow)$; $(\Updownarrow)$ *segment* consists of a marked formula $\langle B[x/\_]\rangle$ that is both the conclusion and the major premise of an application of the $(\Updownarrow)$ rule. The $(\Updownarrow)$; $(\Updownarrow)$ segments can be eliminated by and gluing lateral derivations (after renaming variables properly).

(∇I); (∇E) A (∇I); (∇E) *segment* consists of a marked formula$\nabla v A[_/v]$ which is both the conclusion of an application of the rule (∇I) and the premise of an application of the (∇E) rule. The (∇I); (∇E) segments can be eliminated as follows:

$$\begin{array}{c} \Pi_1 \\ (\nabla I)\dfrac{\langle A[v/\_]\rangle}{(\nabla E)\dfrac{\nabla z A[v/z]}{\langle A[v/z][z/\_]\rangle}} \\ \Pi_2 \end{array} \quad \Rightarrow \quad \begin{array}{c} \Pi_1 \\ \langle A[v/\_]\rangle \\ \Pi_2 \end{array}$$

Note that for a maximal segment $S = (A_1, \ldots, A_n)$, with $n > 1$, in a derivation in $ML(\mathcal{B})$, an application of an introduction rule of a connective or a quantifier c or an application of the rule ($\Updownarrow$) does not immediately precede an application of the c elimination rule or an application of the rule ($\Updownarrow$). Thus, none of the reductions presented above can be applied. In order to eliminate S, we employ the following procedure: first, reverse the order of the applications of the rules, by applying the so called *permutative reductions* [11], to reduce S to a unary maximal segment S'; then, by applying one of the reductions presented above, eliminate S'.

Now, in $ML(\Omega) = ML(\mathcal{B}) \cup \Omega^*$, we can have maximal segments having a marked formula $\langle M \rangle$ as the conclusion of an application of the rule $(\wedge^* I)$ (introduction of $\wedge$ within a marked formula) and the major premise of an application of the rule $\wedge^* E$ (elimination of $\wedge$ within a marked formula):

$(\wedge^* I)$; $(\wedge^* E)$ A $(\wedge^* I)$; $(\wedge^* E)$ *segment* consists of a marked formula $\langle M_1 \wedge M_2 \rangle$ that is both the conclusion of an application of the $(\wedge^* I)$ rule and premise of an application of the $(\wedge^* E)$ rule. One can eliminate such maximal segments with the following reduction (where $i \in \{1, 2\}$):

$$\dfrac{\dfrac{\begin{matrix}\Pi_1 & \Pi_2 \\ \langle M_1 \rangle & \langle M_2 \rangle\end{matrix}}{\langle M_1 \wedge M_2 \rangle}(\wedge^* I)}{\begin{matrix}\langle M_i \rangle \\ \Sigma\end{matrix}}(\wedge^* E) \quad \Rightarrow \quad \begin{matrix}\Pi_i \\ \langle M_i \rangle \\ \Sigma\end{matrix}$$

The reductions involving applications of $\vee$-elimination and $\exists$-elimination rules (with marked or unmarked formulas) turn out to be permutative reductions.

Next, we will consider the case of $IL(\Omega)$ having intuitionistic logic as the underlying logic.

In the natural deduction system for Intuitionistic Logic [15, 24], applications of rule (Abs) can be restricted to those whose conclusions are atomic formulas. In a similar way, in the systems $IL(\mathcal{B})$ and $IL(\Omega)$, applications of (Abs) are restricted to those ones whose conclusions are atomic or generalized formulas.

The kinds of maximal segments that can occur in a derivation in $IL(\mathcal{F})$ are as follows:

(IL^*) Maximal segments of IL^*, namely

- maximal segments of $ML(\mathcal{F})$ (eliminated as before), and
- maximal segments involving applications of the rule (Abs) (which can be eliminated as well).

(Abs); (∇E); ($\Updownarrow$) A (Abs); (∇E); ($\Updownarrow$) *segment* consists of consecutive applications of (Abs), (∇E) and ($\Updownarrow$). One can eliminated such maximal segments as follows:

$$(\Updownarrow)\dfrac{\begin{matrix}[A]^i \\ \Sigma' \\ B\end{matrix} \quad (\nabla E)\dfrac{(Abs)\dfrac{\begin{matrix}\Pi \\ \bot\end{matrix}}{\nabla v A}}{\langle A[v/\_]\rangle} \quad \begin{matrix}[B]^i \\ \Sigma'' \\ A\end{matrix}}{\langle B[v/\_]\rangle} i \quad \Rightarrow \quad (\nabla E)\dfrac{(Abs)\dfrac{\begin{matrix}\Pi \\ \bot\end{matrix}}{\nabla v B}}{\langle B[v/\_]\rangle}$$

A derivation without occurrences of maximal segments and without superfluous applications of $\vee$-elimination or of $\exists$-elimination is said to be in *normal form*.

Thus, we have the normalization result for $ND(\Omega)$ (when ND is ML or IL and Ω^* is as above). Every derivation Π in $ND(\Omega)$ can be normalized: reduced to a derivation Π' in normal form.

5 Minimum Segment in Logics for 'Generally'

We will now present the minimum segment result for our logics $ML(\Omega)$ and $IL(\Omega)$, with operational rules $\Omega^* \subseteq \{(\top^* I), (\bot^* E), (\wedge^* I), (\wedge^* E), (\vee^* I)\}$.

First, we can naturally split a path of a derivation into sub-paths without applications of $(\Updownarrow)$:

$$\rho_1 \ (\Updownarrow) \ \rho_2 \ (\Updownarrow) \ \ldots \ (\Updownarrow) \ \rho_n$$

The next remark follows by a familiar argument.

Remark 2 Let Π be a normal derivation in $ND(\mathcal{F})$. In every path of Π without applications of $(\Updownarrow)$ an application of an introduction rule does not precede an application of an elimination rule.

What about the general situation? To establish that a set belongs to a filter, it suffices to show that it includes a finite intersection of sets in the filter. This natural strategy provides derivations.

Example 3 For instance, one can show that $\nabla v A_1, \nabla v A_2, \forall v[(A_1 \wedge A_2) \to B] \vdash_{ND(\mathcal{F})} \nabla v B$ as follows (where Π' is like the derivation Π in Example 1 **??**):

$$(\nabla I)\dfrac{\begin{matrix}(\wedge^* I)\dfrac{(\nabla E)\dfrac{\nabla v A_1}{\langle A_1[v/\_]\rangle} \quad (\nabla E)\dfrac{\nabla v A_2}{\langle A_2[v/\_]\rangle}}{\langle (A_1 \wedge A_2)[v/\_]\rangle} \\ \Pi' \\ B\end{matrix}}{\nabla v B}$$

The marked region of this derivation goes from $\langle A_1[v/\_]\rangle$ and $\langle A_2[v/\_]\rangle$ to $\langle B[v/\_]\rangle$. It begins with $(\wedge^* I)$ and proceeds with $(\Updownarrow)$ followed by $(\wedge^* E)$, the latter two within Π' (cf. Example 1). So, the structure of each path from $\langle A_i[v/\_]\rangle$ to $\langle B[v/\_]\rangle$ is $(\wedge^* I)$; $(\Updownarrow)$; $(\wedge^* E)$.

We will examine the structure of normal derivations in $ND(\Omega)$ (when ND is ML or IL and Ω^* is as above). We will show that every such derivation (not entirely within FOL) has paths with the following structure:

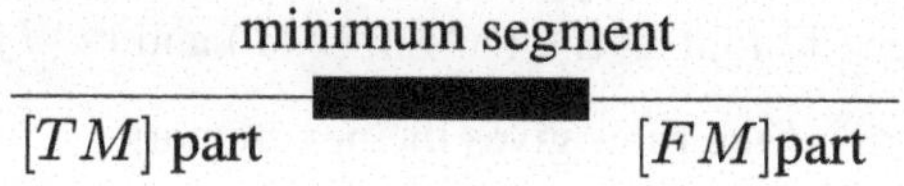

Here, these three regions (each one of them possibly empty) are as follows:

TM: The $[TM]$ part generates marked formulas, entering the marked environment.
MS: The minimum segment $[MS]$ manipulates marked formulas.
FM: The $[FM]$ leaves the marked environment.

Now, a marked region may have several applications of $(\Updownarrow)$. In this case, we can convert the derivation to a normal one where each path has a central marked region much as in Example 3: applications of $(\wedge^* I)$, followed by at most one application of $(\Updownarrow)$ and ending with applications of $(\wedge^* E)$.

In this sequel, we will outline the steps for establishing our minimum segment result for derivations in a logic $ND(\Omega) = ND(\mathcal{B}) \cup \Omega^*$, where $\Omega^* \subseteq \{(\top^* I), (\bot^* E), (\wedge^* I), (\wedge^* E), (\vee^* I)\}$ and ND is ML or IL.

The next result follows from the form of the rules of $ND(\Omega)$.

Lemma 4 *Let Π be a normal derivation in $ND(\Omega)$. In every path of Π an application of an introduction rule of ND^* does not precede an application of an elimination rule of ND^*.*

Now, notice that the conclusion of an application of an introduction rule of ND^* cannot be a marked formula or a generalized formula. Also, applications of the rule $(\top^* I)$ occur always at the top of a marked environment and applications of the rule $(\bot^* E)$ occur always at the bottom of a marked environment. Thus, one has just three kinds of environments (sequences of formulas) in a path of a $ND(\Omega)$ derivation.

(UE) An environment of unmarked formulas that are premises of elimination rules of ND^*.
(GM) An environment with generalized or marked formulas.
(UI) An environment of unmarked formulas that are premises of introduction rules of ND^*.

Furthermore, the path will present the following pattern:

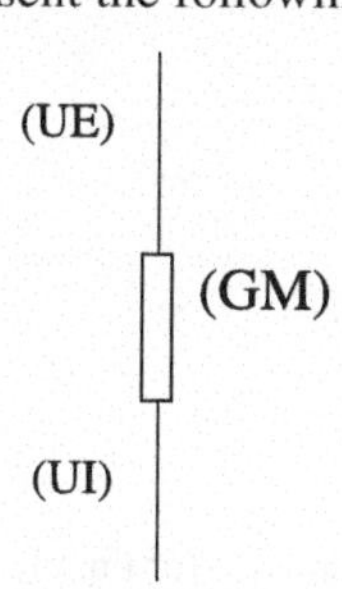

The rules that can occur in the environment (GM) with generalized or marked formulas are

- the basic rules (∇E), (∇I) and $(\Updownarrow)$, as well as
- the operational rules $(\top^* I)$, $(\bot^* E)$, $(\wedge^* E)$, $(\wedge^* I)$ and $(\vee^* I)$.

Once again, the form of the rules gives the next remark.

Remark 5 Let Π be a normal derivation in $ND(\Omega)$. Consider the environment (GM) with generalized or marked formulas in a path of Π.

- An application of (∇E) or $(\bot^* E)$ cannot precede applications of the other rules.
- An application of (∇I) or $(\top^* I)$ cannot be preceded by applications of the other rules.

Thus, the environment (GM) with generalized or marked formulas has the following structure.

1. An application of (∇I) or $(\top^* I)$ introducing the first marked formula.
2. Applications of $(\Updownarrow)$, $(\wedge^* E)$, $(\wedge^* I)$ and $(\vee^* I)$, transforming marked formulas.
3. An application of (∇E) or $(\bot^* E)$ eliminating the last marked formula.

Thus, the structure of a path is as follows.[18]

$$\overbrace{A\ E(ND^*)\ B}^{(UE)}\ \overbrace{\begin{pmatrix}(\nabla E)\\(\top^* I)\end{pmatrix} M \left\{\begin{matrix}(\wedge^* E)\\(\Updownarrow)\\(\wedge^* I),(\vee^* I)\end{matrix}\right\}^{\star} N \begin{pmatrix}(\nabla I)\\(\bot^* E)\end{pmatrix}}^{(GM)}\ \overbrace{C\ I(ND^*)\ D}^{(UI)}$$

Now, we may reorganize this structure as follows:

$$\underbrace{A\ E(ND^*)\ B \begin{pmatrix}(\nabla E)\\(\top^* I)\end{pmatrix}}_{[TM]}\ \underbrace{M \left\{\begin{matrix}(\wedge^* E)\\(\Updownarrow)\\(\wedge^* I),(\vee^* I)\end{matrix}\right\}^{\star} N}_{[MS]}\ \underbrace{\begin{pmatrix}(\nabla I)\\(\bot^* E)\end{pmatrix} C\ I(ND^*)\ D}_{[FM]}$$

[18] We use () for alternative applications and {}* for finitely many applications.

These three regions are as follows:

TM The $[TM]$ part has only applications of elimination rules of ND^*, possibly followed by (∇E) or $(\top^* I)$.

MS The minimum segment $[MS]$ manipulates marked formulas by $(\Updownarrow)$ as well as $(\wedge^* E)$, $(\wedge^* I)$ and $(\vee^* I)$.

FM The $[FM]$ part has possibly applications of (∇E) or $(\perp^* E)$ followed by introduction rules of ND^*.

If a derivation has no marked formulas, then it is a derivation in the underlying logic ND.

Now, we will examine the minimum segment result for derivations in the basic logic $ND(\mathcal{B})$.

Consider a path in a normal derivation in $ND(\mathcal{B})$ with marked formulas.

Imagine that the path has no application of the $(\Updownarrow)$ rule. Then, it is as follows:

$$\underset{[TM]\text{ part}}{\overline{\qquad\qquad}} \overset{(\nabla E)}{\Big|}\ \langle A[v/\_]\rangle\ \overset{(\nabla I)}{\Big|} \underset{[FM]\text{ part}}{\overline{\qquad\qquad}}$$

In this case, the formula $\langle A[v/\_]\rangle$ constitutes the minimum segment of the path.

Now, imagine that the path has an application of $(\Updownarrow)$. Then, it is as follows:

$$\underset{[TM]\text{ part}}{\overline{\qquad\qquad}} \overset{(\nabla E)}{\Big|}\ \langle A[v/\_]\rangle\ \overset{(\Updownarrow)}{\Big|}\ \langle B[v/\_]\rangle\ \overset{(\nabla I)}{\Big|} \underset{[FM]\text{ part}}{\overline{\qquad\qquad}}$$

In this case, the minimum segment of the path consists of the formulas $\langle A[v/\_]\rangle$ and $\langle B[v/\_]\rangle$.

Note that in a path of a normal derivation in $ND(\mathcal{B})$ can have at most one application of the rule $(\Updownarrow)$.[19]

Therefore, a path of a normal derivation in $ND(\mathcal{B})$ has one of the following forms.

(0) No marked formula: $AE(ND^*)B = CI(ND^*)D$, with $[MS] = \emptyset$ and minimum formula B.

(1) One marked formula: $AE(ND^*)B(\nabla E)M = N(\nabla I)CI(ND^*)\ D$, with minimum segment (M).[20]

(2) Two marked formulas: $AE(ND^*)B(\nabla E)M(\Updownarrow)N(\nabla I)CI(ND^*)D$, with minimum segment (M, N).[21]

In general, a minimum segment of $ND(\Omega)$ may have several applications of $(\Updownarrow)$. In this case, we wish to convert the derivation to a normal one where each path has a

[19] Otherwise, the derivation would not be normal.

[20] Here, B, M and C are respectively $\nabla v B'$, $\langle B'[v/\_]\rangle$ and $\nabla u B'[v/u]$.

[21] Notice that M is obtained from A by elimination rules and C is obtained from N by introduction rules.

minimum segment with applications of $(\wedge^* I)$, followed by at most one application of $(\Updownarrow)$ and ending with applications of $(\wedge^* E)$.

To achieve this goal, we move applications of marked introductions $(\wedge^* I)$ and $(\vee^* I)$ toward the beginning of the path and applications of marked elimination $(\wedge^* E)$ toward the end of the path (this can be done by resorting to logical equivalences in the underlying logic). We handle consecutive applications of $(\Updownarrow)$ as before (cf. $(\Updownarrow)$; $(\Updownarrow)$ segments in Sect. 4).

Example 6 Consider the following derivation Π (in $ND(\mathcal{S})$) with 2 applications of $(\Updownarrow)$.

$$(\Updownarrow)\dfrac{\begin{matrix}\Gamma_3,[C]^2\\ \Pi_3\\ D\end{matrix} \qquad (\wedge^* E)\dfrac{(\Updownarrow)\dfrac{\begin{matrix}\Gamma_1,[A]^1\\ \Pi_1\\ B\wedge C\end{matrix} \quad \begin{matrix}\Gamma\\ \Delta\\ \langle A[v/\_]\rangle\end{matrix} \quad \begin{matrix}\Gamma_2,[B\wedge C]^1\\ \Pi_2\\ A\end{matrix}}{\langle (B\wedge C)[v/\_]\rangle}\,1}{\langle C[v/\_]\rangle} \qquad \begin{matrix}\Gamma_4,[D]^2\\ \Pi_4\\ C\end{matrix}}{\begin{matrix}\langle D[v/\_]\rangle\\ \Sigma\end{matrix}}\,2$$

We will eliminate this double application of $(\Updownarrow)$ of derivation Π.

Reversing the applications of $(\wedge^* E)$ and $(\Updownarrow)$, we obtain the following derivation Π':

$$(\wedge^* E)\dfrac{(\Updownarrow)\dfrac{\begin{matrix}\Gamma_3,[B\wedge C]^2\\ \Pi_3'\\ B\wedge D\end{matrix} \qquad (\Updownarrow)\dfrac{\begin{matrix}\Gamma_1,[A]^1\\ \Pi_1\\ B\wedge C\end{matrix} \quad \begin{matrix}\Gamma\\ \Delta\\ \langle A[v/\_]\rangle\end{matrix} \quad \begin{matrix}\Gamma_2,[B\wedge C]^1\\ \Pi_2\\ A\end{matrix}}{\langle (B\wedge C)[v/\_]\rangle}\,1 \qquad \begin{matrix}\Gamma_4,[B\wedge D]^2\\ \Pi_4'\\ B\wedge C\end{matrix}}{\langle (B\wedge D)[v/\_]\rangle}\,2}{\begin{matrix}\langle D[v/\_]\rangle\\ \Sigma\end{matrix}}$$

Here, subderivations Π_3' and Π_4' are as expected.[22] Now, this derivation Π' has 2 consecutive applications of $(\Updownarrow)$, which we can handle as before.

In general, we can proceed as follows (where $\circ$ is $\wedge$ or $\vee$):

$(\wedge^* E)$; $(\circ^* I)$ To commute a marked elimination $(\wedge^* E)$ immediately followed by a marked introduction $(\circ^* I)$, we use the logical equivalence between $(A \wedge B) \circ C$ and $(A \circ C) \wedge (B \circ C)$. We can thus convert $(\wedge^* E)$; $(\circ^* I)$ to $(\circ^* I)$; $(\Updownarrow)$; $(\wedge^* E)$, using FOL derivations Δ', from $(A \wedge B) \circ C$ to $(A \circ C) \wedge (B \circ C)$, and Δ'', from $(A \circ C) \wedge (B \circ C)$ to $(A \wedge B) \circ C$, as follows:

$$(\circ^* I)\dfrac{(\wedge^* E)\dfrac{\begin{matrix}\Pi_1\\ \langle (A\wedge B)[v/\_]\rangle\end{matrix}}{\langle B[v/\_]\rangle} \qquad \begin{matrix}\Pi_2\\ \langle C[v/\_]\rangle\end{matrix}}{\begin{matrix}\langle (B\circ C)[v/\_]\rangle\\ \Sigma\end{matrix}}$$

$$\Downarrow$$

$$(\wedge^* E)\dfrac{(\Updownarrow)\dfrac{\begin{matrix}[(A\wedge B)\circ C]^k\\ \Delta'\\ (A\circ C)\wedge(B\circ C)\end{matrix} \qquad (\circ^* I)\dfrac{\begin{matrix}\Pi_1\\ \langle (A\wedge B)[v/\_]\rangle\end{matrix} \quad \begin{matrix}\Pi_2\\ \langle C[v/\_]\rangle\end{matrix}}{\langle [(A\wedge B)\circ C][v/\_]\rangle} \qquad \begin{matrix}[(A\circ C)\wedge(B\circ C)]^k\\ \Delta''\\ (A\wedge B)\circ C\end{matrix}}{\langle [(A\circ C)\wedge(B\circ C)][v/\_]\rangle}\,k}{\begin{matrix}\langle (B\circ C)[v/\_]\rangle\\ \Sigma\end{matrix}}$$

---

[22] For instance, Π_3' starts with 2 $(\wedge E)$'s, uses Π_3 and ends with $(\wedge I)$.

$(\Updownarrow)$; $(\circ^* I)$ To commute an application of $(\Updownarrow)$ immediately followed by a marked introduction $(\circ^* I)$, we use substitutivity of equivalents $(A \leftrightarrow B \vdash (A \circ C) \leftrightarrow (B \circ C))$. We can thus convert $(\Updownarrow)$; $(\circ^* I)$ to $(\circ^* I)$; $(\Updownarrow)$.

$$
(\circ^* I)\,\dfrac{(\Updownarrow)\,\dfrac{\begin{matrix}\Gamma_1,[A]^i\\ \Delta_1\\ B\end{matrix}\quad \begin{matrix}\Pi_1\\ \langle A[v/\_]\rangle\end{matrix}\quad \begin{matrix}\Gamma_2,[B]^i\\ \Delta_2\\ A\end{matrix}}{\langle B[v/\_]\rangle}\,i \qquad \begin{matrix}\Pi_2\\ \langle C[v/\_]\rangle\end{matrix}}{\begin{matrix}\langle (B\circ C)[v/\_]\rangle\\ \Sigma\end{matrix}}
$$

$$\Downarrow$$

$$
(\Updownarrow)\,\dfrac{\begin{matrix}\Gamma_1,[A\circ C]^j\\ \Delta_1'\\ B\circ C\end{matrix}\quad (\circ^* I)\,\dfrac{\begin{matrix}\Pi_1\\ \langle A[v/\_]\rangle\end{matrix}\ \begin{matrix}\Pi_2\\ \langle C[v/\_]\rangle\end{matrix}}{\langle (A\circ C)[v/\_]\rangle}\quad \begin{matrix}\Gamma_1,[B\circ C]^j\\ \Delta_2'\\ A\circ C\end{matrix}}{\begin{matrix}\langle (B\circ C)[v/\_]\rangle\\ \Sigma\end{matrix}}\,j
$$

Here, derivations Δ_1' and Δ_2' are obtained from Δ_1 and Δ_2 as expected.

$(\wedge^* E)$; $(\Updownarrow)$ To commute a marked elimination $(\wedge^* E)$ immediately followed by an application of $(\Updownarrow)$, we use substitutivity of equivalents. We can thus convert $(\wedge^* E)$; $(\Updownarrow)$ to $(\Updownarrow)$; $(\wedge^* E)$, as follows:

$$
(\Updownarrow)\,\dfrac{\begin{matrix}\Gamma_1,[B]^i\\ \Delta_1\\ C\end{matrix}\quad (\wedge^* E)\,\dfrac{\begin{matrix}\Pi\\ \langle (A\wedge B)[v/\_]\rangle\end{matrix}}{\langle B[v/\_]\rangle}\quad \begin{matrix}\Gamma_1,[C]^i\\ \Delta_2\\ B\end{matrix}}{\begin{matrix}\langle C[v/\_]\rangle\\ \Sigma\end{matrix}}\,i
$$

$$\Downarrow$$

$$
(\wedge^* E)\,\dfrac{(\Updownarrow)\,\dfrac{\begin{matrix}\Gamma_1,[A\wedge B]^j\\ \Delta_1'\\ A\wedge C\end{matrix}\quad \begin{matrix}\Pi\\ \langle (A\wedge B)[v/\_]\rangle\end{matrix}\quad \begin{matrix}\Gamma_2,[A\wedge C]^j i\\ \Delta_2'\\ A\wedge B\end{matrix}}{\langle (A\wedge C)[v/\_]\rangle}\,j}{\begin{matrix}\langle C[v/\_]\rangle\\ \Sigma\end{matrix}}
$$

Here, derivations Δ_1' and Δ_2' are obtained from Δ_1 and Δ_2 as expected.

$(\wedge^* E)$; $(\wedge^* E)$ We can also eliminate consecutive applications of marked elimination $(\wedge^* E)$, by resorting to logical equivalence between $A\wedge(B\wedge C)$ and $(A\wedge B)\wedge C$. Thus, we can convert $(\wedge^* E)$; $(\wedge^* E)$) to $(\Updownarrow)$; $(\wedge^* E)$, using FOL derivations Δ', from $A\wedge(B\wedge C)$ to $(A\wedge B)\wedge C$, and

Δ'', from $(A \wedge B) \wedge C$ to $A \wedge (B \wedge C)$, as follows:

$$(\wedge^*E)\,\dfrac{(\wedge^*E)\,\dfrac{\begin{matrix}\Pi\\ \langle[A\wedge(B\wedge C)][v/\_]\rangle\end{matrix}}{\langle(B\wedge C)[v/\_]\rangle}}{\begin{matrix}\langle C[v/\_]\rangle\\ \Sigma\end{matrix}}$$

$$\Downarrow$$

$$(\wedge^*E)\,\dfrac{(\Updownarrow)\,\dfrac{\begin{matrix}[A\wedge(B\wedge C)]^k\\ \Delta'\\ (A\wedge B)\wedge C\end{matrix}\quad \begin{matrix}\Pi\\ \langle[A\wedge(B\wedge C)][v/\_]\rangle\end{matrix}\quad \begin{matrix}[(A\wedge B)\wedge C]^k\\ \Delta''\\ A\wedge(B\wedge C)\end{matrix}}{\langle[(A\wedge B)\wedge C][v/\_]\rangle}\,k}{\begin{matrix}\langle C[v/\_]\rangle\\ \Sigma\end{matrix}}$$

With these conversions, one can convert a derivation to one where the minimum segments are as desired. Actually, one must be careful, as one may introduce new maximal segments, but these in lateral paths (of higher order).[23] So, one eliminates these inversions from paths of an order and normalizes the resulting derivation, which may introduce new inversions, but of higher order. Paths of maximal order have no application of $(\Updownarrow)$. Thus, one can convert each normal derivation in our $ND(\Omega)$ to a normal one where the minimum segments have at most one application of $(\Updownarrow)$: $M\{(\wedge^*I), (\vee^*I\}^\star\ P\ (\Updownarrow)\ Q\ \{(\wedge^*E)\}^\star N$.[24]

Hence, a normal derivation in $ND(\Omega)$ with nonempty minimum segment has one of the following forms.

(0) No application of $(\Updownarrow)$
minimum segment $[FM] = M\,\{(\wedge^*E)\}^\star\ P\,\{(\wedge^*I), (\vee^*I)\}^\star\ N$, with minimum formula P. path:

$$A\ E(ND^*)\ B\ \begin{pmatrix}(\nabla E)\\(\top^*I)\end{pmatrix}\ [FM]\ \begin{pmatrix}(\nabla I)\\(\bot^*E)\end{pmatrix}\ C\ I(ND^*)\ D$$

(1) One application of $(\Updownarrow)$
minimum segment $[FM] = M\ \{(\wedge^*I), (\vee^*I)\}^\star\ P\ (\Updownarrow) Q\ \{(\wedge^*E)\}^\star\ N$, minimum formulas M and N; possible maximal formulas P and Q. path

$$A\ E(ND^*)\ B\ \begin{pmatrix}(\nabla E)\\(\top^*I)\end{pmatrix}\ [FM]\ \begin{pmatrix}(\nabla I)\\(\bot^*E)\end{pmatrix}\ C\ I(ND^*)\ D$$

The case of one application of $(\Updownarrow)$ has a simpler structure for some logics.

$(\mathcal{B})$ For basic logic $ND(\mathcal{B})$ (as seen above):
minimum segment $[FM] = M\ (\Updownarrow)\ N$, minimum formulas M and N. path:

[23] The definition of order of a path is as usual [15, 24].

[24] See the appendix for more details.

$$A\ E(ND^*)\ B\ (\nabla E)\ [FM]\ (\nabla I)\ C\ I(ND^*)\ D$$

($\mathcal{P}$) For proper logic $ND(\mathcal{P})$
minimum segment $[FM] = M\ (\Updownarrow)\ N$, path:

$$A\ E(ND^*)\ B \begin{pmatrix}(\nabla E)\\(\top^* I)\end{pmatrix} [FM] \begin{pmatrix}(\nabla I)\\(\bot^* E)\end{pmatrix} C\ I(ND^*)\ D$$

($\mathcal{S}$) For proper up-closed logic $ND(\mathcal{S})$
minimum segment $[FM] = M(\Updownarrow)\ Q\ \{(\wedge^* E)\}^\star\ N$, with minimum formulas M and N; possible maximal formula Q. path:

$$A\ E(ND^*)\ B \begin{pmatrix}(\nabla E)\\(\top^* I)\end{pmatrix} [FM] \begin{pmatrix}(\nabla I)\\(\bot^* E)\end{pmatrix} C\ I(ND^*)\ D$$

($\mathcal{L}$) For proper lattice logic $ND(\mathcal{L})$
minimum segment $[FM] = M\ \{(\wedge^* I), (\vee^* I)\}^\star\ P\ (\Updownarrow)\ N$, with minimum formulas M and N; possible maximal formula P. path:

$$A\ E(ND^*)\ B \begin{pmatrix}(\nabla E)\\(\top^* I)\end{pmatrix} [FM] \begin{pmatrix}(\nabla I)\\(\bot^* E)\end{pmatrix} C\ I(ND^*)\ D$$

($\mathcal{F}$) For proper filter logic $ND(\mathcal{F})$
minimum segment $[FM] = M\ \{(\wedge^* I)\}^\star\ P\ (\Updownarrow) Q\ \{(\wedge^* E)\}^\star\ N$, with minimum formulas M and N; possible maximal formulas P and Q. path:

$$A\ E(ND^*)\ B \begin{pmatrix}(\nabla E)\\(\top^* I)\end{pmatrix} [FM] \begin{pmatrix}(\nabla I)\\(\bot^* E)\end{pmatrix} C\ I(ND^*)\ D$$

These considerations can be summarized as follows:

Theorem 7 *Every derivation in a logic $ML(\Omega)$ or $IL(\Omega)$, with $\Omega^* \subseteq \{(\top^* I), (\bot^* E), (\wedge^* I), (\wedge^* E), (\vee^* I)\}$ can be transformed into a normal one where each path has at most one application of the rule $(\Updownarrow)$.*

In general, a path in a normal derivation in our $ND(\Omega)$ has one of the following patterns:

$$- \quad \begin{array}{ccccc} A & & & & C \\ & \searrow & & \nearrow & \\ & & F & & \end{array} \quad \text{with minimum formula } F$$

$$- \quad \begin{array}{ccccccccc} A & & & & P\ (\Updownarrow)\ Q & & & & C \\ & \searrow & & \nearrow & & \searrow & & \nearrow & \\ & & M & & & & N & & \end{array} \quad \text{with} \left|\begin{array}{l}\text{minimum formulas } M \text{ and } N\\ \text{and possibly}\\ \text{maximal formulas } P \text{ and } Q\end{array}\right.$$

6 Conclusion

We now comment on the main features of our approach and on on-going work.

We have examined natural deduction systems for ‘generally.’ We have characterized the structure of normal derivations in such systems, generalizing the minimum formula result in FOL [15].

Logics for ‘generally’ (LGs) were introduced as extensions of FOL for handling assertions with vague notions (e.g., ‘generally,’ ‘most,’ ‘several’) expressed by generalized formulas. Deductive systems have been developed for LGs. With derivation rules that reflect the diverse properties of ‘generally,’ we can modularly construct natural deduction systems for LG’s, which are correct, complete, and normalizable. In our deductive systems, we use marked formulas, to handle more easily the interaction of the ∇ quantifier with the other logical constants, since the behavior of ∇ depends on the logic for ‘generally’ being considered.

The equivalence rule, which characterizes BL, reflects extensionality (equal sets have the same properties) can be seen both as an elimination and an introduction rule. The interaction of this rule with the other ones affects the structure of the derivations, with consequent loss of some usual properties, e.g., subformula property.

In FOL, a path from hypothesis A to conclusion C uses elimination rules ($\searrow$) to reach a minimum formula B and then introduction rules ($\nearrow$) to obtain the conclusion; its structure can be represented as follows:

$$\begin{array}{ccccc} A & & & & C \\ & \searrow & & \nearrow & \\ & & B & & \end{array}$$

Now, LGs also handle generalized information, and the path structure becomes as follows:

$$\begin{array}{ccccc} A & & & & C \\ & \searrow & & \nearrow & \\ & & \boxed{\text{M, …, N}} & & \end{array}$$

The frontier region manipulates marked formulas, possibly with several applications of the equivalence rule ($\Updownarrow$). We have characterized the inner structure of this frontier region, in terms of marked introductions ($\nearrow$) and eliminations ($\searrow$), as well as the equivalence rule ($\Updownarrow$), as follows:

$$\begin{array}{ccc} M & & N \\ & \searrow \nearrow & \end{array} \quad \text{or} \quad \begin{array}{ccccc} & & (\Updownarrow) & & \\ & \nearrow & & \searrow & \\ M & & & & N \end{array}$$

Thus, a path in a normal derivation in our $ND(\Omega)$, e.g. filter logic, has one of the following patterns:

$$\begin{array}{ccccc} & A & & C & \\ - & \searrow & & \nearrow & \text{with minimum formula } F \\ & & F & & \end{array}$$

$$\begin{array}{ccccccccc} & A & & P\ (\Updownarrow)\ Q & & C & \\ - & \searrow & \nearrow & & \searrow & \nearrow & \text{with minimum formulas } M \text{and } N \\ & & M & & N & & \end{array}$$

This result extends the familiar one for FOL derivations [15]. Our result is obtained by controlling and minimizing the number of applications of the equivalence rule.

Here, we have restricted our attention to the cases where the underlying logic is minimal or intuitionisitic logic. We expect to extend our analysis to the case of classical logic by means of strategies as those in [25, 26]. We also have sequent calculi for 'generally' [14]. We plan to employ these ideas to provide a better characterization of such sequent calculi derivations.

In conclusion, we have characterized the structure of derivations involving versions of (nonstandard) generalized quantifiers corresponding to some notions of 'generally'. As such, this approach may be expected to pave the way for proof methods and automatic theorem provers for logics of 'generally' (cf. [3, 8]).

Appendix A

Now, we will present some details of our development.

The notions of path, thread and (maximal) segment in a system $ND(\Omega)$ are natural adaptations of those notions for the system ND [15].

Consider a derivation Π in a system $ND(\Omega)$.

A sequence $(F_1, \ldots, F_n)$ of formula occurrences is a *thread* in Π iff

1. F_1 is a top-formula in Π;
2. F_i stands immediately above F_{i+1} in Π for each $i < n$; and
3. F_n is the end-formula of Π.

A sequence $(F_1, \ldots, F_n)$ is a *path* in Π iff

1. F_1 is a top-formula in Π that is not discharged by an application of $(\vee E)$ or $(\exists E)$;
2. for each $i < n$, F_i is a minor premise of an application of $(\rightarrow E)$, and
   (a) either F_i is not the major premise of $(\vee E)$ or $(\exists E)$, and F_{i+1} is the formula occurrence immediately below F_i,
   (b) or F_i is the major premise of an application α of $(\vee E)$ or $(\exists E)$, and F_{i+1} is an assumption discharged in Π by α;
3. the end-formula F_n is either a minor premise of $(\rightarrow E)$ or a major premise of an application α of $(\vee E)$ or $(\exists E)$ such that α does not discharge any assumption.

A *segment* in Π is a sequence $(F_1, \ldots, F_n)$ of consecutive formula occurrences in a thread in Π such that

1. F_1 is not the consequence of an application of $(\vee E)$ or $(\exists E)$;
2. for each $i < n$, F_i is a minor premise of an application of $(\vee E)$ or $(\exists E)$; and
3. F_n is not the minor premise of an application of $(\vee E)$ or $(\exists E)$.

A segment S in a derivation Π is a *maximal segment* iff

- the first formula in S is the conclusion of an application of an introduction rule, and the last formula in S is the major premise of an application of an elimination rule.
- if $S = (F_1, \ldots, F_n)$ (with $n > 1$), then:
  - F_1 is the conclusion of an application of one of the rules: (Abs), (RaA), $(\Updownarrow)$ and F_n is the major premise of an application of an elimination rule; or
  - F_1 is the conclusion of an application of the $(\Updownarrow)$ rule and F_n is the major premise of an application of the $(\Updownarrow)$ rule.
- if $S = (F)$, then:
  - F is both the conclusion of an application of one of the rules (Abs) or (RaA) and the major premise of an application of an elimination rule of ND^*; or
  - F is both the conclusion and the major premise of an application of the $(\Updownarrow)$ rule.

Let Π be a derivation of G from a set of formulas Γ in $ND(\Omega)$ and let $\rho = (F_1, \ldots, F_n)$ be a path in Π. Then, the *order* of ρ, denoted by $o(\rho)$, is defined as follows:

$$o(\rho) = \begin{cases} 0 & \text{if } F_n = \text{G} \\ o(\rho') + 1 & \text{if } F_n \text{ is a minor premise of an applicationrule of path } \rho' \end{cases}$$

The *order of a derivation* is the maximum order of its paths.

Consider a derivation Π of order M. For each n ($0 \le n \le M$), let $multi_n(\Pi)$ be the set consisting of all paths of order n with more than one application of rule $(\Updownarrow)$. Notice that $multi_M(\Pi) = \emptyset$.

Now, consider a normal derivation Π with some $multi_k(\Pi) \neq \emptyset$ and let n be the least among these. For each path $\rho \in multi_n(\Pi)$, apply the conversions in Sect. 5 (including the reductions for $(\Updownarrow)$; $(\Updownarrow)$ in Sect. 4) to obtain a derivation Π'_n, in which every path of order up to n has at most one application of $(\Updownarrow)$ and no maximal segment. However, a path of order $n + 1 \le M$ may have more than one application of rule $(\Updownarrow)$ as well as maximal segments. Now, normalize Π'_n to obtain Π_n. In Π_n, every $multi_k(\Pi)$, for $0 \le k \le n$ is empty. Applying this process, we construct normal derivations $\Pi_n, \ldots, \Pi_{M-1}$, where in Π_{M-1} every path of order up to $M - 1$ has at most one application of $(\Updownarrow)$ and no maximal segment. Note that $multi_M(\Pi_{M-1}) = \emptyset$.

References

1. Carnielli, W. A., & Veloso, P. A. S. (1997). Ultrafilter logic and generic reasoning. In G. Gottlob, A. Leitsch, & D. Mundici (Eds.), *Computational Logic and Proof Theory (LNCS)* (vol. 1289, pp. 34–53). Berlin: Springer.
2. Veloso, S. R. M., & Veloso, P. A. S. (2001). On a logical framework for 'generally'. In J. M. Abe & J. I. Silva Filho (Eds.), *Logic, Artificial Intelligence and Robotics: Proceedings of LAPTEC* (pp. 279–286). Amsterdam: IOS Press.
3. Veloso, S. R. M., & Veloso, P. A. S. (2002). On special functions and theorem proving in logics for 'generally'. In G. Bittencourt & G. L. Ramalho (Eds.), *Advances in Artificial Intelligence: 16th Brazilian Symposium in Artificial Intelligence: SBIA 2002 (LNAI)* (vol. 2507, pp. 1–10). Berlin: Springer.
4. Rentería, C. J., Haeusler, E. H., & Veloso, P. A. S. (2003). NUL: Natural deduction for ultrafilter logic. *Bulletin of Section of Logic*, *32*(4), 191–199.
5. Veloso, S. R. M., & Veloso, P. A. S. (2004). On modulated logics for 'generally': some metamathematical issues. In J.-Y., Béziau, A. Costa Leite & A. Facchini (Eds.), *Aspects of universal logic* (pp. 146–168). Neuchatel: University of Neuchâtel.
6. Veloso, P. A. S., & Carnielli, W. A. (2004). Logics for qualitative reasoning. In D. Gabbay, S. Rahman, J. Symons, & J. P. van Bendegem (Eds.), *Logic, epistemology and the unity of science* (pp. 487–526). Dordretch: Kluwer Press.
7. Veloso, P. A. S., & Veloso, S. R. M. (2004). On ultrafilter logic and special functions. *Studia Logica*, *78*, 459–477.
8. Veloso, S. R. M., & Veloso, P. A. S. (2005). On logics for 'generally' and their relational interpretations. In *First World Congress on Universal Logic Handbook* (pp. 101–102). Montreux.
9. Veloso, P. A. S., & Veloso, S. R. M. (2005). Functional interpretation of logics for 'generally'. *Logic Journal of the IGPL*, *13*(1), 127–140.
10. Veloso, P. A. S., & Veloso, S. R. M. (2005a). On 'most' and 'representative': Filter logic and special predicates. *Logic Journal of the IGPL*, *13*(6), 717–728.
11. Vana, L. B., Veloso, P. A. S., & Veloso, S. R. M. (1997). Natural deduction for 'generally'. *Logic Journal of the IGPL*, *15*, 775–800.
12. Veloso, P. A. S., Vana, L. B., & Veloso, S. R. M. (2005). Natural deduction strategies for 'generally'. In *Actas de la XI Conferencia de la Asociación Española para la Inteligencia Artificial* (pp. 173–182). Santiago de Compostela.
13. Vana, L. B., Veloso, P. A. S., & Veloso, S. R. M. (2005). Sobre lógicas para 'geralmente' em ambiente de dedução matural. In *XXV Congresso da SBC (ENIA V)* (pp. 622–630).
14. Vana, L. B., Veloso, P. A. S., & Veloso, S. R. M. (2007). Sequent calculi for 'generally'. In *Electronic Notes on Theoretical Computer Science Proceedings of the II Workshop on Logical and Semantic Frameworks with Applications* (vol. 205, pp. 49–65).
15. Prawitz, D. (1957). *Natural deduction: A proof-theoretical study*. Stockholm: Almquist & Wiksdell.
16. Mostowski, A. (1957). On a generalization of quantifiers. *Fundamenta Mathematicae*, *44*, 12–36.
17. Keisler, J. H. (1970). Logic with the quantifier there exist uncountably many. *Annals of Mathematical Logic*, *1*, 1–93.
18. Carnielli, W. A., & Sette, A. M. (1994). Default operators. In *Abstracts of Workshop on Logic, Language, Information and Computation*. Recife.
19. Schlechta, K. (1995). Defaults as generalized quantifiers. *Journal of Logic and Computation*, *5*, 473–494.
20. Barwise, J., & Cooper, R. (1981). Generalized quantifiers and natural language. *Linguistics and Philosophy*, *4*, 159–219.
21. Enderton, H. B. (1972). *A mathematical introductionto logic*. New York: Academic Press.
22. Vana, L. B. (2008). *Dedução natural e cálculo de seqüentes para 'geralmente'*. D. Sc. diss., COPPE-UFRJ, Rio de Janeiro, RJ, Brazil.

23. Gentzen, G. (1969). Investigations into logical deduction. In M. E. Szabo (Ed.), *The collected papers of Gerhard Gentzen*. Amsterdan: North-Holland.
24. van Dalen, D. (1989). *Logic and structure* (2nd edn., 3rd prt.). Berlin: Springer.
25. Seldin, J. (1986). On the proof-theory of the intermediate logic MH. *Jornal of Symbolic Logic*, *5*, 626–647.
26. Medeiros, M. P. N. (2001). *Traduções via Teoria da Prova: aplicações à lógica linear*. D. Sc. diss., PUC-Rio, Rio de Janeiro, RJ, Brazil.

Type Theories from Barendregt's Cube for Theorem Provers

Jonathan P. Seldin

Abstract Anybody using a theorem prover or proof assistant will want to have confidence that the system involved will not permit the derivation of false results. On some occasions, there is more than the usual need for this confidence. This chapter discusses some logical systems based on typed lambda-calculus that can be used for this purpose. The systems are natural deduction systems, and use the propositions-as-types paradigm. Not only are the underlying systems provably consistent, but additional unproved assumptions from which a lot of ordinary mathematics can be derived can also be proved consistent. Finally, the systems have few primitive postulates that need to be programmed separately, so that it is easier for a programmer to see whether the code really does program the systems involved without errors.

Theorem provers and proof assistants are used for a variety of purposes. For some of these purposes, including many cases of formal verification, these theorem provers must be *trusted*. Among the conditions needed for a trusted system are the following:

- *Consistency*. It must not be possible to derive a contradiction in the system.
- *Confidence in the implementation*. The user must have good reason to believe that the coding for the implementation really does accurately program the formal logic involved, and does not accidently include extra principles.

In some cases, it is also necessary for the theorem prover to be small, for example in *proof carrying code* (PCC) [1, 2]. PCC is designed to provide computer users installing new software evidence that the software is safe: the software code is to include a formal proof that the software satisfies certain safety conditions (such as not writing to the wrong memory locations), and the computer on which the software is installed is to have a theorem prover that checks this formal proof. In order not to

J. P. Seldin (✉)
Department of Mathematics and Computer Science,
University of Lethbridge, Lethbridge, AB, Canada
e-mail: jonathan.seldin@uleth.ca
http://www.cs.uleth.ca/~seldin

L. C. Pereira et al. (eds.), *Advances in Natural Deduction*,
Trends in Logic 39, DOI: 10.1007/978-94-007-7548-0_7,

interfere with other computer operations, the theorem prover needs to be small. In order to provide the necessary assurance to the user, it must be trusted.

In this chapter, we look at some formal systems that can be used for theorem provers or proof assistants of this kind.

I would like to thank Roger Hindley for his helpful comments and suggestions.

1 Barendregt's λ-Cube

Definition 1 The λ-cube of Barendregt [3] is a collection of eight systems of type assignment to λ-calculus. The systems all have the same syntax:

$$M \longrightarrow x|c|\mathsf{Prop}|\mathsf{Type}|(MM)|(\lambda x : M\ .\ M)|(\forall x : M)M.$$

Here Prop and Type are special constants called *sorts*; s, s', and s_1, etc., will be used for sorts.[1] Conversion will be β-conversion, generated by

$$(\lambda x : A\ .\ M)N \rhd [N/x]M,$$

where $[N/x]M$ denotes the substitution of N for all free occurrences of x in M, with bound variables being changed to avoid conflicts. Judgements are of the form $\Gamma \vdash M : A$, where Γ is

$$x_1 : A_1, x_2 : A_2, \ldots, x_n : A_n.$$

The systems all have the same axiom, namely

$$\mathsf{Prop} : \mathsf{Type}.$$

They all have the following *general rules* in common:

(start) If $x \notin FV(\Gamma)$

$$\frac{\Gamma \vdash A : s}{\Gamma, x : A \vdash x : A}$$

(weakening) If $x \notin FV(\Gamma)$

$$\frac{\Gamma \vdash A : B \quad \Gamma \vdash C : s}{\Gamma, x : C \vdash A : B}$$

(application)

$$\frac{\Gamma \vdash M : (\forall x : A)B \quad \Gamma \vdash N : A}{\Gamma \vdash MN : [N/x]B}$$

---

[1] It is common to use * for Prop, □ for Type. I formerly referred to sorts as *kinds* [21–25].

(abstraction) If $x \notin FV(\Gamma)$

$$\frac{\Gamma, x : A \vdash M : B \quad \Gamma \vdash (\forall x : A)B : s}{\Gamma \vdash \lambda x : A \,.\, M : (\forall x : A)B}$$

(conversion)

$$\frac{\Gamma \vdash A : B \quad \Gamma \vdash B' : s \quad B =_{\beta} B'}{\Gamma \vdash A : B'}$$

The systems are differentiated by the following *specific rules*:

(ss' rule) If $x \notin FV(\Gamma)$

$$\frac{\Gamma \vdash A : s \quad \Gamma, x : A \vdash B : s'}{\Gamma \vdash (\forall x : A)B : s'} \quad ,$$

where the different systems depend on the possible values of s and s'. Here are some of the specific systems:

- $\lambda \rightarrow$, related to simple type assignment: Both s and s' must be Prop.
- $\lambda 2$, related to second-order typed λ-calculus: s' must be Prop.
- λP, related to AUT-QE and LF: s must be Prop.
- $\lambda\omega$, related to Girard's $F\omega$: If s is Prop, so is s'.
- λC, Calculus of constructions: s and s' can both be of either kind.

HOL, which is Church's simple type theory [9], which is not in the λ-cube, is a subsystem of λC. λC is the strongest system in the λ-cube; all the other systems are subsystems of it. Similarly, $\lambda \rightarrow$ is a subsystem of all other systems in the cube.

These systems all have some advantages for the purposes we consider in this chapter:

- They can all be proved consistent by proving a strong normalization theorem. The proof of this theorem varies from one system to another; the proof for stronger systems is a stronger proof than that for weaker systems. As Gödel's Second Theorem tells us, each of these proofs uses means of proof which cannot be formalized in the system involved.
- They all have a small number of primitive postulates, which means that when they are implemented there are few places for programming errors to occur.[2]

They do have one feature which might be considered as a disadvantage: they are impredicative. However, all the known predicative systems suffer from the disadvantage that they have a large number of primitive postulates. It thus appears that we can have the advantages of a small number of postulates (with the consequence of a relatively small chance of an error in programming) or else predicativity. My own

[2] Compare this to the system Nuprl [10], which has over one hundred primitive postulates, each of which must be programmed separately in implementation.

personal view is that a small number of primitive postulates is more important than predicativity, especially since in practice, most clients for such systems will probably never have heard of predicativity.

The formulation given here is the one given by Barendregt as a Pure Type System (PTS) [3], and, in the case of the calculus of constructions, is much closer to the original formulation of Coquand and Huet [11] than the formulation I used in my previous chapters on this subject [21–25]. The main difference is that before an environment Γ can appear in a deduction in the formulation given here, it must be proved well-formed by proving that

$$\Gamma \vdash \mathsf{Prop} : \mathsf{Type},$$

whereas in the formulation I formerly used, being well-formed was shown to be necessary for the discharge of assumptions by conditions on the rules corresponding to (abstraction) and the (ss' rule). For

$$x_1 : A_1, x_2 : A_2, \ldots, x_n : A_n$$

to be well-formed means that

- The variable x_i does not occur free in $A_1, A_2, \ldots, A_i$ (but it may occur free in $A_{i+1}, \ldots, A_n$), and
- $x_1 : A_1, x_2 : A_2, \ldots x_{i-1} : A_{i-1} \vdash A_i : s$ for some sort s.

From now on, all environments will be assumed to be well-formed.

2 Representing Logic with Equality

Logic is represented in these typed systems by the *propositions-as-types* interpretation, which is also known as the *Curry-Howard isomorphism* [14]. The idea is that the types are interpretated as formulas or propositions, and the terms are interpreted as proofs or deductions. It is not hard to see that deductions in any of these systems using the rules for (application) and (abstraction) follow the constructions of the terms involved, and the other rules play the auxiliary role of determining when these two rules, and especially the rule for (abstraction), can be legally applied.

The notation $A \to B$ is used for the type $(\forall x : A)B$ when x does not occur free in B. As a type, $A \to B$ is the type of functions whose arguments are in type A and whose values are in type B. When $A : \mathsf{Prop}$ and $B : \mathsf{Prop}$, the type $A \to B$ will be interpreted as the implication from A to B, and the notation $A \supset B$ will often be used for this. The rule (application) gives us modus ponens, and the rule (abstraction) gives us implication introduction in the usual sense of natural deduction.

The other connectives and quantifiers are defined to the extent that they can be defined in the various systems of the λ-cube[3]:

Definition 2 The connectives and quantifiers are defined as follows:

- If $A : \mathsf{Prop}$ and $B : \mathsf{Prop}$, use $A \wedge B$ for $(\forall w : \mathsf{Prop})((A \supset B \supset w) \supset w)$. The terms of type $A \wedge B$ are pairs with projections, and their types give us the usual properties of conjunction in natural deduction.
- If $A : \mathsf{Prop}$ and $B : \mathsf{Prop}$ use $A \vee B$ for

$$(\forall w : \mathsf{Prop})((A \supset w) \supset ((B \supset w) \supset w))$$

  Terms of type $A \vee B$ are disjoint unions with injections, and their types give us the usual properties of disjunction in natural deduction.
- Define $\mathsf{void} \equiv \perp \equiv (\forall x : \mathsf{Prop})x$. If $A : \mathsf{Prop}$, use $\neg A$ for $A \supset \perp$. This gives us the usual properties of intuitionistic negation. Furthermore, it follows from the strong normalization theorem that there is no closed term M such that $\vdash M : \mathsf{void}$.
- If $A : \mathsf{Prop}$ and $x : A \vdash B : \mathsf{Prop}$, then use $(\exists x : A)B$ for

$$(\forall w : \mathsf{Prop})((\forall x : A)(B \to w) \to w).$$

  Terms of type $(\exists x : A)B$ are pairs (differently typed from those of type $A \wedge B$) with a left projection but (for technical reasons) no right projection. The types involved give us the usual natural deduction rules for the existential quantifier.
- If $A : s$, $M : A$ and $N : A$, use $M =_A N$ for

$$(\forall z : A \to \mathsf{Prop})(zM \supset zN).$$

  This is called *Leibniz equality*. It is easy to prove that it satisfies the reflexive, symmetric, and transitive laws and that equals under this equality can always be replaced by equals.

With these definitions, we have a higher-order intuitionistic logic.
We can also define a *Boolean* type:

$$\mathsf{Bool} \equiv (\forall u : \mathsf{Prop})(u \to u \to u),$$

$$\mathsf{T} \equiv \lambda u : \mathsf{Prop} \,.\, \lambda x : u \,.\, \lambda y : u \,.\, x,$$

$$\mathsf{F} \equiv \lambda u : \mathsf{Prop} \,.\, \lambda x : u \,.\, \lambda y : u \,.\, y.$$

Here T and F have distinct normal forms. The usual truth functions are easy to define, but none of them are identified with the connectives and quantifiers defined above.

---

[3] See [22, Sect. 6].

3 Adding Unproved Assumptions

The higher-order intuitionistic logic with equality that we have seen above is not enough for practical theorem provers and proof assistants. We will also need to allow for new postulates in the form of unproved assumptions. Adding such assumptions is easy: to make A an assumption, just add $c : A$ as a new assumption, where c is a new atomic constant. New atomic constants can be obtained from variables by simply deciding not to make substitutions for them. In an implementation, variables and constants are just identifiers anyway, and any computer implementation allows for any number of those.

However, adding unproved assumptions in this way can cause problems with consistency. The basic system is consistent, as noted above. However, it is easy to come up with a set of new assumptions which leads to a contradiction:

$$c_1 : \mathsf{Prop}, c_2 : c_1, c_3 : \neg c_1.$$

The contradiction that follows from these assumptions does not in any way negate the consistency that follows from the strong normalization theorem. Nevertheless, it is important to avoid them in any practical application.

Definition 3 A set Γ of assumptions is *consistent* if there is no term M such that $\Gamma \vdash M : \bot$. This is equivalent to: there is no term N such that $\Gamma, x : \mathsf{Prop} \vdash N : x$, where $x \notin \mathrm{FV}(\Gamma)$.

We now want to prove consistent sets of assumptions that are useful in practical applications of theorem-proving. The best way to do this is to prove these consistency results for the strongest system we are considering, namely the calculus of constructions. The results so obtained will then follow for all those weaker systems in which the assumptions can be typed. In some weaker systems, the results can be used to justify assuming new assumptions. For example, in $\lambda\rightarrow$, our definition of conjunction cannot be typed. However, we can add a new constant Λ, take $A \wedge B$ as an abbreviation for ΛAB, and then add the following assumptions:

- $\Lambda : \mathsf{Prop} \rightarrow \mathsf{Prop} \rightarrow \mathsf{Prop}$.
- $\mathsf{and.in} : \lambda x : \mathsf{Prop} . \lambda y : \mathsf{Prop} . x \rightarrow y \rightarrow \Lambda xy$.
- $\mathsf{and.left} : \lambda x : \mathsf{Prop} . \lambda y : \mathsf{Prop} . \Lambda xy \rightarrow x$.
- $\mathsf{and.right} : \lambda x : \mathsf{Prop} . \lambda y : \mathsf{Prop} . \Lambda xy \rightarrow y$.

The fact that the resulting system can be interpreted in a stronger system (the calculus of constructions) in which these assumptions are interpreted as provable results shows that adding them to $\lambda\rightarrow$ will not lead to contradiction.

4 The Consistency of Unproved Assumptions

In order to obtain the consistency results we desire, we need to consider normalizing deductions as well as terms.

Deduction Normalization A deduction of the form

$$\dfrac{\dfrac{\dfrac{\Gamma, x : A \vdash M : B \quad \Gamma \vdash (\forall x : A)B : s}{\Gamma \vdash \lambda x : A \,.\, M : (\forall x : A)B}\text{(abstraction)}}{\Gamma \vdash \lambda x : A \,.\, M : (\forall x : C)D}\text{(conversion)} \qquad \Gamma \vdash N : C}{\Gamma \vdash (\lambda x : A \,.\, M)N : [N/x]D}\text{(application)}$$

where $x \notin \mathrm{FV}(\Gamma, A)$, $A =_\beta C$, and $B =_\beta D$, reduces to

$$\dfrac{\dfrac{\Gamma, x : A \vdash M : B \quad \dfrac{\Gamma \vdash N : C}{\Gamma \vdash N : A}\text{(conversion)}}{\Gamma \vdash [N/x]M : [N/x]B}\text{(substitution lemma)}}{\Gamma \vdash [N/x]M : [N/x]D.}\text{(conversion)}$$

The substitution lemma is proved for all systems of the λ-cube in [3, Lemma 5.1.11].

Strong normalization holds for deductions with this reduction relation.

Note that this kind of reduction step is related to the kind of implication reduction step considered by Prawitz in [20]:

$$\dfrac{\dfrac{\begin{matrix} 1 \\ [A] \\ \mathcal{D}_1 \\ B \end{matrix}}{A \supset B} \supset \mathrm{I} - 1 \quad \begin{matrix} \mathcal{D}_2 \\ A \end{matrix}}{\begin{matrix} B \\ \mathcal{D}_3 \\ C \end{matrix}} \supset \mathrm{E}$$

reduces to

$$\begin{matrix} \mathcal{D}_2 \\ A \\ \mathcal{D}_1 \\ B \\ \mathcal{D}_3 \\ C. \end{matrix}$$

Here, as part of the reduction step, the deduction

$$\begin{matrix} \mathcal{D}_2 \\ A \end{matrix}$$

is placed above all occurrences of A as an undischarged assumption in

$$\begin{array}{c} A \\ \mathcal{D}_1 \\ B. \end{array}$$

This part of the reduction step corresponds to the use of the Substitution Lemma above. But the use of the substitution lemma is more complicated. This is because in order to make use of an assumption in a deduction, it must be made the conclusion of the (start) rule, and so, unlike ordinary natural deduction, the assumption does not occur at a leaf of the tree. In fact, the only way to make as assumption occur at the top of a branch is to *define* a branch to start at the conclusion of an inference by rule (start).

One of the main features of normalized deductions used in [20] is that if the deduction ends in an elimination rule, then the only inferences which can occur in the main branch are other inferences by elimination rules, and so the assumption at the top of the main branch is not discharged. This form of reasoning applies here as well, provided that we understand that the assumption at the top of the main branch is on the right side of the conclusion of an inference by (start). The rules are written here so that the main branch is always the leftmost branch, so the main branch will sometimes be referred to as the "left branch".

There are certain kinds of inferences that we want to avoid in the main branch of a deduction if we are to have consistency. Let Γ be

$$A : \mathsf{Prop}, w : (\forall z : A \to \mathsf{Prop})(zN), x : \mathsf{Prop}.$$

What we want to avoid is the following:

$$\dfrac{\dfrac{\Gamma \vdash w : (\forall z : A \to \mathsf{Prop})(zN) \quad \begin{array}{c}\vdots \\ \Gamma \vdash \lambda y : A \,.\, x : A \to \mathsf{Prop}\end{array}}{\Gamma \vdash w(\lambda y : A \,.\, x) : (\lambda y : A \,.\, x)N}\text{(application)}}{\Gamma \vdash w(\lambda y : A \,.\, x) : x.}\text{(conversion)}$$

In order to avoid this, we define a class of assumptions that cannot occur undischarged at the top of the main branch in which such an inference occurs:

Definition 4 Let Γ be an environment of the form

$$x_1 : A_1, x_2 : A_2, \ldots, x_n : A_n.$$

For each i, let A_i convert to

$$(\forall y_{i1} : B_{i1})(\forall y_{i2} : B_{i2}) \ldots (\forall y_{im_i} : B_{im_i})S_i,$$

where S_i, the *tail* of A_i, does not convert to a term of the form $(\forall z : C)D$.[4] Then Γ is *strongly consistent* if for each A_i for which $m_i > 0$ and S_i converts to $z_i M_{i1} M_{i2} \ldots M_{il}$, if z_i is a variable, then the tail of its type does not convert to Prop.

A strongly consistent environment is consistent.

This definition is very weak. A strongly consistent environment cannot contain any types of the form $A \wedge B$, $A \vee B$, $\bot$, $\neg A$, $(\exists x : A)B$, or $M =_A N$.

Although a strongly consistent environment cannot contain negations of types, there are consistent environments that can. See [22, Definition 27].

Theorem 5 *Let Γ_1 be a well-formed environment in which each type is the negation of an equation between terms with distinct normal forms, and let Γ_2 be strongly consistent. Then if, for $B : s$ and a closed term R,*

$$\Gamma_1, \Gamma_2 \vdash R : M =_B N,$$

then $M =_\beta N$.

This is [22, Theorem 20]. The idea behind the proof is to show that the deduction ending in $R : M =_B N$ must have the form

$$\cfrac{\cfrac{\cfrac{\Gamma_1, \Gamma_2, z : B \to \mathsf{Prop}, u : zM \vdash R_1 : zM \quad \vdots \;\vdots}{\Gamma_1, \Gamma_2, z : B \to \mathsf{Prop}, u : zM \vdash R_1 : zN}\,(\text{conversion}) \quad \vdots}{\Gamma_1, \Gamma_2, z : B \to \mathsf{Prop} \vdash \lambda u : zM \,.\, R_1 : zM \supset zN}\,(\text{abstraction}) \quad \vdots}{\Gamma_1, \Gamma_2 \vdash \lambda z : B \to \mathsf{Prop} \,.\, \lambda u : zM \,.\, R_1 : (\forall z : B \to \mathsf{Prop})(zM \supset zN).}\,(\text{abstraction})$$

In this deduction, the inference by (conversion) is only valid if $M =_* N$ and R_1 is u. In showing that the deduction is of this form, it is necessary to show in each case that the formula in question cannot be the conclusion of an inference by (application), and this is shown by reasoning about the assumption at the top of the main branch if it is: the assumptions $z : B \to \mathsf{Prop}$ and $u : zM$ can both be in a strongly consistent environment, and it cannot be any of those, and if it is in Γ_1, then the minor premise would have to have a conclusion of the form $M' =_{B'} N'$, contrary to the hypothesis that we are dealing with the shortest deduction of that kind.

Note that a consequence of this theorem is the following corollary:

Corollary 1 *If Γ is a strongly consistent environment, then*

$$\Gamma, c : \neg(\mathsf{T} =_{\mathsf{Bool}} \mathsf{F})$$

is consistent.

Note also that the theorem identifies Leibniz equality with conversion.

Berardi [5] assumes that Leibniz equality is extensional in the sense that, for terms M and N of type $(\forall x_1 : A_1)(\forall x_2 : A_2) \ldots (\forall x_m : A_m)\mathsf{Prop}$,

---

[4] It can be shown that every type in a well-formed environment converts a term of this form.

$$(\forall x_1 : A_1)(\forall x_2 : A_2)\ldots(\forall x_m : A_m)(Mx_1x_2\ldots x_m \leftrightarrow Nx_1x_2\ldots x_m) \supset$$

$$M =_B N,$$

where B is $(\forall x_1 : A_1)(\forall x_2 : A_2)\ldots(\forall x_m : A_m)\mathsf{Prop}$ and $A \leftrightarrow B$ is $(A \supset B)\wedge(B \supset A)$. This assumption has unusual consequences when $m = 0$. For let A be any inhabited type. Then A is provable. But so is $A \rightarrow A$. It follows from this assumption that $A =_{\mathsf{Prop}} (A \rightarrow A)$ is provable. It follows from this that A is a model of the untyped λ-calculus, and so any function of type $A \rightarrow A$ has a fixed point. Since the type N defined in the next section is inhabited and since the successor function $\boldsymbol{\sigma}$ has type $\mathsf{N} \rightarrow \mathsf{N}$, $\boldsymbol{\sigma}$ has a fixed point! This is a highly unusual and counterintuitive result, and it is undesirable in small trusted theorem provers and proof assistants. For this reason, this assumption of extensionality will not be made here.

Classical logic can be obtained by adding the assumption

$$\mathsf{cl} : (\forall u : \mathsf{Prop})(\neg\ \neg\ u \supset u),$$

where cl is a new constant. This assumption can be proved consistent with strongly consistent assumptions plus negations of equations between distinct normal forms by means of a variation of the double-negation translation. See [22, Corollary 22.1].

5 Representing Arithmetic

It is standard to represent arithmetic in this system with the following definitions:

- $\mathsf{N} \equiv (\forall A : \mathsf{Prop})((A \rightarrow A) \rightarrow (A \rightarrow A))$
- $\mathbf{0} \equiv \lambda A : \mathsf{Prop}\ .\ \lambda x : A \rightarrow A\ .\ \lambda y : A\ .\ y$
- $\boldsymbol{\sigma} \equiv \lambda u : \mathsf{N}\ .\ \lambda A : \mathsf{Prop}\ .\ \lambda x : A \rightarrow A\ .\ \lambda y : A\ .\ x(uAxy)$

Here, N is the type of the natural numbers, $\mathbf{0}$ is the number zero, and $\boldsymbol{\sigma}$ is the successor function. Then a natural number n is represented by

$$\mathsf{n} =_\beta \lambda A : \mathsf{Prop}\ .\ \lambda x : A \rightarrow A\ .\ \lambda y : A\ .\ \underbrace{x(x(\ldots(x}_{n}\ y)\ldots))$$

It is possible to define $\boldsymbol{\pi}$ so that

$$\boldsymbol{\pi}\mathbf{0} =_\beta \mathbf{0}$$
$$\boldsymbol{\pi}(\boldsymbol{\sigma}\mathsf{n}) =_\beta \mathsf{n}$$

Using this $\boldsymbol{\pi}$, it is possible to define R so that if $A : \mathsf{Prop}$, $M : A$, and $N : \mathsf{N} \rightarrow A \rightarrow A$,

$$\mathsf{R}MN\mathbf{0} =_\beta \quad M$$
$$\mathsf{R}MN(\sigma\mathsf{n}) =_\beta \quad N\mathsf{n}(\mathsf{R}MN\mathsf{n})$$

We can prove

$$\vdash \quad \mathsf{N} : \mathsf{Prop}$$
$$\vdash \quad \mathbf{0} : \mathsf{N}$$
$$\vdash \quad \sigma : \mathsf{N} \to \mathsf{N}$$

This is an example of an inductively defined datatype. This particular representation is an example of the kind of datatype treated in [6, 8].

We would naturally like to prove mathematical induction for N, which is

$$(\forall A : \mathsf{N} \to \mathsf{Prop})((\forall u : \mathsf{N})(Au \supset A(\sigma u)) \supset A\mathbf{0} \supset (\forall x : \mathsf{N})(Ax)),$$

but there is a major problem about this. It says that every term of type N is Leibniz equal to a term representing a natural number. However, the theorem we proved above identifies Leibniz equality with conversion, which in these systems is β-conversion. But there is a term in type N that does not β-convert to a term representing a natural number: $\lambda A : \mathsf{Prop} . \lambda x : A . x$. The term does η-convert to a term representing a natural number, but we are not using η-conversion here.[5] Pfenning and Paulin-Mohring [19] give an example of a recursive datatype represented this way in which there is a term in the type which does not β- or η-convert to anything constructed from the constructors of the datatype. For this reason, it appears that this form of mathematical induction is false in these systems and that it would be a mistake to assume it.

Instead, we can get the principle of mathematical induction by using Dedekind's definition of the natural numbers [13]:

$$\mathcal{N} \equiv \lambda n : \mathsf{N} . (\forall A : \mathsf{N} \to \mathsf{Prop})((\forall m : \mathsf{N})(Am \supset A(\sigma m)) \supset A\mathbf{0} \supset An)$$

We can prove

$$\vdash \quad \mathcal{N} : \mathsf{N} \to \mathsf{Prop},$$
$$\vdash \quad \mathcal{N}\mathbf{0},$$
$$\vdash \quad (\forall n : \mathsf{N})(\mathcal{N}n \supset \mathcal{N}(\sigma n)),$$
$$\vdash \quad (\forall A : \mathsf{N} \to \mathsf{Prop})((\forall m : \mathsf{N})(Am \supset A(\sigma m)) \supset A\mathbf{0} \supset, \quad (\forall n : \mathsf{N})(\mathcal{N}n \supset An)).$$

Thus, by relativizing the quantifiers to $\mathcal{N}$, we can obtain the use of mathematical induction. Furthermore, we have used only definitions; there are no new unproved

[5] The rule for η-reduction is that $\lambda x : A . Ux \rhd U$ if x is not free in U.

assumptions. Thus, if we are working within an environment known to be consistent, we retain consistency.

We can extend these results to the other Peano axioms. For example, using π we can prove

$$\vdash (\forall n : \mathsf{N})(\forall m : \mathsf{N})(\mathcal{N} n \supset \mathcal{N} m \supset \sigma n =_{\mathsf{N}} \sigma m \supset n =_{\mathsf{N}} m).$$

Also, using $\mathsf{Bool} : \mathsf{Prop}, \mathsf{T} : \mathsf{Bool}$, and $\lambda n : \mathsf{N} . \lambda x : \mathsf{Bool} . \mathsf{F} : \mathsf{N} \rightarrow \mathsf{Bool} \rightarrow \mathsf{Bool}$, all of which are provable, we can define

$$\mathsf{Iszero} \equiv \mathsf{RT}(\lambda n : \mathsf{N} . \lambda x : \mathsf{Bool} . \mathsf{F})$$

Then for $\mathsf{n} : \mathsf{N}$, since $\vdash \mathcal{N}\mathsf{n}$,

$$\begin{aligned} \mathsf{Iszero}\ \mathbf{0} &=_{\beta} \mathsf{T}, \\ \mathsf{Iszero}\ (\sigma \mathsf{n}) &=_{\beta} \mathsf{F}. \end{aligned}$$

Hence, we can prove

$$\mathsf{Bool} : \neg\ \mathsf{T} =_{\mathsf{Bool}} \mathsf{F} \vdash (\forall n : N)(\mathcal{N} n \supset \neg\ \sigma n =_{\mathsf{N}} \mathbf{0}).$$

This means that *arithmetic in a typed system is consistent.*

In a recent chapter [24], this is extended to a large class of abstract recursively defined data types. But not all: only those for which the type of each constructor has the form

$$A_1 \rightarrow (A_2 \rightarrow \cdots (A_n \rightarrow D) \cdots)$$

where D is the type of the database and each A_i is either D or is the type of another such database or is a variable of type Prop. All of these datatypes are covariant in the sense of [1]. A method of representing non-covariant datatypes is given by Appel and McAllester in [2]. I believe that a more natural method is to extend Dedekind's definition to arbitrary datatypes by defining a predicate for any set of constructors which says that an object (of a suitable type) is an object of the datatype if it satisfies every predicate which is closed under the constructors.[6] To deal with recursive definitions (i.e., definitions of recursive functions), I propose to use the logic to prove that valid definitions do, indeed, define functions. For the case of definitions corresponding to primitive recursive functions of natural numbers, I propose to adapt a proof that primitive recursive definitions define total functions of natural numbers that was given by Lorenzen [18] and Kalmár [17].[7]

---

[6] As I write this, I have not yet found a general way of writing this predicate, but I have succeeded for an example which is not covariant.

[7] The adaptation of this proof seems to work in the case of the example mentioned in the preceding footnote.

This approach to inductive definitions, as well as the approach of [1, 2], has the advantage that it is carried out by making definitions, and not by adding new rules to the type theory as is done in [7, 12, 26]. When extra rules are added, there are existential committments that follow from them, and this means that restrictions must be placed on the rules in advance in order to avoid inconsistency. When new definitions are used without new unproved assumptions, there is more flexibility. And with the approach I am advocating, merely extending Dedekind's approach to defining the natural numbers does not, in itself involve any existential committments. The worst that can happen if a definition is given that does not correspond to a real inductive structure is that it will be impossible to prove that anything satisfies the definition, and this causes no problems with consistency. Furthermore, when the proof of Lorenzen and Kalmár that primitive recursive definitions define total functions is adapted to this kind of theory, the theorem that is proved has an existential hypothesis, so this theorem only proves that something exists if it can be proved that there is an object of the right type satisfying the inductive definition. With this approach, the logic itself will take care of these problems, and this, in my opinion, makes the approach more general than the others.

6 Representing Sets as Predicates

In [15, Chap. 5] and [16], Huet proposed to represent a significant part of elementary set theory by means of predicates. The idea is to take a type $U : \mathsf{Prop}$[8] as a universe and to define Set_U to be $U \to \mathsf{Prop}$. Then if $A : \mathsf{Set}_U, x \in A$ is just Ax. Furthermore, for $P : \mathsf{Prop}$, the set $\{x : U | P\}$ is $\lambda x : U \,.\, P$. In addition, we have the following definitions:

$$
\begin{aligned}
A \subseteq B &\equiv (\forall x : U)(x \in A \supset x \in B), \\
A =_{\mathrm{ex}} B &\equiv (A \subseteq B) \wedge (B \subseteq A), \\
\emptyset &\equiv \{x : U \mid \bot\}, \\
\{x\} &\equiv \{y : U | y =_U x\}, \\
A \cap B &\equiv \{x : U \mid x \in A \wedge x \in B\}, \\
A \cup B &\equiv \{x : U \mid x \in A \vee x \in B\}, \\
\sim A &\equiv \{x : U \mid \neg\, x \in A\}, \\
\mathcal{P}A &\equiv \lambda B : \mathsf{Set}_U \,.\, B \subseteq A, \\
\mathsf{Class}_U &\equiv \mathsf{Set}_U \to \mathsf{Prop}.
\end{aligned}
$$

Note that if $A : \mathsf{Set}_U$, then $\mathcal{P}A : \mathsf{Class}_U$. Other definitions involving classes are

[8] The definition will also work if $U : \mathsf{Type}$, but this is not needed here.

$$\bigcap C \equiv \{x : U | (\forall A : \mathsf{Set}_U)(CA \to x \in A)\},$$
$$\bigcup C \equiv \{x : U | (\exists A : \mathsf{Set}_U)(CA \wedge x \in A)\}.$$

If $A, B : \mathsf{Set}_U$, we can represent the collection of functions from A to B by

$$\lambda f : U \to U \,.\, (\forall x : U)(x \in A \supset fx \in B).$$

This is a lot of set theory. In [22, Remark 17], it is shown that all the axioms of the constructive set theory IZF ([4], p. 164) except for $\in$-induction and power set are provable in this representation. The axiom of $\in$-induction is the constructive replacement for the axiom of foundation, whose role is to prevent infinite descending $\in$-chains, but here such chains are prevented by the type structure. So the important axiom that cannot be proved here is the axiom of power set. But power sets may be taken any finite number of times, and this is enough for many practical purposes.

7 Conclusion

We have seen that with only two unproved assumptions, namely

$$c_1 : \neg(\mathsf{T} =_{\mathsf{Bool}} \mathsf{F}), \quad (1)$$
$$\mathsf{cl} : (\forall u : \mathsf{Prop})(\neg\neg u \supset u), \quad (2)$$

we can obtain classical arithmetic and a good deal of set theory. Furthermore, the combination of these assumptions has been proved consistent. This shows that we have systems which

- are small;
- have few primitive postulates;
- are provably consistent; and
- are sufficient for a lot of mathematical reasoning.

These systems should be useful for a number of applications of theorem proving.

Acknowledgments This work was supported in part by grant RGP-23391-98 from the Natural Sciences and Engineering Research Council of Canada.

References

1. Appel, A. W., & Felty, A. P. (2000). A semantic model of types and machine instructions for proof-carrying code. In *Proceedings of the 27th ACM symposium on the principles of programming languages* (pp. 243–253).

2. Appel, A. W., & McAllester, D. (2000). An indexed model of recursive types for foundational proof-carrying code. Technical report TR-629-00, Princeton University.
3. Barendregt, H. P. (1992). Lambda calculi with types. In S. Abramsky, D. M. Gabbay, & T. S. E. Maibaum (Eds.), *Handbook of logic in computer science* (Vol. 2, pp. 117–309). Oxford: Oxford University Press.
4. Beeson, M. (1985). *Foundations of constructive mathematics*. Berlin: Springer.
5. Berardi, S. (1989). Type dependence and constructive mathematics. Ph.D. thesis, Universita di Torino.
6. Berardi, S. (1993). Encoding of data types in pure construction calculus: a semantic justification. In G. Huet, & G. Plotkin (Eds.), *Logical Environments* (pp. 30–60). Cambridge: Cambridge University Press.
7. Blanqui, F. (1998). The calculus of algebraic and inductive constructions. Technical report, DEA Sémantique, preuve et Programmation.
8. Böhm, C., & Berarducci, A. (1985). Automatic synthesis of typed Λ-programs on term algebras. *Theoretical Computer Science*, *39*(2–3), 135–154.
9. Church, A. (1940). A formulation of the simple theory of types. *Journal of Symbolic Logic*, *5*, 56–68.
10. Constable, R., et al. (1986). *Implementing mathematics with the Nuprl proof development system*. Englewood Cliffs: Prentice-Hall.
11. Coquand, T., & Huet, G. (1988). The calculus of constructions. *Information and Computation*, *76*, 95–120.
12. Coquand, T., & Paulin, C. (1990). Inductively defined types. In P. Martin-Löf, & G. Mints (Eds.), *COLOG-88: Proceedings of the international conference on computer logic held in Tallinn*, December 12–16, 1988, Lecture notes in computer science (Vol. 417, pp. 50–66). Springer.
13. Dedekind, R. (1887). *Was sind und was sollen die Zahen?* (10th ed., 1965). Braunschweig: Friedr. Vieweg & Sohn.
14. Howard, W. A. (1980). The formulae-as-types notion of construction. In J. Roger Hindley & J. P. Seldin (Eds.), *To H. B. Curry: Essays on combinatory logic, lambda calculus and formalism* (pp. 479–490). New York: Academic. A version of this paper was privately circulated in 1969.
15. Huet, G. (1986). *Formal structures for computation and deduction* (1st ed.). Course notes, Carnegie-Mellon University.
16. Huet, G. (1987). Induction principles formalized in the calculus of constructions. In H. Ehrig, R. Kowalski, G. Levi, & U. Montanari (Eds.), *TAPSOFT '87: Proceedings of the international joint conference on theory and practice of software development*, Pisa, Italy, March 23–27, 1987. Advanced seminar on foundations of innovative software development I and colloquium on trees in algebra and programming (CAAP '87) (Vol. 1), Lecture notes in computer science (Vol. 249, pp. 276–286). Berlin: Springer.
17. Kalmár, L. (1940). On the possibility of definition by recursion. *Acta Szeged*, *9*(4), 227–232.
18. Lorenzen, P. (1939). Die Definition durch vollständige Induktion. *Monatshefte für Mathematik und Physik*, *47*, 356–358.
19. Pfenning, F. & Paulin-Mohring, C. (1989). Inductively defined types in the calculus of constructions. In M. Main, A. Melton, M. Mislove, & D. Schmidt (Eds.), *Proceedings of the 5th international conference on mathematical foundations of programming semantics*, Tulane University, New Orleans, Louisiana, USA, March 29–April 1, 1989, Lecture notes in computer science, (Vol. 442, pp. 209–228). Springer.
20. Prawitz, D. (1965). *Natural deduction*. Stockholm: Almqvist & Wiksell.
21. Seldin, J. P. (1992). Coquand's calculus of constructions: a mathematical foundation for a proof development system. *Formal Aspects of Computing*, *4*, 425–441.
22. Seldin, J. P. (1997). On the proof theory of Coquand's calculus of constructions. *Annals of Pure and Applied Logic*, *83*, 23–101.
23. Seldin, J. P. (2000). A Gentzen-style sequent calculus of constructions with expansion rules. *Theoretical Computer Science*, *243*, 199–215.

24. Seldin, J. P. (2000). On lists and other abstract data types in the calculus of constructions. *Mathematical Structures in Computer Science, 10*, 261–276. Special issue in honor of J. Lambek.
25. Seldin, J. P. (2001). Extensional set equality in the calculus of constructions. *Journal of Logic and Computation, 11*(3), 483–493. Presented at festival workshop in foundations and computations held at Heriot-Watt University, Edinburgh, July 16–18, 2000.
26. Werner, B. (1994). Une théorie des constructions inductives. Ph.D. thesis, Université Paris 7.

What is Propositional Logic a Theory of, if Anything?

Oswaldo Chateaubriand

Abstract In this chapter I discuss some traditional philosophical questions relating to propositional logic, among which are the following: (1) Must the objects of propositional logic (propositions, sentences, thoughts, judgments, etc.) have structure? (2) What is the nature of quantification in propositional logic? (3) What is the connection between material implication and the material conditional? (4) What is the role of the material conditional in propositional logic? (5) What is the role of truth-values?

1 Motivation

In Chap. 11 of Mates' *Elementary Logic* he introduces first-order theories, and gives as one of his examples the theory of Aristotelian syllogistic. He formulates Aristotelian syllogistic by means of four binary relations (A, E, I, O) and axiomatizes it with seven axioms, including counterparts of the well-known syllogisms of the first figure. This is quite interesting because we normally do not present logical theories as theories, but present them in a different way. Also, it emphasizes something which I think is quite correct, namely, that Aristotelian logic is a theory of four logical relations: universal subordination, exclusion, partial subordination, and partial exclusion. These are basically second-order relations that apply to first-order properties. Thus, we have a very nice view of the character of Aristotle's logic. Evidently, that does not take into account the character of Aristotle's logic as a deductive system, which has to be developed in a different way. Another interesting thing about Mates' formulation is that if one looks at the (natural) models of the theory, the individual variables need not range over properties. They may range over sets, or predicates, or mental concepts, or any other entities that may be said to be instantiated.

O. Chateaubriand (✉)
PUC-Rio/CNPq, Rio de Janeiro, Brazil
e-mail: oswaldo@puc-rio.br

L. C. Pereira et al. (eds.), *Advances in Natural Deduction*,
Trends in Logic 39, DOI: 10.1007/978-94-007-7548-0_8,

In my view, Frege generalized this idea of logic, and I think his notation is very interesting in this respect. For, whereas Aristotle had four second-order relations—which could also be higher-order—Frege had infinitely many logical relations, which are all those properties and relations expressed by the combinations of the universal quantifier, the conditional, and negation which appear as the purely logical part of his formulas. And one could actually axiomatize Frege's logic as a first-order theory in the same spirit in which Mates axiomatizes Aristotelian logic.

What I want to discuss in this chapter is the analogous question for propositional logic. If propositional logic is presented as a theory, what is it a theory of? What kind of theory is it? I will now mention some arguments and puzzles that were part of my motivation for pursuing this issue.

One argument that has been insistently presented in the literature, especially by Quine, is the argument that one should not confuse the conditional with a notion of implication. For, says Quine, whereas the conditional connects sentences into a sentence, implication is a relation between sentences, and one should use names of these sentences as terms of the relation. An interesting thing, however, is that if we look at the truth-conditions for the material conditional and for material implication, they are exactly the same. In both cases, the whole is true if the left sentence is false or the right sentence is true. So, what is the role of the conditional? Why do we need it? Why not simply work with material implication? Thus, a question that worried me was how to account for the role of the material conditional in propositional logic.

Another argument that puzzled me as I lectured in elementary logic courses, is an argument Quine uses, and Mates also develops to some extent at the beginning of his book, which maintains that in order to do propositional logic we must have "things" that have structure. And since they don't see that propositions have any clear structure, one should work with something like sentences, where one can see the negations, conjunctions, disjunctions, etc. In other words, the claim is that in order to do propositional logic it is essential that the stuff we are dealing with has a structure. (I was convinced by this argument for about 30 years, I think.) Of course, there are also other arguments used by Quine and by Mates against propositions, thoughts, judgments, etc., which are of a metaphysical character; namely, questions about individuation, abstractness, and so on. In this way propositional logic became sentential logic and people don't talk about propositions anymore. The question I want to raise is whether we really need to have structure in order to do propositional logic. Do the things we talk about in propositional logic need to have any kind of structure?

An important puzzle that also comes up is the use of quantification in propositional logic. There is something very odd when one writes something like

$$\forall pp$$

which does not seem to make any sense. For, as Frege taught us, quantification is a higher order predication, and thus we must have a predicate in order to quantify. What happens is that people have an implicit predicate like "holds," or "is true," or something like that. In fact, it is interesting that one of the people Quine considers responsible for the "confusion" between the conditional and implication is Russell, whereas Russell is quite rightly very explicit about using "is true," "is false," "implies," etc. in connection with quantified formulas. In the reading of the formula

$$\forall p \forall q((p \wedge q) \rightarrow (p \vee q))$$

the conjunction and disjunction are read as term builders (connectives) whereas the arrow is read as the relation of material implication.

Another interesting issue is the use of truth-values in propositional logic. Frege introduced truth-values as two objects, but whereas everybody uses truth-values in propositional logic, almost nobody agrees with Frege that they are objects. So, truth-values are elliptical for something like truth and falsity as predicates, but how is that to be interpreted?

One more question that motivated me while teaching logic courses is the large variety of non-equivalent ways in which propositional logic is presented in textbooks, both elementary and advanced. In each case the logical system is the same, but the interpretation of the symbols and formulas is quite different. In what follows I hope to shed some light on these and other issues that seem to me relevant for understanding the character of propositional logic.

2 Introduction

Mathematicians and mathematician-logicians often read the propositional '$\rightarrow$' as 'implies', and have been chastened by philosophers for doing so. Thus Quine claims that this reading confuses the object language conditional 'if-then' with the metalogical notion of implication—a form of use-mention confusion.[1] There is a confusion,

---

[1] Quine's explanation for the confusion is as follows ([17], p. 43):

> Now when we say that one statement or schema implies another... we are not to write 'implies' between the statements or schemata concerned, but between their names. In this way we mention the schemata or statements, we talk about them, but use their names. These names are usually formed by adding quotation marks....
> When on the other hand we compound a statement or schema from two others by means of 'if-then,' or '$\supset$,' we use the statements or schemata themselves and not their names. Here we do not mention the statements or schemata. There is no reference to them; they merely occur as parts of a longer statement or schema.

According to Quine the main culprits for the confusion were Whitehead and Russell (Ibid., p. 44):

> The distinction that was stressed just now was woefully neglected by Whitehead and Russell, who accorded '$p \supset q$' the readings 'if p then q' and 'p implies q' indifferently. The old controversy over the material conditional... was, in consequence, aggravated. The truth-function '$-(p\overline{q})$' meets some opposition as a rendering of 'if-then'; it meets more, and rightly, as a rendering of 'implies'. Certainly implication must be preserved as a strong relation, dependent upon the structure of the related statements and not just the truth-values.

as a matter of fact, and it runs deep, but it is a confusion in propositional logic itself, and the mathematician's reading is a rather sensible one.

Let us consider the language of propositional logic in one of its usual presentations.[2] We have:

- Propositional variables: $p, q, r, s, t, p_1, q_1, \ldots$;
- Connectives: $\neg$, &, $\vee$, $\rightarrow$, $\leftrightarrow$;
- Parentheses: (,).

The connectives are treated as functional operators that build up complex propositions from simpler ones, and the notion of formula is defined as usual with the parentheses as punctuation marks to avoid ambiguities. To economize on parentheses we may use Polish notation, which being a form of functional notation also emphasizes the functional character of the connectives.[3] If p is a proposition, then $\neg p$ is the negation of that proposition; if p and q are two propositions, then $p \rightarrow q$ is the conditional proposition with antecedent p and consequent q; and so on. Each propositional formula is a *term* that refers to or describes a proposition with a certain amount of (propositional) structure.

An arithmetical analog of this propositional language is the following:

- Numeral variables: $i, j, k, l, m, n, i_1, j_1, \ldots$;
- Operations: $'$, $+$, $\cdot$;
- Parentheses: (,).

The notion of formula is defined as before, and Polish notation can also be used to economize on parentheses. And again each formula is a term.

How should we interpret this language? By analogy with the sentential interpretation of propositional logic, we can interpret the numeral variables as ranging over such English expressions as 'zero', 'one', 'two', 'three', and so on, or over the actual numerals '0', '1', '2', '3', and so on. If 'i' is interpreted as 'two' and 'j' is interpreted as 'three', then 'i'' is interpreted as 'the successor of two', and '$i+j$' is interpreted as

---

[2] I will use 'proposition' throughout but I could use 'sentence' (or 'statement') instead. When I want to talk about propositions as opposed to sentences, or about sentences as opposed to propositions, I will indicate this explicitly.

[3] Everybody agrees that the connectives are functions, although there are disagreements as to whether they are functions of propositions, or sentences, or statements, or whatever. Under the heading "The fundamental functions of propositions," ([23], p. 6) Whitehead and Russell say:

> An aggregation of propositions, considered as wholes not necessarily unambiguously determined, into a single proposition more complex than its constituents, is a function with propositions as arguments. The general idea of such an aggregation of propositions, or of variables representing propositions, will not be employed in this work. But there are four special cases that are of fundamental importance, since all the aggregations of subordinate propositions into one complex proposition which occur in the sequel are formed out of them step by step.

They are (1) the Contradictory Function, (2) the Logical Sum, or Disjunctive Function, (3) the Logical Product, or Conjunctive Function, (4) the Implicative Function.

'two plus three'. We can also use the numerical symbols themselves as interpretation of the operations and write 'two′' and 'two + three'; and if we are using the actual numerals as interpretation, we will have expressions '2′' and '2+3'. If we also have numerical constants as part of our language, then it may become difficult to discern whether we are talking about expressions or about numbers. We could start using this arithmetical language to describe the building up of arithmetical terms and then find ourselves talking about numbers and functions.

This is a standard practice in logic of not distinguishing use and mention—or, as Church puts it, of using logical symbols autonymously. Which means that when we describe the language of propositional logic as object language, in our metalanguage we use certain variables—say 'φ', 'ψ', etc.—to vary over expressions of the object language, but we use '$\neg$', '$\rightarrow$', etc., as names of themselves.[4] Thus, the metalinguistic expression '$\varphi \rightarrow \psi$' may describe the expression '$p \rightarrow q$' of the object language. This does not need to lead to any problems, but since the object language is also supposed to be describing linguistic constructions, the same idea is also applied (implicitly) to the object language in relation to what it describes—so that the object language expression '$p \rightarrow q$' can describe something like 'John went to the movies $\rightarrow$ Mary went to the movies'. This is where the trouble comes, because we now say that this sentence is true if and only if the antecedent 'John went to the movies' is false or the consequent 'Mary went to the movies' is true. And this is explained as the standard truth-functional analysis of the connectives.

We categorize this move as a semantic interpretation of the language of propositional logic, but what it involves is a *re-interpretation* of the language of propositional logic (the object language) as follows:

- truth-value variables: $p, q, r, s, t, p_1, q_1, \ldots$;
- truth-functions: $\neg$, &, $\vee$, $\rightarrow$, $\leftrightarrow$;
- parentheses: (,).

Since we have only changed the *names* of the categories, our formulas stay the same and we can switch from one version to the other without seeming to. The truth-value variables range over truth-values, which are referred to as 'T' and 'F', and are usually conceptualized as the very letters 'T' and 'F'—or as some arbitrary objects such as 1 and 0. The truth-functions are operations on the set {T, F}. Our language is still a language of terms though, except that instead of these terms referring to propositions they refer to truth-values.

This could be a coherent interpretation of the language of propositional logic if we gave an account of the truth-values. In Fregean terms we can say that we are describing certain operations on the objects the True and the False. Mathematically

[4] Church formulates the following practical principle ([5], p. 63):

> The primitive symbols of the object language will be used in the syntax language as names of themselves, and juxtaposition will be used for juxtaposition.

Church ([5], pp. 61–63) discusses Frege's use-mention distinction but decides to disregard it, and argues for his principle on the grounds that it accords with mathematical practice and that it does not need to lead to use-mention confusions if one observes certain precautions.

minded philosophers may say that we are developing the Boolean algebra of truth and falsity. This is a mistake though, because what we have is not a Boolean algebra; there are no relations. The problem to which I referred above is related to this.

We have two different understandings of the propositional language: one as a description of a structure of propositions, and the other as certain operations on so-called truth-values. When we decide to call the second an interpretation of the first we give room for misinterpretation.

We now have an interpretation that assigns T to 'p' and T to 'q', say, and hence in which '$p \rightarrow q$' is also assigned T due to the interpretation of '$\rightarrow$' as a truth function. And we say that the *formula* '$p \rightarrow q$' is *true* in that interpretation. This switch seems innocent enough, but it identifies the object T (or 'T', or 1) with the property of *being true in an interpretation*, which is an entirely unjustified and illegitimate move. It *seems* all right because we are now taking the formula '$p \rightarrow q$' to be a *conditional* of two *propositions* that is true when both propositions are *true*. As a matter of fact, we are taking the conditional as a *relation* between two propositions; a relation that *holds* in all cases except when the antecedent is true and the consequent is false. This is part of the answer as to why mathematicians talk about *implication*; it is a relation that can be asserted or denied of two propositions.

The confusion essentially derives from Frege's introduction of truth-values, although Frege himself was not confused, because for him conditionality is a relation. In *Begriffsschrift* the relation relates judgment contents and is characterized in terms of the affirmation and denial of those contents [9]. Even if we interpret affirmation and denial as truth and falsity, it is still a relation and not a propositional connective. The same goes for negation and for the other "connectives" that Frege defines. In *Grundgesetze* conditionality is introduced as a function of objects whose values are the truth-values the True and the False, and it seems to be a generalization of what we are talking about when we talk of '$\rightarrow$' as a truth-function. We must remember, however, that in *Grundgesetze* there are only functions and objects; *everything* that is not an object is a function, including relations. Concepts and relations are functions whose value is always a truth-value [8]. Thus, even in *Grundgesetze*, conditionality is a relation, not a connective. Conditionality *holds* (or does not hold) for any two objects, including thoughts, and this holding is analyzed functionally in terms of the objects the True and the False.

Going back to the modern version, we can see the confusion in action in a rather pristine form when we consider the extended propositional calculus, now quite out of fashion. Let us consider Church's presentation in ([5], p. 151) where he begins to discuss quantification in propositional logic. He adds the quantifiers '$\forall$' and '$\exists$' to the propositional language and we are told that the result of prefixing a quantifier followed by a variable to a formula is a formula. Thus we can write such formulas as

$$\forall pp,$$

which Church then uses as a definition of the constant '**f**' (for falsity). This is actually quite extraordinary, if one thinks about it.

Suppose that someone introduces quantification in the arithmetical language and writes

$$\forall ii.$$

Suppose, moreover, that this formula is used to define a constant '**f**.' We would surely complain that quantification can only be applied to predicates and not to names or to variables for objects. We would argue, reasonably, that it does not make sense to say such things as 'for every numeral i, i', or 'for every number i, i'. 'Nonsense,' we are told, 'what it means is that for any numeral i, i denotes the number 1; and since this is false, we can define the constant '**f**' (for falsity).'

Church's quantification

$$\forall pp,$$

is supposed to mean that for every proposition p, p is the case; and since this is false, we can define '**f**' from it. We have variables that vary over propositions (or sentences) but when we write them down by themselves, as in

$$p$$

we are writing down something that means

it is the case that p.

An invisible predicate manifests itself.[5] How can this be?

One explanation is that Frege was wrong when he maintained that variables are mere letters; rather, propositional variables are variable propositions. Thus when we write 'p' we are writing down a variable proposition that is or is not the case.

---

[5] The classic paper on these questions is [12], where the quantifier is introduced as follows (p. 54):

> For the universal quantifier Łukasiewicz uses the sign '$\prod$' which was introduced by Peirce. With this notation the formula '$\prod pq$' is the symbolic expression of the sentence 'for all p, q (holds).'

Church describes the calculus but does not see anything odd about it. In fact, Church is consistent in reading '→' as '(materially) implies' and in treating it as a relation between truth-values. In this connection he remarks ([5], note 89, p. 38):

> As a matter of fact, the words "if...then" and "implies" as used in ordinary non-technical English often seem to denote a relation between propositions rather than between truth-values. Their possible meanings when employed in this way are difficult to fix precisely and we shall make no attempt to do so. But we select the one use of the words "if...then" (or "implies") – their material use, we shall call it – in which they may be construed as denoting a relation between truth-values, and we assign this relation as the associated function for the connective [⊃].

This is fine, except that if 'p' ranges over truth-values, which is Church's ([5], p. 73) principal interpretation of propositional logic, then the qualification 'it is the case that' makes no sense—and, therefore, quantification makes no sense.

Moreover, we can quantify it by interpreting the quantifier as a kind of metalinguistic qualification. What we are saying with

$$\forall pp$$

is:

$$p \text{ (is the case)},$$

for every proposition p.

Another explanation—the correct one, I think—is that we are confusing different interpretations of the language of propositional logic.

Suppose I write

(1) $\forall p \exists q(p \rightarrow q)$.

What does this mean? It means that for any proposition p there is a proposition q such that p implies q. It is quite clear here that '$\rightarrow$' is being interpreted as a relation between propositions. If we do not like the word 'implies,' or if we want to avoid confusion with logical implication, then we can read '$(p \rightarrow q)$' as 'p materially implies q'—or even as 'if p is the case, then q is the case.' The point though, is that on these alternative readings we still have a relation between propositions.

Suppose now that we want to justify (1), and that we do it by arguing that we can pick 'p' itself as a value for 'q' because

(2) $p \rightarrow p$

always holds; i.e., because the conditional proposition

$$p \rightarrow p$$

is a tautology (a logical truth). We are now switching to the interpretation of '$\rightarrow$' as a propositional connective, rather than as a relation. Of course, we can interpret '$\rightarrow$' in (2) as a relation, but then our argument is not by instantiation to a particular formula that the truth-value analysis has shown to be tautologous. Rather, we are identifying the value of 'p' with the value of 'q' and it is not so clear that we have established

(3) $\forall p(p \rightarrow p)$

for the relation of material implication.

If we are honest about it we begin to realize that the interpretation of the connectives as propositional connectives that result in terms describing complex propositions, and the interpretation of the connectives as relations between propositions are becoming inextricably mixed up. We see this very clearly when we use conjunction and disjunction together with conditional (or implication). If we write

(4) $\forall p \forall q((p\&q) \rightarrow (p \vee q))$,

the natural tendency is to read it as asserting that the proposition that is the conjunction of p and q implies the proposition that is the disjunction of p and q; so that '&' and '$\vee$' are treated as functions (term builders) and '$\rightarrow$' is treated as a relation.

Let us go back now to the quantification

$$\forall ii.$$

We can make good the earlier explanation of the meaning of this quantification by explicitly introducing a predicate 'D1( )'—to be read as '( ) denotes the number 1'—into the language, and rewriting the quantification as

$$\forall i \mathrm{D1}(i).$$

Similarly, if we try to understand what we are doing in the extended propositional logic, we may introduce a unary predicate 'H( )' into the language; where 'H(p)' is to be read as 'p holds', or as 'p is the case', or simply as 'p is true'.[6] For any interpretation, as an assignment of truth-values to the propositional variables, the extension of 'H( )' will be the set of formulas that receive the value T in that interpretation. But this only solves one confusion by exposing another: the use-mention confusion between language and metalanguage. What happens is that we start thinking of the very formulas of the object language, which are the subject matter of the metalanguage, as being what the object language is about. Thus, propositional logic is about those 'p''s and 'q''s and '$\neg$''s and '&''s, and their combinations. But if we do this, then we do not have propositional variables anymore and we certainly cannot quantify without further ado. What we have to do is to clear the air by making explicit what we are talking about. I list some standard options below.

Option 1 We really want to stick to truth-values. Then our variables range over truth-values and our connectives are truth-functions. In order to avoid ambiguities let us use Polish notation with the letters '$\mathcal{N}$', '$\mathcal{K}$,' '$\mathcal{A}$,' '$\mathcal{C}$,' and '$\mathcal{E}$' for the connectives and keep the symbols introduced earlier for use

---

[6] This is Russell's reading in his early paper "The Theory of Implication". Russell is quite consistent in reading '$\supset$' as 'implies', and this serves him well when he introduces quantification in Section *7. Thus, in p. 193 he reads

$$\vdash :. \sim p. \supset : p. \supset .(q).q$$

as "For any value of p, not-p implies that p implies that, for all values of q, q is true." And a little later in the page he says:

> Note that $(q).q$ is an absolute constant, meaning "everything is true". Thus the formula
>
> $$\vdash :. \sim p. \supset : p. \supset .(q).q$$
>
> may be read: "If p is false, then p implies that everything is true."

I shall comment on these readings of Russell's later on.

in the metalanguage. We have many complex terms for truth-values, but if we want to make any assertions about them, we must introduce predicates that we can apply to truth-values. We can introduce a relation 'Imp(,)', for instance. As a matter of fact, we can introduce a predicate corresponding to each connective: False( ), Conj(,), Disj(,), Imp(,), Eq(,). In addition we can still have the predicate 'True( )'. These predicates have the following sets as their extension: {F}, {<T, T>}, {<T, T>, <T, F>, <F, T>}, {<T, T>, <F, T>, <F, F>}, {<T, T>, <F, F>}, and {T}.

According to this option propositional logic treats of a domain {T, F} over which vary the propositional variables and where the truth functions and the truth predicates are interpreted in a fixed way—i.e., are logical notions. An interpretation is a specific assignment of truth-values to the propositional variables, making it possible thereby to treat their free occurrences as constants.[7] We are actually embedding propositional logic into a first-order logic (say) where the formulas of the object language are terms, the various truth predicates are the atomic predicates, and where we can use the quantifiers and connectives of the metalanguage to build up more complex expressions. Given that we have restricted our interpretation in the ways indicated above, certain formulas of our first-order logic are clearly valid. For instance:

(5) $\forall p \mathrm{True}(\mathcal{C}pp)$,
(6) $\forall p \mathrm{Imp}(p, p)$,
(7) $\forall p \forall q (\mathrm{Imp}(p, q) \leftrightarrow \forall r (\mathrm{Imp}(q, r) \rightarrow \mathrm{Imp}(p, r)))$,
(8) $\forall p \forall q \mathrm{Imp}(\mathcal{K}pq, \mathcal{A}pq)$.

When mathematicians develop the Boolean algebra of {T, F} they are working essentially within this option, and they manage to clear things up to some extent by an appropriate choice of notation. Suppose that we use the arithmetical language with the variables interpreted as ranging over {0, 1}, and that we add constants '**t**' and '**f**' (denoting 1 and 0, respectively) and a binary predicate 'Eq(,)'—which we can also call '='. We can then define Imp(i, j) as Eq$(i + j, j)$—and we can call it '$\leq$'. We see now that we have negation ($'$), conjunction ($\cdot$), and disjunction (+) as truth functions, we have the constants as names of the truth-values, and we have equivalence (=) and implication ($\leq$) as truth relations. We leave out the conditional and the biconditional as truth functions so as not to mix things up, but we can define them if we wish.[8]

[7] That is the role of interpretations in general, as a matter of fact. We have certain variables that we want to treat as if they were constants—we may even call them 'constants'—so we assign a fixed meaning to them (an object as denotation, a set as extension, etc.) and call this assignment an 'interpretation.'

[8] Usually people do not work with the pure Boolean algebra of {T, F} (or {0, 1}), but combine this with propositional logic treated as a language of formulas or of sentences of a certain sort. See, for instance, [14], or [20] See also Option 4.

Option 2 We leave out the connectives as functions and only have the predicates. The rest is exactly the same. We cannot say such things as (5) or (8), but (6) and (7) are still valid. In fact, for any tautology we have something like (6). For instance, corresponding to the tautology '$\mathcal{CK}pq\mathcal{A}pq$' we have

(9) $\forall p \forall q \forall r \forall s(((\text{Conj}(p, q) \leftrightarrow \text{True}(r))\&(\text{Disj}(p, q) \leftrightarrow \text{True}(s))) \rightarrow \text{Imp}(r, s))$.

This is analogous to doing arithmetic with relations S(,), A(, ,) and M(, ,) instead of using the operations $'$, $+$ and $\cdot$.

Option 3 In this option we leave out the truth-values and decide to take propositions (or sentences, or statements, etc.) as the range of the propositional variables. An interpretation consists of a domain of propositions for which the operations of negation, conjunction, disjunction, conditional, and biconditional are always defined. We keep within classical two-valued logic by requiring that each proposition in the domain must be either true or false, and not both, and we characterize the extensions of the truth predicates as sets of propositions, or of ordered pairs of propositions, that satisfy the standard conditions. We can do everything as in the first option; in fact, if we collapse all true propositions into one proposition (denoted by the constant '**t**') and all false propositions into one proposition (denoted by the constant '**f**'), then we get the first option as a special case. Since the choice of the objects T and F is supposed to be arbitrary, they might as well be two propositions.

When mathematicians apply logic to mathematics in the ordinary course of events, they are working within this option. They have the propositions of arithmetic as an interpretation, for instance, and they talk about a proposition implying another, being true, etc. What is confusing is that the notations for the propositional connectives and for the truth predicates are not clearly distinguished—but they are kept separate by mathematical horse sense. We saw this in the treatment of the Boolean algebra of the truth-values, and another indication of it is the reading of '$\rightarrow$' as 'implies.' The reason that this seems to be a confusion is because mathematical truths are necessary truths and, therefore, if a mathematical proposition materially implies another, then it necessarily implies it. This suggests that mathematicians are confusing material implication with necessary implication. Maybe they are, but with their subject matter it does not make any difference. They are definitely not confusing this implication with logical implication (or with logical consequence), however.

Option 4 Instead of treating 'p', 'q', 'r', etc., as variables we treat them as dummy symbols, or schematic letters, or non-logical constants, or place-holders, etc. The basic idea is that we are developing linguistic structures that are built up from certain sentences (ordinary or formal) that we can put in place of the place-holders 'p', 'q', 'r', etc., using the symbols '$\neg$', '&', '$\vee$', etc. (or, actually, '$\mathcal{N}$', '$\mathcal{K}$', '$\mathcal{A}$', etc.). So these symbols are being used autonymously (i.e., as names of themselves) when we describe the language of propositional logic. This is really a language schema, therefore, that becomes an actual language when the place-holders are filled with actual something-or-others. In other words, given a stock of something-or-others to take the place of 'p', 'q', 'r',..., then we get a (propositional) language.

A better way of looking at this option is to see the letters 'p', 'q', 'r',..., as variables that vary only over certain sentences of a specific language, or over formulas of a specific formal language, and not have any variables that vary over sentences, propositions, etc., in general. Our propositional language then describes certain more or less artificial languages that can be built from those atoms (or atomic sentences). There is no need for the connectives to be used autonymously, for this just confuses the issue. The connectives will be names of certain words, or word combinations, or symbols that can be chosen in each specific case. For instance, if the variables vary over a certain range of English sentences (finite or infinite), then the word combinations denoted by the connectives may be 'not', 'and', 'or', 'only if' and 'if and only if'. If 'Jane went to the movies' and 'Peter went to the movies' are in the range of the variables, then we can get a construction such as '(not(Jane went to the movies) only if (Peter went to the movies))'. This is not an English sentence, but it bears a certain relation to English.

In whichever way we conceive it, the important point is that we are using the language of propositional logic to describe the construction of certain grammatical structures. It is essentially what we do when we build the numerals from a symbol '0' and an accentuation construction. How we do the interpretations and the semantics will vary, but we will need some truth predicates in order to do it. The usual confusion tends to set in, however, because when dealing with the syntax the dummy letters (or non-logical constants) stand for statements or sentences, whereas when doing the semantics they stand for truth-values.[9]

The options I have listed are not the only possibilities, of course, and if one looks at various books on logic (especially books that go beyond the most elementary level) one finds a number of other alternatives and variations. Thus in [6], Chap. 1 Enderton proceeds as follows. He lists

$$\neg, \&, \vee, \rightarrow, \leftrightarrow, (,),$$

[9] This is the option that fits most treatments nowadays. For example, [13] (non-logical constants), [17] (dummy schematic letters), [14] (atomic sentences).

and

$$A_1, A_2, A_3, \ldots$$

as names of symbols—I have changed his notation slightly. So, '$\neg$' is a name of the symbol $\neg$, which is not specified, and similarly for the other names. He characterizes expressions as finite sequences of symbols in the set-theoretic sense, and he uses juxtaposition of names of symbols to denote expressions—i.e., '$((\neg A_2)\&A_2)$' denotes $<(, (, \neg, A_2,), \&, A_2,)>$. The only assumption he needs is that no symbol is a finite sequence of other symbols. He then introduces certain (formula-building) operations, $\mathscr{E}_\neg, \mathscr{E}_\&, \mathscr{E}_\vee, \mathscr{E}_\rightarrow, \mathscr{E}_\leftrightarrow$, on expressions such that, e.g., for any expressions $\alpha =< \alpha_1, \ldots, \alpha_n >$ and $\beta =< \beta_1, \ldots, \beta_m >$,

$$\mathscr{E}_\&(\alpha, \beta) =< (, \alpha_1, \ldots, \alpha_n, \&, \beta_1, \ldots, \beta_m,) > .$$

Thus,

$$\mathscr{E}_\&((\neg A_2), A_2) = ((\neg A_2)\&A_2) .$$

The set of well-formed formulas is the set inductively generated from the sentence symbols A_1, A_2, A_3,...by the formula-building operations. Since the symbols are not specified, this amounts to introducing an abstract algebra. This abstract algebra can be characterized in several (more or less) equivalent ways, and even the set-theoretic treatment of expressions can be left aside in favor of certain assumptions that are necessary for applying the Recursion Theorem to the set of well-formed formulas—which is used, among other things, to justify mathematically the truth-value interpretation. The explicit use of the Recursion Theorem shows very clearly the distinction between the formula building operations and the Boolean functions (truth-functions).[10]

There are many reasons, historical and other, for the differences in the interpretation of propositional logic. One reason is that there is no real advantage in taking the third option over the first, because we never make any logical use of the features of propositions that are additional to their truth-value. This was actually Frege's fault, because his arguments in [7] and his treatment of logic in *Grundgesetze* began the replacement of propositions by truth-values in logic. Just as the simple theory of types collapsed all the definability distinctions that Russell built into the ramified hierarchy, two-valued extensional logic collapsed all the distinctions of sense that Frege built into his account of logic.

---

[10] Aside from this algebraic treatment, Enderton also introduces a notion of "translation" from English to the formal language. Thus, the English sentence

(a) The evidence obtained is admissible, and the suspect need not be released from custody, can be translated as
(b) $(A_1\&(\neg A_2))$.

This is problematic because even if we have a function from a certain set of English sentences to sentence symbols, it is not at all clear how to extend this function so that a "complex" English sentence is assigned a well-formed formula. What is the relation between (a) and (b)?

3 Material Implication and the Material Conditional

Let us now go back to the beginning. Is there really a confusion about the conditional? A typical student question is: 'Why do you call the conditional 'material'? Material as opposed to what?' The standard answer is that the conditional is material because it is analyzed in terms of truth and falsity, and that this is opposed to a meaning analysis, or a "contentual" analysis, or a causal analysis, or something like that. One usually appeals to the subjunctive conditional to make the point. If on top of that one brings in logical implication, then one feels like walking on quicksand. Why should anyone confuse the conditional with logical implication? Because, goes the standard answer, logical implication amounts to validity of a conditional. But a conditional of the ordinary variety is (almost) never logically valid; it must contain logical structure in order to be valid. In what sense is logical implication not material? It is not because it is causal, or contentual, but because it is formal. Yet, if we look up 'formal implication,' we are told that it is something like

$$\forall x(Fx \rightarrow Gx),$$

which is not valid.[11] It seems right that there should be a contrast between material and formal, though not in these terms.

[11] This is how Church introduces formal implication in ([5], p. 44). Of course, if 'F' and 'G' were to range over formulas (or forms) rather than over classes (which is how Church takes them), then it would be more appropriate to talk about formal implication—as Church himself remarks in note 104:

> The names formal implication and formal equivalence are those used by Whitehead and Russell in [23], and have become sufficiently well-established that it seems best not to change them—though the adjective formal is perhaps not very well chosen, and must not be understood here in the same sense that we shall give it elsewhere.

In [23], Whitehead and Russell justify their use of 'formal implication' as follows (pp. 20–21):

> When an implication, say $\varphi x.\supset.\psi x$, is said to hold always, i.e., when $(x)\ :\varphi x.\supset.\psi x$, we shall say that φx *formally implies* ψx; and propositions of the form "$(x) : \varphi x.\supset.\psi x$" will be said to state formal implications. In the usual instances of implication, such as "'Socrates is a man' implies 'Socrates is mortal'", we have a proposition of the form "$\varphi x.\supset.\psi x$" in a case in which "$(x)\ :\varphi x.\supset.\psi x$" is true. In such a case, we feel the implication as a particular case of a formal implication. Thus it has come about that implications which are not particular cases of formal implications have not been regarded as implications at all. There is also a practical ground for the neglect of such implications, for, speaking generally, they can only be *known* when it is already known either that their hypothesis is false or that their conclusion is true; and in neither of these cases do they serve to make us know the conclusion, since in the first case the conclusion need not be true, and in the second it is known already. Thus such implications do not serve the purpose for which implications are chiefly useful, namely that of making us know, by deduction, conclusions of which we were previously ignorant. *Formal* implications, on the contrary, do serve this purpose, owing to the psychological fact that we often know "$(x) : \varphi x.\supset.\psi x$" and φy, in cases where ψy (which follows from these premises) cannot easily be known directly.

I always felt that I did not know what I was talking about when I answered that question about the conditional. It does not sound right. And the reason that it does not sound right is because the distinction is not between the conditional and something else, but it is a distinction between various kinds of implication relations. Material implication is a relation that holds between propositions in terms of their truth-values; formal implication is a relation that holds between propositions in terms of their logical form; and besides these there are also generalized implication relations, necessary implication relations, and so on.[12] I think that what is confused, and confusing, is Quine's account of this issue.

Quine discusses it at some length in Sect. 5 of *Mathematical Logic*, and makes two main points about material implication. One is that as a relation it "is so broad as not to deserve the name of implication at all except by analogy" (p. 29). The other is that as a relation that connects sentences it should be stated of specific sentences by naming them—in order to accompany Quine, I shall now switch from 'proposition' to 'sentence.' That is, one should say

(10) 'John is tall' materially implies 'Mary is blonde',

and not

(Footnote 11 continued)
The distinction between formal and material implication is also discussed in ([21], pp. 36–41) and in ([22], pp. 161–162, 194).

[12] Bochenski discusses this question in several sections of [1]. In Sections 20.5–20.13 he discusses implication in Megarian-Stoic logic. He distinguishes Philonian implication, which is material implication as I characterized it in the text, from Diodorean implication, which he formulates as follows (p. 118):

> If p, then q, if and only if, for every time t it is not the case that p is true at t and q is false at t.

There is also connexive implication, which is a form of strict implication. Bochenski quotes Diogenes Laertius (p. 118):

> A connected (proposition) is true in which the opposite of the consequent is incompatible with the antecedent, e.g. 'if it is day, it is light'. This is true, since 'it is not light', the opposite of the consequent, is incompatible with 'it is day'. A connected (proposition) is false in which the opposite of the consequent is not incompatible with the antecedent, e.g. 'if it is day, Dion walks about'; for 'Dion is not walking about' is not incompatible with 'it is day'.

In Sections 30.12–30.16 Bochenski discusses various scholastic authors who distinguish material from formal implication. An interesting passage is the following from Albert of Saxony (p. 193):

> Of consequences, one kind is formal, another material. That is said to be a formal consequence to which every proposition which, if it were to be formed, would be a valid consequence, is similar in form, e.g. 'b is a, therefore some a is b'. But a material consequence is one such that not every proposition similar in form to it is a valid consequence, or, as is commonly said, which does not hold in all terms when the form is kept the same; e.g. 'a man runs, therefore an animal runs'. But in these (other) terms the consequence is not valid: 'a man runs, therefore a log runs'.

(11) John is tall materially implies Mary is blonde.

Therefore, he concludes, one cannot equate (10) with

(12) If John is tall, then Mary is blonde.

In other words, as a grammatical construction the conditional connects the sentences themselves and not their names. And from this it follows that if the logical symbol '→' can be written between sentences, as in

(13) John is tall → Mary is blonde,

then '→' should be equated to the conditional and not to material implication. Quine is very consistent about this, and it is one of the reasons that he treats propositional logic in terms of dummy letters—as in Option 4.

The problem comes in the semantics. If one holds, with Frege, that sentences literally denote truth-values, then there is a clear sense in which '→,' as analyzed by Quine, is a truth-function. If the individual sentences in (13) denote the True, then (13) denotes the True. And the same holds for other combinations. But Quine, like most philosophers and logicians, does not hold that sentences *literally* denote truth-values—i.e., they hold that there are no such things as the objects the True and the False. Talk of truth-values is supposed to be elliptical for something else. What is it elliptical for? Presumably, it is elliptical for talk of truth and falsity, using 'true' and 'false' as predicates. So one should say that (12) is true if and only if either 'John is tall' is false or 'Mary is blonde' is true. But, given this analysis, we should say that to assert (12) is to assert

(14) 'John is tall' is false or 'Mary is blonde' is true,

which suggests that what we are asserting is that the relation of material implication holds between 'John is tall' and 'Mary is blonde.'

In fact, to take the general case, how do we give the truth conditions for a conditional using 'true' and 'false' as predicates? We say

(Footnote 12 continued)
Another interesting passage he quotes is from Ockham (p. 192):

> Of consequences, one kind is formal, another material. Formal consequence is twofold, since one holds by an extrinsic medium concerning the form of the proposition, such as these rules: 'from an exclusive to a universal (proposition) with the terms interchanged is a correct consequence', 'from a necessary major and an assertoric minor (premiss) there follows a necessary (conclusion)', etc. The other kind holds directly through an intrinsic medium and indirectly through an extrinsic one concerning the general conditions of the proposition, not its truth, falsity, necessity or impossibility. Of this kind is the following: 'Socrates does not run, therefore some man does not run'. The consequence is called 'material' since it holds precisely in virtue of the terms, not in virtue of some extrinsic medium concerning the general conditions of the proposition. Such are the following: 'If a man runs, God exists', 'man is an ass, therefore God does not exist' etc.

The first kind of formal implication is what I meant in the text by 'formal implication'; the second kind of formal implication is what I meant by 'generalized implication,' that we find in Russell as formal implication.

(15) $p \rightarrow q$ is true if and only if p is false or q is true,

which again suggests that 'p' and 'q' are variables that range over sentences and that '$\rightarrow$' is either a relation among sentences or is an operator which applied to two sentences yields a sentence as result. If 'p' and 'q' were to stand for sentences, i.e., be place-holders for sentences, then we should use quotation marks; or, rather, quasi-quotation marks.[13]

One of the great difficulties in getting through the material analysis of the conditional to logic students, and not only to logic students, is that it seems hardly credible that the ordinary English conditional is truth-functional. One forces the truth-functional analysis down their throats, and eventually gets to the point of saying that by (13) one "just means" (14)—"take it as a convention, if you wish," is the usual tack. The intuition that people have about the English conditional is that it operates on "contents"—i.e., on meanings, or on senses, or on something like that—and not on truth-values. This suggests that the ordinary conditional is really an intensional operator; and, in fact, it does not pass the usual substitutional test for extensionality. But material implication is also a relation between contents (i.e., propositions); only, it is material. That is, we can formulate (13) as

(16) That John is tall materially implies that Mary is blonde.

Of course, if we think of sentences as denoting truth-values, and expressing senses, then we can formulate (16) as (13) and substitutivity salva veritate will always hold—which does not mean that the sense will be preserved, however. This is Frege's analysis of conditionality.

If in (16) we substitute 'necessarily' or 'logically' for 'materially,' there will be no (grammatical) problem. But if we switch to a formulation like (13) substitutivity salva veritate will not hold anymore. Although Quine is right that it would be odd to say in English

(17) If John is tall, then logically Mary is blonde,

it is perfectly correct to say in English

(18) If John is tall, then necessarily Mary is blonde.

Why should Quine's reformulation of (17) as

(19) 'John is tall' logically implies 'Mary is blonde'

---

[13] See [15, Sect. 6]. One would say:
(Footnote 13 continued)

$$\ulcorner p \rightarrow q \urcorner \text{ is true if and only if } \ulcorner p \urcorner \text{ is false or } \ulcorner q \urcorner \text{ is true,}$$

which, for a specific choice of sentences for the place-holders 'p' and 'q,' would become:

'John is tall $\rightarrow$ Mary is blonde' is true if and only if 'John is tall' is false or 'Mary is blonde'is true.

count in favor of his analysis of the conditional, and the reformulation of (18) as

(20) 'John is tall' necessarily implies 'Mary is blonde'

not count in favor of an analysis of the conditional as expressing various forms of implication?

I think there is an important confusion about the conditional, but that it is not the confusion Quine emphasizes. What is one talking about in sentence (12)? I believe there are two main options. Either one is talking about John and Mary, or one is talking about the sentences 'John is tall' and 'Mary is blonde.'[14] If the former, then one is predicating something of John and Mary; namely the predicate '[x is not tall or y is blonde](x, y)'. In this case (12) should be interpreted as

(12') [x is not tall or y is blonde](John, Mary).

If the latter, then one is predicating something of the sentences 'John is tall' and 'Mary is blonde'; namely, the predicate '[p is false or q is true](p, q)'. In this case (12) should be interpreted as

(12") [p is false or q is true]('John is tall','Mary is blonde').

What is Quine's analysis of the conditional? What is his so-called "connection"?

The distinction I just made can be reformulated in terms of the distinction between *de re* and *de dicto*. Is Quine's analysis of the conditional *de re* or is it *de dicto*? His truth-functional analysis suggests that it is *de dicto*, yet he wants to claim that the conditional is not *de dicto*. But if it is *de re*, then it cannot be analyzed truth-functionally—and Quine gives no other analysis. In terms of the distinction between *de re* and *de dicto*, we can also see why (17) is incorrect and (18) is correct. Logical implication is a *de dicto* relation (between sentences, or propositions, or whatever), and therefore (17) cannot be interpreted as being about John and Mary—i.e., it cannot be interpreted *de re*. Necessary connections, on the other hand, have both a *de re* and a *de dicto* interpretation; hence the distinction between (18) and (20).

[14] This is Boole's distinction between primary and secondary propositions. In ([2], pp. 52–53) he says:

> Every assertion that we make may be referred to one or the other of the two following kinds. Either it expresses a relation among *things*, or it expresses, or is equivalent to the expression of, a relation among *propositions*. An assertion respecting the properties of things, or the phaenomena which they manifest, or the circumstances in which they are placed, is, properly speaking, the assertion of a relation among things. To say that "snow is white," is for the ends of logic equivalent to saying, that "snow is a white thing." An assertion respecting facts and events, their mutual connexion and dependence, is, for the same ends, generally equivalent to the assertion, that such and such propositions concerning those events have a certain relation to each other as respects their mutual truth or falsehood. The former class of propositions, relating to things, I call "Primary;" the latter class, relating to propositions, I call "Secondary." The distinction is in practice nearly but not quite co-extensive with the common logical distinction of propositions as categorical or hypothetical.

See also the discussion in ([4], pp. 203–204).

Although the fundamental issue seems to me to be an issue about logic, I do not think that the situation in English is as Quine thinks it is. The kind of sentential analysis Quine favors is mostly based on the rejection of propositions and on the view that, therefore, logic reflects grammatical and semantic features of language. But the analysis of sentences as denoting truth-values, even if not literally, and the truth functional analysis of the conditional that depends upon it, do not seem to reflect the most significant features of conditionality in English—as shown by the reactions of elementary logic students, among others. Quine's first objection that material implication would be a trivial implication relation (too broad), is precisely what people feel about the material interpretation of the English conditional. In fact, I think that there is an analogical process there, but that the bad analogy is to interpret the conditional by analogy to material implication. And the main reason for this is the misunderstanding of Frege's analysis of conditionality.

Far from being trivial, material implication is one of the fundamental relations upon which logic is based in its attempt to formulate general laws of truth and of truth preservation. The basic idea of material implication is that it is a relation of truth preservation: if a proposition p materially implies a proposition q, then if p is true, q is also true. This is what we want to guarantee. Hence, if p is false, the relation holds independently of q; and if q is true, the relation holds independently of p. Classical propositional logic is an attempt to analyze the most general laws of truth preservation taking into account only the truth and falsity of propositions, and assuming that each proposition is either true or false and not both. What we may call 'pure propositional logic' is the most general theory of material implication. It involves no analysis whatever of the structure of propositions, or any assumptions as to the relations that propositions may have to each other. All it assumes, in the classical formulation, is that propositions are either true or false and not both. On this basis it formulates laws of material implication.

In order to carry out the propositional analysis of the laws of truth and of truth preservation it is not necessary to have *any* analysis of the structure of propositions. That is what gives propositional logic its generality. All we need do is suppose that propositions can have certain *truth relations* to each other. We can consider a relation $\mathrm{N}(p, q)$ which holds if p is true when and only when q is false; a relation $\mathrm{K}(p, q, r)$ which holds if p is true when and only when q and r are both true; and so on. As I shall argue below, this is what classical propositional logic is about. And the laws are formulated in terms of material implication precisely because what we want to figure out are the *broadest laws of truth preservation*.

There are no propositional logical forms in this sense—and, therefore, there is no notion of logical implication either in the strict sense of a proposition implying another in virtue of its *propositional logical form*. There are only *forms of truth*. Of course, if we have a certain realm of propositions that are structured in some way, and such that this structure determines certain truth relations among them, then we can talk about logical implication among these propositions. And to the extent that sentences express propositions we can apply this to sentences. In fact, as long as we have the notion of truth for sentences, we can do everything directly in terms of sentences, and it gives a better account of propositional logic also in that case.

The account of propositions I gave in [4] includes propositions that are neither true nor false, and it seems to me that this will be a feature of any general account of truth either for propositions or for sentences. It seems best, therefore, to formulate propositional logic both on the assumption that bivalence holds and without that assumption. Since my view of propositional logic is that it is a theory about truth, I will formulate it as a first-order theory with the propositional variables as individual variables. The basic predicates are 'T' (for 'is true') and 'F' (for 'is false).

4 Classical Propositional Logic as First-Order Theory

The most general case is the case where we do not assume anything about the structure of propositions. The two fundamental laws of truth, which are axioms of classical two-valued propositional logic, are the principles of excluded middle (bivalence) and of non-contradiction. Excluded middle can be stated as

(21) $\forall p(\mathrm{T}p \vee \mathrm{F}p)$,

or, equivalently, as

(22) $\forall p(\neg\mathrm{T}p \rightarrow \mathrm{F}p)$.

Non-contradiction can be stated as

(23) $\forall p\neg(\mathrm{T}p \,\&\, \mathrm{F}p)$,

or, equivalently, as

(24) $\forall p(\mathrm{T}p \rightarrow \neg\mathrm{F}p)$.

If we take them together, they can be formulated as the axiom

(A1) $\forall p(\mathrm{T}p \leftrightarrow \neg\mathrm{F}p)$.

Even if we assume absolutely nothing about the structure of propositions, we can define the following predicates:

(D1) $\forall p\forall q(\mathrm{N}pq \leftrightarrow (\mathrm{T}p \leftrightarrow Fq))$
(D2) $\forall p\forall q\forall r(\mathrm{K}pqr \leftrightarrow (\mathrm{T}p \leftrightarrow (\mathrm{T}q \,\&\, \mathrm{T}r)))$
(D3) $\forall p\forall q\forall r(\mathrm{A}pqr \leftrightarrow (\mathrm{T}p \leftrightarrow (\mathrm{T}q \vee \mathrm{T}r)))$
(D4) $\forall p\forall q\forall r(\mathrm{C}pqr \leftrightarrow (\mathrm{T}p \leftrightarrow (\mathrm{T}q \rightarrow \mathrm{T}r)))$
(D5) $\forall p\forall q\forall r(\mathrm{E}pqr \leftrightarrow (\mathrm{T}p \leftrightarrow (\mathrm{T}q \leftrightarrow \mathrm{T}r)))$.

These definitions amount to the truth table characterization of the standard connectives. Since we do not have connectives, we can read the predicates as follows (in order): p negates q, p conjoins q and r, p disjoins q and r, p conditionalizes r on q, p biconditionalizes q and r. On the basis of these definitions and axiom (A1) we can develop the theory of classical propositional logic.[15] This is the usual semantic analysis of propositional logic, and we can prove all the "tautologies." For example:

(25) $\forall p \forall q(\mathrm{C}pqq \rightarrow \mathrm{T}p)$
(26) $\forall p \forall p_1 \forall q \forall q_1 \forall q_2 \forall r \forall s \forall t((\mathrm{C}qrs \,\&\, \mathrm{C}q_1st \,\&\, \mathrm{C}q_2rt \,\&\, \mathrm{K}p_1qq_1 \,\&\, \mathrm{C}pp_1q_2) \rightarrow \mathrm{T}p)$
(27) $\forall p \forall q \forall r((\mathrm{N}qr \,\&\, \mathrm{A}pqr) \rightarrow \mathrm{T}p)$
(28) $\forall p \forall q \forall q_1 \forall r \forall s((\mathrm{K}qrs \,\&\, \mathrm{A}q_1rs \,\&\, \mathrm{C}pqq_1) \rightarrow \mathrm{T}p)$.

We can do more, in fact, because we can not only show that if a proposition has certain truth relations to others then it "must" be true, but we can also do the same thing for falsity. For example:

(29) $\forall p \forall q \forall r((\mathrm{C}qrr \,\&\, \mathrm{N}pq) \rightarrow \mathrm{F}p)$
(30) $\forall p \forall q \forall r((\mathrm{N}qr \,\&\, \mathrm{K}pqr) \rightarrow \mathrm{F}p)$.

We cannot define tautologous structure in general, but the totality of theorems like (25)–(28) gives us a certain characterization of tautologous structure, and the totality of theorems like (29)–(30) gives us a certain characterization of contradictory structure (in the propositional sense).

Suppose now that we do have some operations on propositions such as negation, conjunction, disjunction, etc. We can then add function symbols to our theory—for example, the script Polish notation I used earlier on—and axiomatize it in the same spirit. We can have the following axioms—I will use parentheses with the predicates 'T' and 'F' applied to terms to avoid confusions:

(A2) $\forall p(\mathrm{T}(\mathcal{N}p) \leftrightarrow \mathrm{F}p)$
(A3) $\forall p \forall q(\mathrm{T}(\mathcal{K}pq) \leftrightarrow (\mathrm{T}p \,\&\, \mathrm{T}q))$
(A4) $\forall p \forall q(\mathrm{T}(\mathcal{A}pq) \leftrightarrow (\mathrm{T}p \vee \mathrm{T}q))$
(A5) $\forall p \forall q(\mathrm{T}(\mathcal{C}pq) \leftrightarrow (\mathrm{F}p \vee \mathrm{T}q))$
(A6) $\forall p \forall q(\mathrm{T}(\mathcal{E}pq) \leftrightarrow (\mathrm{T}p \leftrightarrow \mathrm{T}q))$.

These are the truth tables again, and on the basis of these axioms we can prove all the tautologies. In fact, the tautologies are the terms τ for which we can prove $\mathrm{T}(\tau)$; the

[15] From (A1) we cannot prove that there is at least one true proposition and at least one false proposition, but this does not affect the theory. If we add

(E) $\exists p \mathrm{T}p \,\&\, \exists p \mathrm{F}p$

as a further axiom, then using (A1) we can prove that there are at least two propositions. We can also prove:

(E1) $\forall q \exists p \mathrm{N}pq$
(E2) $\forall q \forall r \exists p \mathrm{K}pqr$
(E3) $\forall q \forall r \exists p \mathrm{A}pqr$
(E4) $\forall q \forall r \exists p \mathrm{C}pqr$
(E5) $\forall q \forall r \exists p \mathrm{E}pqr$.

contradictions are the terms τ for which we can prove $\mathrm{F}(\tau)$; the term τ tautologically implies τ' if we can prove $\mathrm{T}(\mathcal{C}\tau\tau')$.[16]

Since we can define the truth predicates just as we did before, we can see that for everything we can prove in this theory there is an exact analog (or translation) we can prove in the earlier theory. From which it follows that it would be a mistake to suppose that by introducing the operations we are analyzing the *structure* of propositions in a more explicit way than we did before. In other words, in a model of this theory the proposition denoted by a term of the form $\mathcal{C}\tau\tau'$ need not have any kind of recognizable "conditional" structure aside from the truth relations it has to the propositions denoted by τ and by τ'.

It does seem, however, that if we interpret the theory with sentences rather than with propositions, then we get something we can sink our teeth into. But this is really an illusion that derives from a very peculiar interpretation that has no universality at all. This is Option 4 above, which consists of starting with a stock of English sentences, say, and then defining a structure of sentences, which are not generally English anymore, through certain constructions that mimic the syntax of the logical theory we have been discussing. Thus, if p and q are part of our stock of atomic sentences, we get such things as

(31) not-not-not-p,
(32) (not-p only if q) or (not-q and p),

and so on. As far as English is concerned, however, and language more generally, this gets us nowhere unless we can relate these structures to the actual sentences of the language. Even a sentence like 'Quine is not a dentist,' which is the standard negation of 'Quine is a dentist' in English, does not fit our format. Unless we can map sentences of English onto those forms, we cannot talk about the logical form of English sentences; thus the problem of translation, symbolization, regimentation, etc. And if we can so map them, then we do not need those phony structures because the mapping will give us an interpretation of the logical theory. Thus, either this nominalistic attempt to conceive of logic through language does not get us anywhere, or is a very misleading way of presenting a theory of the truth relations of propositions (or sentences) as if it were a theory of linguistic forms.[17]

One can complain against my formulation that it does not make much sense to present propositional logic as a first-order theory. I shall discuss this later, but

---

[16] Since N is an operation(totally defined), from (A1) and (A2) we can prove (E) in the previous note.

[17] That Quine's motivation was ontological can be seen from his early paper [19], which contains all the fundamental ideas of his interpretation of propositional logic. Thus (pp. 267–268):

> In the theory of deduction the signs '*p*,' '*q*', etc., are customarily construed as proposition variables, i.e., as signs ambiguously denotative of propositions, i.e., as signs ambiguously denotative of the things that sentences denote. We now cancel this circuit through denoted entities, and explain the signs '*p*', '*q*', etc., directly as ambiguously abbreviated sentences—which comes to the same thing as before except that the existence of denoted entities, propositions, is no longer presupposed.

to clear the air temporarily I would point out that this is what everybody does, except that generally the theory is not made explicit but left in the background (meta)language. All I did was to formulate the usual truth analysis explicitly as a first-order theory. What people generally use is a second order-theory (or set theory), which is indispensable for characterizing the syntax of propositional languages, and also for characterizing the linguistic forms I discussed in the previous paragraph. But let us go back to the issue about material implication.

5 Material Implication (Continued)

Just as I introduced the truth relations in (D1)–(D5), we can introduce the earlier truth relations: Conj, Disj, Imp, and Eq. The definitions are:

(D6) $\forall p \forall q(\mathrm{Conj}(p, q) \leftrightarrow (\mathrm{T}p \,\&\, \mathrm{T}q))$
(D7) $\forall p \forall q(\mathrm{Disj}(p, q) \leftrightarrow (\mathrm{T}p \vee \mathrm{T}q))$
(D8) $\forall p \forall q(\mathrm{Imp}(p, q) \leftrightarrow (\mathrm{T}p \rightarrow \mathrm{T}q))$
(D9) $\forall p \forall q(\mathrm{Eq}(p, q) \leftrightarrow (\mathrm{T}p \leftrightarrow \mathrm{T}q))$.

Of these, the last two correspond to material implication and to material equivalence. We can still do everything we could do before. Instead of (25) and (26), for example, we can state

(33) $\forall q \mathrm{Imp}(q, q)$
and
(34) $\forall r \forall s \forall t((\mathrm{Imp}(r, s) \,\&\, \mathrm{Imp}(s, t)) \rightarrow \mathrm{Imp}(r, t))$.
(27) can be stated as
(35) $\forall q \forall r((\mathrm{T}q \leftrightarrow \mathrm{F}r) \rightarrow \mathrm{Disj}(q, r))$,
and (28) as
(36) $\forall r \forall s(\mathrm{Conj}(r, s) \rightarrow \mathrm{Disj}(r, s))$.

In fact, since (D1)–(D9) are definitions, anything we can do by means of them we can also do by using directly the two truth predicates in the primitive notation. In (35) I used the truth predicates to "help out" in stating the principle of excluded middle.

Suppose, however, that we only have the truth relations introduced in (D6)–(D9) and do not have the truth predicates; can we state everything we could before? The answer is no; we cannot state (35). And even if we had the relation of negation introduced in (D1), the answer would still be no—although we can now state something like (35) as

(37) $\forall q \forall r(\mathrm{N}(q, r) \rightarrow \mathrm{Disj}(q, r))$.

Take the usual law of propositional logic

(38) $p \rightarrow (q \rightarrow p)$.

There is no way to state this just using material implication. And we should not fool ourselves that we can state it by using the operations (connectives) of the second formulation. If we do not have the truth predicates, then in that formulation we cannot state *anything at all*. We can write down

(39) $CpCqp$,

but unless we take this writing down as involving an implicit qualification 'is true,' all we are writing down is a term. It is for this reason that the usual formulation of quantified propositional logic is confused; it combines, without telling us straight, connectives and relations. The confusion derives from propositional logic itself, however, not from the introduction of quantification. The Boolean algebra treatment is not confused because it explicitly combines operations (connectives) and relations (material implication, material equivalence).

But why can we not axiomatize the relation of material implication directly as a relation among propositions and define the truth predicates by

(D10) $\forall p(\mathrm{T}p \leftrightarrow \forall q \mathrm{Imp}(q, p))$
and
(D11) $\forall p(\mathrm{F}p \leftrightarrow \forall q \mathrm{Imp}(p, q))$?

We can, actually, although neither (D10) nor (D11) are good definitions unless we make certain existential assumptions about propositions. If all propositions are truth-valueless, then what (D10) and (D11) define is 'not false' and 'not true.' If some propositions are true and some not true, then (D10) defines 'true' but (D11) still defines 'not true.' If all propositions are either true or false, then (D10) and (D11) define 'true' and 'false.'

As a propositional relation, material implication has two basic features that are characteristic of *all* implication relations; i.e., it is reflexive and transitive. In other words, it satisfies the two principles

(I1) $\forall p \mathrm{Imp}(p, p)$
(I2) $\forall p \forall q \forall r((\mathrm{Imp}(p, q) \,\&\, \mathrm{Imp}(q, r)) \rightarrow \mathrm{Imp}(p, r))$.

Although these principles state necessary features of any implication relation, they are not sufficient to characterize an implication relation. What characterizes an implication relation is that, in addition, it must be truth-preserving; i.e.,

(I3) $\forall p \forall q(\mathrm{Imp}(p, q) \rightarrow \neg(\mathrm{T}p \,\&\, \neg\mathrm{T}q))$.[18]

Material implication is the broadest implication relation (as Quine says) that satisfies these conditions. That is, we can state (I3) as the biconditional

(MI) $\forall p \forall q(\mathrm{Imp}(p, q) \leftrightarrow \neg(\mathrm{T}p \,\&\, \neg\mathrm{T}q))$.

Given the usual classical assumptions about the truth and falsity of propositions (stated in (A1)), we can formulate (I3) and (MI) as

[18] The rule of Modus Ponens (from p and $\mathrm{Imp}(p, q)$, infer q), which holds for any implication relation, is a way of stating (I3) without appealing to the truth predicate.

(I3') $\forall p \forall q(\mathrm{Imp}(p,q) \rightarrow \neg(\mathrm{T}p \,\&\, \mathrm{F}q))$

and

(MI') $\forall p \forall q(\mathrm{Imp}(p,q) \leftrightarrow \neg(\mathrm{T}p \,\&\, \mathrm{F}q))$.

The latter is Frege's characterization of conditionality, and implies that material implication has the features (I1) and (I2) as well.

We can avoid the use of the 'T' predicate by means of the following axioms:

(I4) $\forall p(\forall q \mathrm{Imp}(q,p) \vee \forall q \mathrm{Imp}(p,q))$,

and

(I5) $\exists p \exists q \neg \mathrm{Imp}(p,q)$.

(I5) implies that there is at least one true proposition and at least one proposition that is not true. This is important because (I1), (I2), and (I4) hold if all propositions are true or if all propositions are truth-valueless, as can be seen by the characterization (MI) of material implication. We can then use (D10) as definition of 'T.' (I4) implies (I1) directly, and using (D10) it implies the converse of (I3) as well. And, using (D10), (I2) implies (I3). Thus, (I2), (I4), (I5), and (D10) axiomatize material implication as characterized by (MI). But since (D11) still defines 'not true' rather than 'false,' this is not an axiomatization of (MI'). In fact, there is no way to define falsity and axiomatize (MI') unless we exclude truth-valueless propositions. But let me go back to the conditional now.

6 The Material Conditional (Continued)

That the material interpretation of the conditional is simply an analogy to material implication follows from the fact that the only assignment of a truth-value for the material conditional one can actually justify is the assignment of falsity for the combination of true antecedent and false consequent. And the reason one can justify that categorically is because it holds *for every implication relation*, and because the conditional "represents" an implication relation. In fact, the ordinary conditional can represent *any* implication relation; that seems to be its fundamental function in the language. It is precisely this feature of the conditional that leads to the common intuition that one should make it truth-valueless for every combination except the combination of true antecedent and false consequent. Since this obviously will not do, for several reasons, one hits upon the idea of making it true for every other combination. Why will it not do? Well, for one thing, if a sentence is true (false), then the conditional of the sentence and itself should be true. That is because of (I1). But from this it does not follow that the conditional of any two true (false) sentences should be true. Similarly, the conditional of a false sentence with a true sentence that the former unquestionably implies should be true, without the truth-value of the combination of false antecedent and true consequent being always true.

I think the usual idea that the proper analysis of the truth conditions of an ordinary conditional depends on "meaning" (in some sense) is essentially correct, except there is not a unique correct analysis because the meaning of an ordinary conditional is very strongly context dependent. The context may be such that a material analysis is appropriate, or a causal analysis, or a necessity analysis, or a logical analysis, etc.[19] It is for this reason I said that the function of the conditional is to "represent" (or express) any implication relation. What is this representational feature of the conditional?

Consider the formulation of propositional logic purely in terms of connectives. By the bit of cheating (or interpretation) that consists of viewing the writing down of something like (39) as asserting the truth of a (complex) proposition, we gain a tremendous amount of power. We cannot match that using the relation of material implication, but we can if we appeal to the construction 'that.' Leaving the qualification 'materially' implicit, we can state (39) as

(40) p implies that q implies p.

Something more complicated, such as

(41) $(p \to (q \to r)) \to ((p \to q) \to (p \to r))$,

can be stated as

(42) that p implies that q implies r, implies that p implies q implies that p implies r.

This is a little messy, but readable; it would be a lot messier if we tried to use the 'if-then' construction or the 'only if' construction. For example:

(43) If if p then if q then r, then if if p then q, then if p then r
(44) p only if, q only if r; only if p only if q, only if p only if r.[20]

---

[19] The context dependence of the conditional is seen very clearly when we switch to the subjunctive mood. Goodman's idea that a counterfactual conditional is true if the consequent follows by law from the antecedent and an appropriate set of relevant conditions is an expression of this contextual dependence—see [10]. But the idea that 'law' here refers to something like physical laws seems to me quite mistaken. The "laws" in question are also context dependent.

[20] Quine actually thinks that part of the reason for the mathematician's reading of '→' as 'implies' has to do with word order. He says ([17], p. 44):

> The habit of pronouncing '→' as 'implies' still persists, and is to be deplored. Partly it is encouraged by the trivial circumstance that 'if' breaks the word order, while 'implies' falls into the position of '→'. Anyone thus tempted should observe that '→' can be read without change of position as 'only if'.

(44) seems to show that the problem has nothing to do with word order but, rather, with the fact that *we want to state implications that involve other implications*. 'Implies' combined with 'that' is rather good for that; 'only if' is a total loss unless we use parentheses—my use of semicolon and commas was a poor attempt to avoid the ambiguities. Parentheses are what we use combined with '→,' and *their function is precisely the function of the* 'that.'

The idea that parentheses are "mere" punctuation marks is a rather misleading idea, because combined with other features of the notation they can serve various (hidden) functions. It is just like

In fact, the most natural way to state (42) is:

(45) If p implies that q implies r, then if p implies q, p implies r,

which is essentially Russell's reading.

In any case, the point of the 'that' construction can be seen as follows. Suppose that in our language we can not only assert that a proposition materially implies another, but we can assert that material implication holds or does not hold between two propositions, and we can iterate this any number of times. In other words, suppose we have a term building operator 'that' (which we can also read 'it is the case that') that applied to any formula gives us a term—i.e., gives us something that behaves like the propositional variables, which are the only terms we have. Then, we can state (40) as

(46) $\forall p \forall q \mathrm{Imp}(p, \mathrm{thatImp}(q, p))$;

and we can state (42) as

(47) $\forall p \forall q \forall r (\mathrm{Imp}(p, \mathrm{thatImp}(q, r)) \rightarrow (\mathrm{Imp}(p, q) \rightarrow \mathrm{Imp}(p, r)))$.

This is essentially what we want, and we can achieve precisely the same effect if we have a term building operator that applied to any two propositions gives us a proposition whose truth amounts to the holding or not holding of the material implication of these propositions. Something like '$\mathcal{C}$' fits the bill, and with it (40) and (42) look like this

(48) $\forall p \forall q \mathrm{Imp}(p, \mathcal{C}pq)$
(49) $\forall p \forall q \forall r (\mathrm{Imp}(p, \mathcal{C}qr) \rightarrow (\mathrm{Imp}(p, q) \rightarrow \mathrm{Imp}(p, r)))$.

The point is that what we get from all these constructions are *iterations* of the truth relations; and this amounts to iterations of the truth predicates 'T' and 'F.' In the first formulation of propositional logic I gave above, I used the logical machinery of first-order logic to iterate the truth relations at will via the definitions (D1)–(D5). But I did not need to; I could have done the same thing by repeated use of the truth predicates.

The usual formulation of propositional logic in terms of sentential structures is essentially a cumulative theory of the truth relations via the truth-functional interpretation of the connectives. If we use sentences, then we have to use quotation

(Footnote 20 continued)
the idea that juxtaposition is "merely" placing symbols next to each other. The juxtaposition of 'F' and 'x' in 'Fx' serves to indicate predication—i.e., a logical operation—and it does this no matter what the arity of the predicate variable (or constant) 'F'; so, in fact, we get an infinity of logical operations subsumed under one form of notation. Syntactically all we are doing is juxtaposing, but semantically we interpret some juxtapositions in one way and others in other ways. Similarly, when we juxtapose quantifiers and read their dependence from right to left (through the definition of truth), there is a lot more there than meets the eye. If all quantifiers were totally independent of each other, we would not be able to do much logic because we would not be able to distinguish between '$\forall x \exists y Rxy$' and '$\exists y \forall x Rxy$', for instance. If we introduced distinctions of scope for the quantifiers in relation to other quantifiers (to indicate their dependence), as we do for the variables they bind, then the "punctuation" symbols we may use would not be "mere" punctuation marks. What is a punctuation mark, in fact, if not the expression of some sort of operation?

marks throughout both with the truth predicates and with the truth relations. Using quasi-quotation, we can state (40) as

(50) $\forall p \forall q(\ulcorner p \urcorner \text{implies} \ulcorner\ulcorner q \urcorner \text{implies} \ulcorner p \urcorner\urcorner)$,

which in a particular case reads

(51) ‘John is tall’ implies ‘‘Mary is blonde’ implies ‘John is tall’’.

This means that we will have to iterate these quotation constructions any finite number of times. If we use the connectives and the truth predicates—i.e., do the usual semantic analysis—then we can avoid all but the first use of quotation and truth predicates by treating each sentential formula as a complex construction out of a stock of given sentences. The truth functional interpretation of negation, conjunction, and disjunction as sentential connectives is quite natural and easy to justify. But the truth functional interpretation of the conditional is completely unnatural and can only be justified as some sort of convention. By making that convention we get the conditional to represent material implication but we get stuck with a two-valued interpretation of the basic stock of sentences and we lose the power of the conditional to express other implication relations.[21] To see this let us go back to the logic.

7 Truth-Value Gaps

Let us formulate it now for the case where there may be propositions (or sentences) that are neither true nor false. Let us assume we have the function symbols as above. Axiom (A1) must be weakened to

(B1) $\forall p(\mathrm{T}(p) \rightarrow \neg\mathrm{F}(p))$,

because we do not have excluded middle (or bivalence) anymore. The other axioms are:

(B2) $\forall p(\mathrm{T}(\mathcal{N}p) \leftrightarrow \mathrm{F}p)$
(B3) $\forall p(\mathrm{F}(\mathcal{N}p) \leftrightarrow \mathrm{T}p)$
(B4) $\forall p\forall q(\mathrm{T}(\mathcal{K}pq) \leftrightarrow (\mathrm{T}p \,\&\, \mathrm{T}q))$
(B5) $\forall p\forall q(\mathrm{F}(\mathcal{K}pq) \leftrightarrow ((\mathrm{T}p \,\&\, \mathrm{F}q) \vee (\mathrm{F}p \,\&\, \mathrm{T}q) \vee (\mathrm{F}p \,\&\, \mathrm{F}q)))$
(B6) $\forall p\forall q(\mathrm{T}(\mathcal{A}pq) \leftrightarrow ((\mathrm{T}p \,\&\, \mathrm{T}q) \vee (\mathrm{T}p \,\&\, \mathrm{F}q) \vee (\mathrm{F}p \,\&\, \mathrm{T}q)))$
(B7) $\forall p\forall q(\mathrm{F}(\mathcal{A}pq) \leftrightarrow (\mathrm{F}p \,\&\, \mathrm{F}q))$
(B8) $\forall p\forall q(\mathrm{T}(\mathcal{C}pq) \leftrightarrow ((\mathrm{T}p \,\&\, \mathrm{T}q) \vee (\mathrm{F}p \,\&\, \mathrm{T}q) \vee (\mathrm{F}p \,\&\, \mathrm{F}q)))$
(B9) $\forall p\forall q(\mathrm{F}(\mathcal{C}pq) \leftrightarrow (\mathrm{T}p \,\&\, \mathrm{F}q))$
(B10) $\forall p\forall q(\mathrm{T}(\mathcal{E}pq) \leftrightarrow ((\mathrm{T}p \,\&\, \mathrm{T}q) \vee (\mathrm{F}p \,\&\, \mathrm{F}q)))$
(B11) $\forall p\forall q(\mathrm{F}(\mathcal{E}pq) \leftrightarrow ((\mathrm{T}p \,\&\, \mathrm{F}q) \vee (\mathrm{F}p \,\&\, \mathrm{T}q)))$.

[21] The sentential logic of order ω at the end of Chap. 6 of [4] was built up like that to avoid the two-valued interpretation. It amounted to an embedding of a three valued logic (the sentences could be neither true nor false) into a two-valued predicate logic with the predicate ‘T’.

These conditions amount to saying that a "complex" proposition is either true or false only when all its "component" propositions are either true or false—this is Kleene's *weak* interpretation of three-valued logic.[22] Material implication should be defined by (MI), however; i.e., material implication holds or does not hold for any propositions, including truth-valueless propositions.

Although in this theory the conditional has been characterized materially, it does not express material implication anymore. In other words, we do not have

(52) $\forall p \forall q(\mathrm{T}(\mathcal{C}pq) \leftrightarrow \mathrm{Imp}(p, q))$.

On the contrary, we can prove that, for example,

(53) $\forall p \forall q((\mathrm{F}p \,\&\, \neg \mathrm{T}q \,\&\, \neg \mathrm{F}q) \rightarrow (\mathrm{Imp}(p, q) \,\&\, \neg \mathrm{T}(\mathcal{C}pq)))$.

Of course, for any term τ for which we can prove $\mathrm{T}(\tau)$ or $\mathrm{F}(\tau)$ in the earlier formulation, we can now prove $\neg\mathrm{F}(\tau)$ or $\neg\mathrm{T}(\tau)$. And if for every propositional variable ρ that appears in τ we relativize our theorems by adding the qualification '$\mathrm{T}\rho \vee \mathrm{F}\rho$', then we can prove $\mathrm{T}(\tau)$ or $\mathrm{F}(\tau)$ just as before. Thus, although we cannot prove

(54) $\forall p \mathrm{T}(\mathcal{C}pp)$,

we can prove

(55) $\forall p \neg \mathrm{F}(\mathcal{C}pp)$

and

(56) $\forall p((\mathrm{T}p \vee \mathrm{F}p) \rightarrow \mathrm{T}(\mathcal{C}pp))$.

Therefore we lose nothing we had before.

What we lose is the connection between the conditional as connective and material implication, because material implication continues to behave exactly as before. In fact, the laws of material implication are axiomatized in the implicational propositional calculus by means of the principles (38), (41), and Peirce's law

(57) $((p \rightarrow q) \rightarrow p) \rightarrow p$,

which can be interpreted as expressing the idea that a false proposition (or a non-true proposition) materially implies any proposition. If we introduce the operations on propositions as in (D1)–(D5), then the theory of material implication is the same in both formulations. That is, we can prove the analogs of (38), (41) and (57). For example, (38) becomes

(58) $\forall p \forall q \forall r \forall s((\mathrm{C}rqp \,\&\, \mathrm{C}spr) \rightarrow \mathrm{T}s)$.

But we cannot prove the analogs of the fundamental principles of negation, such as

[22] See ([11], p. 334). I emphasize that the first-order logic I am using to describe all of this is just ordinary first-order logic, however.

(59) $(\neg q \rightarrow \neg p) \rightarrow (p \rightarrow q)$,

which would give us the usual formulations of classical two-valued propositional logic.

As a theory of truth relations between propositions classical propositional logic is a theory of extreme generality one of whose aims can be characterized as the formulation of laws of truth preservation. It is actually independent of any assumptions as to the nature of the bearers of truth, since it applies equally well to propositions, sentences, thoughts, beliefs, etc. The only assumption is that these things can be true or false and cannot be both. Of course, as we add assumptions as to the structure that the bearers of truth may have and we add operations on them, then our theory becomes more specific and more complex. The usual idea that propositional logic has to do with sentences by analyzing their structure in certain ways seems to me to confuse the general aims of propositional logic with one of its important domains of application. This confusion goes hand in hand with the rejection of abstract analyses of reality and of its fundamental traits, and at the end one even loses sight of what the actual aims are.

Suppose we want to make sure that our domain of application of propositional logic consists of sentences, or of things that keep the same distinctions as the imaginary sentences molded upon the actual formalism of propositional logic. As is well known we will not be able to guarantee this if our theory of propositional logic is formulated as a first-order theory. Many nominalists and quasi-nominalists believe theirs is an ontologically simpler account on the grounds that sentences are good old material objects one can see and kick around. This is basically an illusion, however, resulting from lack of care in the formulation of their theory.

For the second-order formulation, corresponding to Option 4 above, we add a unary predicate 'At' to pick out the atomic sentences. We then have the following axioms that correspond to the usual inductive definitions (using '$\oplus$' for exclusive disjunction):

(S1) $\forall p(\mathrm{At}(p) \oplus \exists q(p = \mathcal{N}q) \oplus \exists q \exists r(p = \mathcal{K}qr \oplus p = \mathcal{A}qr \oplus p = \mathcal{C}qr \oplus p = \mathcal{E}qr))$

(S2) $\forall p \forall q(\mathcal{N}p = \mathcal{N}q \rightarrow p = q)$

(S3) $\forall p \forall q \forall r \forall s(\mathcal{K}pq = \mathcal{K}rs \rightarrow (p = r \,\&\, q = s))$

(S4) $\forall p \forall q \forall r \forall s(\mathcal{A}pq = \mathcal{A}rs \rightarrow (p = r \,\&\, q = s))$

(S5) $\forall p \forall q \forall r \forall s(\mathcal{C}pq = \mathcal{C}rs \rightarrow (p = r \,\&\, q = s))$

(S6) $\forall p \forall q \forall r \forall s(\mathcal{E}pq = \mathcal{E}rs \rightarrow (p = r \,\&\, q = s))$

(S7) $\forall Z((\forall p(\mathrm{At}(p) \rightarrow Z(p)) \,\&\, \forall p(Z(p) \rightarrow Z(\mathcal{N}p)) \,\&\, \forall p \forall q((Z(p) \,\&\, Z(q)) \rightarrow (Z(\mathcal{K}pq) \,\&\, Z(\mathcal{A}pq) \,\&\, Z(\mathcal{C}pq) \,\&\, Z(\mathcal{E}pq))) \rightarrow \forall p Z(p)))$.

The last axiom is the usual axiom of induction for propositional "formulas." It is a second-order axiom that cannot be replaced by first-order axioms. We can join these axioms either to (A1)–(A6) or to (B1)–(B11), depending on whether we want to exclude truth-valueless sentences or not.

Quine's formulation of propositional logic is a model of this, but English (even idealized English) is not. But, we may ask, what is the advantage of restricting propositional logic to the models of this theory? And, as a matter of fact, as far as good old material objects are concerned, those we can get our teeth into, there aren't any models at all. What there are, are certain rough finite approximations, which are better approximations as models of the earlier theories.

8 Philosophical Foundations

Let us go back now to the idea that presenting propositional logic in these ways—whether as first-order theories or as second-order theories—inverts the natural order of things because these theories "presuppose" propositional logic. The idea behind the 'presuppose' is that we start out from scratch building up the systems of logic in a more and more inclusive way. At the bottom is propositional logic; the most superficial analysis of logical form in terms of connectives. Then we start digging deeper and deeper into the structure of sentences to reveal subjects and predicates, quantifiers, identity, etc., and in this way we get richer and richer analyses of logical form and, consequently, we get richer and richer logical systems. This is what the foundations of logic are supposed to look like.

This picture is challenged by the beginner student's question: "How can you use 'not' to characterize the properties of '$\neg$'?" (And the same for the other connectives.) "Aren't you *presupposing* logic in order to *characterize* logic?"[23] Although these are good questions, the students do not get a good answer. Either they do not get an answer at all, or else they get something like: "Well, I am not *making up* logic, I am just characterizing it; and for that I use language, including such words as 'not', 'and', 'or', and so forth."

Very well, we can start from there. What does it mean to characterize something, and what is it we claim to be characterizing? When mathematicians characterize topological spaces, say, they are building a theory about them, or they are defining them as being certain kinds of mathematical structures as part of building a theory about them. But in either case, as logicians and philosophers always insist, the mathematician *uses* logic. Similarly, when logicians or philosophers characterize propositional logic, or first-order logic, or any other kind of logic, they are either building a theory about something or they are defining something, and in both cases

[23] Quine raises the question in ([16], p. 40) in connection with Tarski's definition of truth:

> A reasonable way of explaining an expression is by saying what conditions make its various contexts true. Hence one is perhaps tempted to see the above satisfaction conditions as explaining negation, conjunction, and existential quantification. However, this view is untenable; it involves a vicious circle. The given satisfaction conditions for negation, conjunction, and quantification presuppose an understanding of the very signs they would explain, or of others to the same effect... If we are prepared to avail ourselves thus of 'not', 'and', and 'some' in the course of explaining negation, conjunction, and existential quantification, why not proceed more directly and just offer these words as direct translations?

they are also using logic. In fact, they generally use a lot of logic—or set theory, which for some reason seems less problematic.

At this point there are two main lines that can be taken; or, at any rate, variations on two main themes. The adepts of the formalist line claim they are using informal logic and language to build up formal languages and systems of logic. They claim, moreover, that this is essential for precision and rigor, and that this precision and rigor are essential for mathematics, science, and philosophical understanding. They should not object, therefore, if one brings their more or less informal process of characterization of logic up to their standards of precision and rigor.

But, anyway, what is this informal logic? Is it propositional logic? First-order logic? Set theory? A combination of everything? There are many possibilities here, but the most popular among mathematical logicians seems to be that since logic is part of mathematics one can use any mathematical techniques in building up logical systems. This is quite natural because logic is mostly conceived as syntactic and semantic structures, and these are mathematical structures. Formal constructivists will balk at this, of course, and will restrict themselves to various inductive techniques they consider more primitive than set theory.

The linguistic analysis line claims that logic is something that is part of our language—it derives from grammar, for example, or from our use of language, or both. One approach is that 'not' has a certain meaning, and so do 'and,' 'or,' etc. Part of the meaning of 'not' is that it transforms truths into falsehoods, and falsehoods into truths. That is just how the word 'not' is used in our language, and logic goes on to analyze this use systematically. So 'syntax' and 'semantics' are really straightforward words in the logical vocabulary. Logic builds up certain theories of linguistic use, both syntactic and semantic. Linguists, however, have no qualms about using logic in formulating their theories. So why should logicians and philosophers of this persuasion object to a formal presentation of their theory of informal logical use?

Is this avoidance merely a didactic technique? In some cases it is, especially in elementary presentations. In more or less advanced presentations mathematicians are generally not afraid to be explicit about the logic and set theory they are using—and this even if they accept the basic claims of this linguistic view. But in philosophical presentations, people do try to hide any insinuations that logic is presupposed by logic. What are they afraid of?

They smell a rat, I think, a circle that may be vicious and will disturb an epistemological structure built upon language that sees logic as its greatest initial achievement. This is a view put together in the beginning decades of the twentieth century by the logical positivists, among others, and that is sometimes even attributed to Frege.[24]

Knowledge and science are founded on logic, which is founded on language, which is founded on meaning, which is founded on use, and practice, and behavior. This epistemological structure received a few hard blows from Quine and others,

[24] It is really ironic that after fighting so hard against the view that logic studies or formulates the laws of thinking, Frege should get landed with the view that logic formulates the laws of talking. In fact, all the same arguments that he used against the former view, and more, can be used against the latter view.

but in one way or another it has kept together, at least as far as the connection between logic and language is concerned. And, ironically, Quine has become the major proponent of a revised version of it (for logic) where meaning has been replaced by syntax, or grammar—which, in effect, amounts to switching to a version of the earlier formalist line.[25]

In spite of claims to the contrary, this line reflects the quest of an epistemology based on first principles. They are not the kind of absolute first principles philosophers of old were looking for, but they are still the philosopher's stone. Our essence is to talk, therefore the laws of logic are the laws of talking. There is a certain grain of truth in this view, of course, because language does reflect logic in many fundamental ways. But the attempt to build logic from language is very much like the attempt to lift oneself by one's own bootstraps. To use logic to characterize logic would be cheating on this view; so one leaves it in the nebulous background language hidden behind a lot of double-talk.

It is quite obvious that if logic is universal in the way that it is usually assumed, then it must be necessary to formulate one's theory of *anything*, including one's theory of logic itself. Since language does indeed embody logic in fundamental ways, the mere use of language in describing logic does involve logic; we can make it explicit or not, but it does not change anything. The student's question is whether this process involves some sort of vicious circularity. Trying to make logic *result* from language, in one way or another, does seem circular in this way. That is the reason for the appeal to practices, uses, etc., which would make logic result from linguistic behavior—but the practices, uses, etc., are quickly idealized to fit a prior idea of what should or should not be logical.

Or else the appeal to a formalist account that is ultimately a conventional account. As happened with geometry, logic becomes meaningful only relative to some interpretation, formal or informal. This leads to a nice cozy relationship between all sorts of people who for quite a while thought they were disagreeing about such things as the nature of reality. They all gang up on the metaphysician and hold the view that just as Beauty failed to lead the way for aesthetics, Truth (i.e., Reality) failed to lead the way for logic.

If logic is about some of the broadest traits of reality, then there is no circularity in describing these traits and theorizing about them by means of linguistic theories that involve logic. If what one is describing is not linguistic at all and if the way

---

[25] This is Quine's conclusion to the remarks I quoted in note 20 (Ibid., p. 40):

> Tarski, to whom the three satisfaction conditions are due, saw their purpose the other way around; not as explaining negation, conjunction, and quantification, which would be untenable, but as contributing to a definition of satisfaction itself and so, derivatively, of truth.

But what about negation, conjunction, and quantification, keep wondering the beginning students—and not only them.

If the idea is that a definition of satisfaction somehow "justifies" having those logical notions, then a definition of satisfaction would justify having *anything* as a logical notion—as Quine objected to Carnap in ([18], p. 125). The formalism comes in by appealing to a logical grammar given by *fiat*, that includes what one wants and excludes what one does not want.

in which language involves logic is by reflecting and expressing these features of reality, then language is a perfectly legitimate tool for the description of logic. The logician need not fear the logical features of language just as the physicist need not fear the physical features of language and the biologist need not fear the biological features of language.

9 Propositional Logic as Theory

In my earlier discussion I sought to separate pure propositional logic from applied propositional logic. Pure propositional logic is an absolute theory about reality, given a certain conception of it, which is completely independent of language. It is a theory about truth and truth preservation based on a certain analysis of truth and on the principles of non-contradiction and of excluded middle.

We can distinguish among propositions those that determine a state of affairs, and call them 'true,' from those that do not. If we say that a true proposition *denotes* the state of affairs it determines, then truth preservation amounts to the preservation of denotation. The point of the relation of material implication is to assure us that we do not lose the connection with reality. In other words, if we have some propositions that are hooked up to reality, and they materially imply another proposition, then we are still hooked up to reality. Material implication is quite important, therefore, because it prevents us from starting up with our feet on the ground and wandering off. This is one of the true foundations of logic.

In fact, if we think of propositional logic in this way, then we see how we can formulate it in terms of states of affairs, although involving propositions as a second component. Aside from propositional variables we would have variables '**s**', '**u**', '**v**', etc. over states of affairs and one fundamental binary predicate 'Den(,)' to express the relation of denotation. Classical propositional logic will involve in addition a propositional operator '$\mathcal{N}$' for negation. Our axioms are

(C1) $\forall p \forall \mathbf{u} \forall \mathbf{v}((\mathrm{Den}(p, \mathbf{u}) \,\&\, \mathrm{Den}(p, \mathbf{v})) \rightarrow \mathbf{u} = \mathbf{v})$,

which expresses the uniqueness of denotation (or propositional instantiation), and the principle of non-contradiction

(C2) $\forall p \neg(\exists \mathbf{u} \mathrm{Den}(p, \mathbf{u}) \,\&\, \exists \mathbf{u} \mathrm{Den}(\mathcal{N} p, \mathbf{u}))$.

We can define truth as

(C-D1) $\forall p(\mathrm{T}p \leftrightarrow \exists \mathbf{u} \mathrm{Den}(p, \mathbf{u}))$,

and falsity as

(C-D2) $\forall p(\mathrm{F}p \leftrightarrow \mathrm{T}(\mathcal{N} p))$.

Material implication can be characterized by

(C-D3) $\forall p \forall q(\mathrm{Imp}(p, q) \leftrightarrow \neg(\exists \mathbf{u} \mathit{Den}(p, \mathbf{u}) \,\&\, \neg \exists \mathbf{u} \mathrm{Den}(q, \mathbf{u})))$.

This is classical propositional logic in that it involves a classical conception of truth as what is real, a classical conception of discourse (*logos*) as propositions, and a classical conception of falsity (as opposed to what is not true) through the operation of negation. It does not require, however, that every proposition be either true or false.

The theory of material implication is the most fundamental part of pure propositional logic. But we can study *all* truth relations among propositions that we can define purely in terms of truth and falsity. It is not a question of *choosing* alternative definitions of the truth relations that may be more to our taste, but of studying them all. And we can study how they behave for propositions that are either true or false, and how they behave for propositions more generally. We can study infinitary truth relations as well, of course; for example a relation of material implication that holds between denumerably many propositions and a proposition. The same for disjunction relations, conjunction relations, etc. We can introduce operations from propositions to propositions, finitary or infinitary, as long as they are characterized in terms of truth and falsity. All of this is pure propositional logic and, by considering one or another aspect (or part) of it, we get practically everything that logicians have done so far in their analysis of truth.

In this sense of pure propositional logic there are no really alternative systems of propositional logic.[26] It is truly universal and can reasonably be said to study some of the broadest features of reality. Where we get particularity is at the level of applications.

There are many applications of propositional logic, both to abstract entities such as linguistic propositions (in one conception or another), and to more or less concrete entities such as sentences, statements, utterances, inscriptions, etc.—also in one conception or another. There are also applications to formal or informal theories, including logical theories, of one or another aspect of reality. We can apply it to English, for example, and to various simulations of English, both at the level of meaning and at the level of syntax; we can apply it to mathematics; we can apply it to first order logic; we can apply it everywhere where we have some sort of discourse. The specific features of one or another application will vary widely, of course, depending on the structure of the things which we are considering, on the operations that we can actually define for these structures, etc. But these applications are not arbitrary; they have to tie up to the absolute theory.

Pure propositional logic, as just characterized, belongs to ontological logic and it does not include a theory of deduction as a human activity. This would be part of epistemological logic and is more closely connected to the applications of pure propositional logic. Systems of deduction can also be seen as systematizations of the laws of material implication and of the other truth relations, however, and in this sense deduction is also a part of ontological logic.[27]

---

[26] I am not referring here to intuitionistic logic or to propositional logics such as modal logic that study other implication relations, of course.

[27] I was very happy to be able to participate in the Rio 2000 Natural Deduction conference in honor of Dag Prawitz. The talk I gave at the conference (on logical deduction and logical consequence)

References

1. Bochenski, I. (1961). *A history of formal logic*. Notre Dame, IN: University of Notre Dame Press.
2. Boole, G. (1854). *An investigation of the laws of thought*. London (Reprinted, New York): Dover.
3. Chateaubriand, O. (1999). Proof and logical deduction. In H. H. Hauesler & L. C. Pereira (Eds.), *Pratica: Proofs, types and categories*. Rio de Janeiro: Editora PUC.
4. Chateaubriand, O. (2001). *Logical forms. Part I: Truth and description*. Campinas: Unicamp.
5. Church, A. (1956). *Introduction to mathematical logic* (Vol. 1). Princeton, NJ: Princeton University Press.
6. Enderton, H. B. (1972). *A mathematical introduction to logic*. New York: Academic Press.
7. Frege, G. (1960). On sense and reference. In P. Geach & M. Black (Eds.), *Translations from the philosophical writings of Gottlob Frege*. Oxford: Blackwell.
8. Frege, G. (1964). *The basic laws of arithmetic: Exposition of the system* (M. Furth, Trans.). Los Angeles, Berkeley: University of California.
9. Frege, G. (1967). Begriffsschrift, a formula language, modelled upon that of arithmetic, for pure thought. In J. van Heijenoort (Ed.), *From Frege to Gödel*. Cambridge, MA: Harvard University Press.
10. Goodman, N. (1947). The problem of counterfactual conditionals. *The Journal of Philosophy*, *44*, 113–128.
11. Kleene, S. C. (1952). *Introduction to Metamathematics*. New York: Van Nostrand.
12. Lukasiewicz, J., & Tarski, A. (1956). Investigations into the sentential calculus. In A. Tarski (Ed.), *Logic, semantics, metamathematics*. Oxford: Clarendon Press.
13. Mates, B. (1972). *Elementary logic* (2nd ed.). New York: Oxford University Press.
14. Monk, J. D. (1976). *Mathematical logic*. New York: Springer.
15. Quine, W. V. (1951). *Mathematical logic* (revised edition). Cambridge, MA: Harvard University Press.
16. Quine, W. V. (1970). *Philosophy of logic*. Englewood Cliffs, NJ: Prentice-Hall.
17. Quine, W. V. (1972). *Methods of logic* (3rd ed.). New York: Holt, Rinehart and Winston.
18. Quine, W. V. (1976a). Carnap on logical truth. In W. V Quine (Ed.), *The ways of paradox and other essays* (revised edition). Cambridge, MA: Harvard University Press.
19. Quine, W. V. (1976b). Ontological remarks on the propositional calculus. In W. V. Quine (Ed.), *The ways of paradox and other essays* (revised edition). Cambridge, MA: Harvard University Press.
20. Rasiowa, H., & Sikorski, R. (1963). *The mathematics of metamathematics*. Warzawa: Panstwowe Wydawnictwo Naukowe.
21. Russell, B. (1903). *The principles of mathematics*. Cambridge: Cambridge University Press.
22. Russell, B. (1906). The theory of implication. *American Journal of Mathematics*, *28*, 159–202.
23. Whitehead, A. N., & Russell, B. (1925). *Principia Mathematica* (Vol. 1, 2nd ed.). Cambridge: Cambridge University Press.

(Footnote 27 continued)
involved a combination of ideas from [3] and from the introduction of [4]. The present chapter is a version of a chapter of the unpublished Part II of *Logical Forms*, and although it is not on natural deduction, it seems to me appropriate for this volume.

Categorical Semantics of Linear Logic for All

Valeria de Paiva

Abstract This note compares several notions of categorical model of intuitionistic linear logic in the literature. The emphasis is on explaining why choices can be made and what they amount to. My conclusion is that despite being an older and more complicated notion, linear categories are still the best way to describe models of linear logic, if one wants the correspondence between syntax and semantics to be as tight as possible.

1 Introduction

This chapter is a survey of results on categorical modeling of linear logic, oriented toward logicians, interested in proof theory, category theory, and linear logic, but not experts in either of these. There are other such surveys available [12, 14, 16] and hence there is a need to explain why another such. The explanation is simple: I find the other surveys too hard to read, going too deeply into the details of definitions and proofs, when an interested, but not committed reader, would prefer to have intuitive explanations of which (and why) choices were made. Of course what is an intuitive explanation for some, will be considered glib glossing over by others. But if it convinces the reader that the categorical modeling of the linear modality '!" is an interesting, albeit subtle, problem, this note will not have been a pointless exercise.

V. de Paiva (✉)
Nuance Communications, 1198 Arques Ave, Sunnyvale, CA 94085, USA
e-mail: valeria.depaiva@gmail.com

L. C. Pereira et al. (eds.), *Advances in Natural Deduction*,
Trends in Logic 39, DOI: 10.1007/978-94-007-7548-0_9,

2 The Problem

What are the requirements for a categorical model of linear logic? There are several solutions to this question in the literature. How do we compare and relate these solutions? That is, how do we compare and relate the several possible notions of 'categorical model of linear logic' available? This is the problem we want to address in this chapter.

To solve this problem, first we must discuss what are categorical models, why should we bother with them and what are the requirements for a categorical model of *linear logic* in particular. Then we must explain why there are several solutions to this question and what these solutions are. Finally we must compare these solutions, pointing out the trade-offs between approaches.

The question "what is a modal of linear logic?" has been around for a while and there are many proposals available. For a while the best answer was the very clear chapter of Bierman entitled exactly "What is a categorical model of intuitionistic linear logic?" [5], but more work ensued and there is a need to incorporate the new work into a coherent framework. This has been done in a series of recent articles [12, 14, 16] which aim at a categorically sophisticated audience. This chapter attempts to explain to a practicing logician, in simple terms, using as little mathematics as possible, what is known about the categorical semantics of linear logic: as such it is a simplification of the work in [12, 14, 16], which have the same goals, but are more technical, presuppose more knowledge of category theory, and leave many of their conclusions to be worked out by the reader. Practicing logicians may brush up on their basic category theory in "Category Theory for Linear Logicians," [6] which has similar goals, and defines all categorical notions necessary for this chapter. Note however that [6] avoids the discussion of what is required to deal the linear logic *exponentials* (cf. page 31), the main object of this work simply remarking that "the structure seems less canonical."

3 Categorical Modeling: What and Why

There are two main uses of "categorical modeling": The first one considers the notion of a structure valued in a category C, the analogue of the classical model-theoretical notion of structure, except that one generalizes from (the category of) sets and functions to an abstract category C with morphisms. This leads to what may be called "categorical model theory."

The second use, the one that concerns us here, is when we want to model more than theorems, we want to model different proofs of a given theorem. The idea, which has been called the "extended Curry-Howard isomorphism," is that since we can associate terms in a typed lambda calculus with derivations in Natural Deduction and with morphisms in certain appropriate classes of categories, we can model these terms or "proofs" as morphisms in the categories and we can then use equality of morphisms to induce a notion of equality of proofs in Natural Deduction for the logic in question. This is sometimes called "categorical proof theory." Applications of categorical proof theory include foundations of functional programming, proofs of correctness of compiler optimizations, new insights into expressive type systems, etc.

However, while the situation for categorical proof theory of (a Natural Deduction formulation of) intuitionistic propositional logic can be summarized in one sentence "categorical models of intuitionistic propositional logic are *cartesian closed categories* with coproducts," the situation for categorical proof theory of propositional *linear logic* is more complicated. In an ideal world one would define a special class of categories, perhaps called *linear* categories (some sort of analogue of cartesian closed categories), prove that linear categories are sound and complete for linear logic, and have one or two corollaries showing that other notions of model of linear logic in the literature are equivalent to (or subsumed by) the definition of a linear category. This kind of situation obtains for the modality-free (or exponential-free) fragment of intuitionistic linear logic, where the appropriate class of categories are symmetric monoidal closed categories (with products and coproducts to model additives). But the modalities pose some real problems. For the modality "!" (which Girard calls the exponential "of course!"), it is possible to prove soundness and completeness for several, in principle very different, notions of categorical model and the corollaries relating these different notions of model are confusing. Hence this note.

Before dealing with the mathematics of the modality, let us first settle some smaller issues. First, it is clear that even if one is interested in *classical* linear logic, producing the appropriate requirements for modeling *intuitionistic* linear logic is sufficient, as categorical duality will produce a model of classical linear logic from one of intuitionistic linear logic, automatically.

Second, the logic in question, presented as a sequent calculus by Girard and Lafont [8] is pretty much uncontroversial. Part of the problem here is an adequate Natural Deduction formulation of the sequent calculus for intuitionistic linear logic plus a term assignment for this Natural Deduction formulation satisfying some natural type-theoretical properties, like 'subject reduction', normalization, confluence,[1] etc. The rest of the problem is how to phrase the categorical concepts in the simplest, more direct way possible.

Third, one could perhaps be tempted to 'ditch' the modality "!" (pronounced 'plink', 'shriek', or 'of course!') given the mathematical difficulties associated with it. This is not sensible, as it is the modalities that make Linear Logic so useful as a logical system. While Linear Logic is a resource sensitive logic, where one can pay attention to how many times resources like premises are used, Girard's translation of usual (intuitionistic or classical) logic into Linear Logic allows us to forget this resource-consciousness, if desired. Since "!" is the way to recover classical logic expressivity in a linear setting, we must deal with it. Not modeling the exponential is not an option. The question is how to do it as pain-free as possible.

[1] While logicians may not be worried about confluence or subject reduction, they should care that the Natural Deduction produced satisfies at least normalization.

The fundamental idea of a categorical treatment of proof theory is that propositions should be interpreted as the objects of an appropriate category and (natural deduction) proofs of propositions should be interpreted as morphisms of that category. We use the traditional notational simplification of using the same name for a proof/proposition and its representation in the category. Fixing notation we say:

Definition 1 A category C is said to be a *categorical model* of a given logic $\mathcal{L}$, iff

1. For all proofs $\Gamma \vdash_{\mathcal{L}} M : A$ in $\mathcal{L}$ there is a morphim $[[M]] : [[\Gamma]] \rightarrow [[A]]$ in C.
2. For all equalities $\Gamma \vdash_{\mathcal{L}} M = N : A$ it is the case that $[[M]] =_{\mathsf{C}} [[N]]$, where $=_{\mathsf{C}}$ refers to equality of morphims in the category C.

We say a notion of categorical model is *complete* if for any signature of the logic $\mathcal{L}$ there is a category C and an interpretation of the logic in the category such that:

> If $\Gamma \vdash M : A$ and $\Gamma \vdash N : A$ are derivable in the system then M and N are interpreted as the same map $\Gamma \rightarrow A$ in the category C just when $M = N : A$ is provable from the equations of the typed equational logic defining $\mathcal{L}$.

As a reminder, recall the categorical modeling of the fragment of intuitionistic linear logic consisting only of linear implications and tensor products, plus their identity, the constant I. (This fragment is sometimes called rudimentary linear logic.) For this fragment, the natural deduction formulation of the logic is uncontroversial. The rules for linear implication are like the rules for implication in intuitionistic logic (with the understanding that variables are always used a single time). The rules for tensor are slightly different from the rules for usual cartesian product, as tensor products $A \otimes B$, unlike cartesian products ($A \times B$), do not have projections ($A \otimes B$ does not prove A or B) and do not allow for duplication (A does not prove $A \otimes A$). While such structures are less common than usual cartesian products, they are not uncommon in mathematics. Moreover, structures consisting of linear-like implications and tensor-like products had been named and investigated by category theorists where many years before the introduction of Linear Logic. They are called *symmetric monoidal closed* categories or smccs. Symmetric monoidal closed categories are sound and complete for rudimentary linear logic.

4 Modeling the Modality

The original sequent rules for the modality '!' are intuitive. One should think of $!A$ in linear logic as a formula which one is allowed to duplicate and erase at will, unlike a plain formula A. Duplicating and erasing formulae prefixed by "!" correspond to the usual structural rules of weakening and contraction:

$$\frac{\Delta \vdash B}{\Delta, !A \vdash B} \qquad \frac{\Delta, !A, !A \vdash B}{\Delta, !A \vdash B}$$

But to make '!" a proper connective we must also introduce it to the right and to left of the turnstile. These rules are more complicated, but familiar from Prawitz's work on S4.

$$\frac{!\Delta \vdash B}{!\Delta \vdash !B} \qquad \frac{\Delta, A \vdash B}{\Delta, !A \vdash B}$$

(Note that $!\Delta$ means that every formula in Δ starts with a ! operator.)

To transform the rules above into Natural Deduction ones with a sensible term assignment is not hard [3]. We obtain rules like:

$$\frac{\Delta \vdash M\colon !A}{\Delta \vdash \mathsf{derelict}(M)\colon A} \qquad \frac{\Delta_1 \vdash M\colon !A \quad \Delta_2 \vdash N\colon B}{\Delta_1, \Delta_2 \vdash \mathsf{discard}\ M\ \mathsf{in}\ N\colon B}$$

$$\frac{\Delta_1 \vdash M\colon !A \qquad \Delta_2, a\colon !A, b\colon !A \vdash N\colon B}{\Delta_1, \Delta_2 \vdash \mathsf{copy}\ M\ \mathsf{as}\ a, b\ \mathsf{in} N\colon B}$$

$$\frac{\Delta_1 \vdash M_1\colon !A_1, \ldots, \Delta_k \vdash M_k\colon !A_k \qquad a_1\colon !A_1, \ldots, a_k\colon !A_k \vdash N\colon B}{\Delta_1, \Delta_2, \ldots, \Delta_k \vdash \mathsf{promote}\ M_i\ \mathsf{for}\ a_i\ \mathsf{in}\ N\colon !B}$$

The upshot of these rules is that each object $!A$ has morphisms of the form $\mathsf{er}\colon !A \to I$ and $\mathsf{dupl}\colon !A \to !A \otimes !A$, which allow us to erase and duplicate the object $!A$. These morphisms give $!A$ the structure of a (commutative) *comonoid* (a comonoid is the dual of a monoid, intuitively like a set with a co-multiplication and co-unit).

Also each object $!A$ has morphisms of the form $\mathsf{eps}\colon !A \to A$ and $\mathsf{delta}\colon !A \to !!A$ that provide it with a *coalgebra* structure, induced by a *comonad*. How should the comonad structure interact with the comonoid structure? This is where the picture becomes complicated and different choices can be made.

Lafont's Models

Even before Girard's original article in Linear Logic appeared, Lafont had come up with a very elegant suggestion for a categorical model of intuitionistic linear logic. He suggested that one should model $!A$ via *free* comonoids.

Definition 2 A *Lafont category* consists of

1. A symmetric monoidal closed category C with finite products,
2. For each object A of C, the object $!A$ is the free commutative comonoid generated by A.

Freeness (and cofreeness) of algebraic structures gives very elegant mathematics, but concrete models satisfying cofreeness are very hard to come by. None of the original models of Linear Logic satisfies this strong requirement.[2] Moreover, as remarked by Mellies, in categories of games, researchers are interested in co-existing notions of '!', to model particular memory management paradigms. But if we use Lafont's notion of model there is only one cofree comonoid. Luckily there was

[2] The notable exception being *dialectica categories* [7].

already another proposed set of requirements, that came to be known as the Seely model.

Seely's Models

Seely's notion of model [17] is much more encompassing, most models of linear logic (with additive conjunction) satisfy it. Instead of giving comonoids the prominent position they had in Lafont's definition, it takes the view, originally put forward by Girard, that linear logic is the basic logic and intuitionistic logic is to be considered a derived logic. Thus, because intuitionistic implication $A \to B$ can be decomposed into an use of the modality and one of linear implication, $A \to B =_{\mathsf{def}} (!A) \multimap B$ via Girard's translation we must model the interaction between linear logic and intuitionistic logic via the notion of a comonad ! relating these systems.

Seely's definition of a categorical model of intuitionistic linear logic requires the presence of additive conjunctions in the logic and it depends both on named natural isomorphims

$$\mathsf{m}\colon !A \otimes !B \cong !(A \& B) \text{ and } \mathsf{p}\colon 1 \cong !\mathsf{T}$$

and on the requirement that the functor part of the comonad '!' take the comonoid structure of the cartesian product to the comonoid structure of the tensor product. This notion of model had been in use for a few years, when, surprisingly, Bierman pointed out that Seely's definition missed one crucial condition. This meant that the comonoid structure of $!A$ and its comonad structure did not interact as well as needed to make it a *sound* categorical model. The condition missing can be simply added via what Bierman called a *new-Seely* category.

Definition 3 (*Bierman*) A *new-Seely category*, C, consists of

1. A symmetric monoidal closed category C, with finite products, together with
2. A comonad $(!, \varepsilon, \delta)$ to model the modality, and
3. Two natural isomorphism, $\mathsf{n}\colon !A \otimes !B \cong !(A \& B)$ and $\mathsf{p}\colon I \cong !\mathsf{T}$,

such that the adjunction between C and its co-Kleisli category is a *monoidal* adjunction.

Soundness of new-Seely categories is proved by showing that every new-Seely category is a *linear category*.

Linear Categories

Linear categories were introduced by Benton, Bierman, de Paiva, and Hyland in [2] and carefully investigated by Bierman in his dissertation [4]. The original definition of a linear category is a bit of a mouthful, as it spells out the relationship between the comonoid and the comonad structure of $!A$ as fully as possible. It also unpacks into quite a collection of diagrams that need to be checked. But it does what it is meant to do and the soundness of the model is fully proved in [4, 5].

Another ways of describing linear categories, championed by Hyland and Schalk [9] is to say that

Definition 4 *(Hyland/Schalk)* A *linear category* is a symmetric monoidal closed category S with products and equipped with a *linear exponential comonad.*

This is certainly easier to say, but to prove that it is indeed equivalent to linear categories as originally defined requires some work, done by Schalk in [16].

Similarly, the following reformulation of the original definition was given by Maneggia in her thesis [13], where it is also proved equivalent to the original definition. This is the simplest formulation of the concept I know of, but it does require introducing the category of Eilenberg-Moore coalgebras. This definition was used by Maietti et al. [12] which we discuss later.

Definition 5 *(Maneggia)* A *linear category* is a symmetric monoidal closed category S together with a symmetric *monoidal* comonad such that the monoidal structure induced on the associated category of Eilenberg-Moore coalgebras is a finite product structure.

All the notions of model that we have seen so far agree on having a basis, which is a symmetric monoidal closed category S modeling the linear propositions. They also agree that there must be a functor $! \colon \mathsf{S} \to \mathsf{S}$ modeling the modality 'of course!' They differ from each other on how they model the action of the modality functor. Basically linear categories (and new-Seely models) consider minimal conditions on some given monoidal comonad to make sure that the conditions guarantee the existence of an associated cartesian closed category modeling intuitionistic logic.

A different proposal came from ideas discussed independently by Benton, Plotkin, and Barber. They suggested modeling the system more symmetrically, giving linear logic and intuitionistic logic the same status. This required introducing a new kind of term calculus, somewhat similar to Girard's *Logic of Unity*, where sequents have different kinds of contexts. But as far as the models are considered this is not such a big step. Having a monoidal comonad on a category C means that this comonad induces a spectrum of monoidal adjunctions spanning from the category of Eilenberg-Moore coalgebras to the co-Kleisli category. (This is basic category theory, cf. MacLane's textbook—[10], 144.) Now putting linear logic and intuitionistic logic on the same footing, corresponds to saying that we decide to consider as the model of the system the *monoidal adjunction* itself, instead of the inducing comonad.

Monoidal Adjunctions

Benton first suggested taking the monoidal adjuction (between a cartesian closed category C and a symmetric monoidal closed category S) as the mathematical definition of model of the interaction between linear logic and intuitionistic logic. He called these kind of monoidal adjunctions *LNL models* (for linear–nonlinear models) and designed a term calculus corresponding to these models that he proved sound and complete.

This, more than simply a simplification of the notion of model of linear logic, corresponds to a shift of paradigms. While before one was trying to model a system of linear logic and it so happened that to model one of its connectives we needed a bridge to a different logical system, intuitionistic logic, now we are talking about modeling *pairs* of logical systems, with some important relationships between them. Moreover, since the whole of intuitionistic logic can be coded up inside linear logic via Girard's translation, in principle by modeling the systems side-by-side we have a wholesale duplication of the intuitionistic proofs. This can be seen both as a good or a bad thing. In the one hand, one of the hopes for linear logic was that it would point out where duplication and erasing were necessary in specific constructive proofs. In the LNL system this ability to pinpoint where duplication and erasing are required, disappears. But on the other hand, thinking of programming in the type system, it allows you to keep and use all the libraries and optimizations that you may have created for intuitionistic logic previously. Moreover, the definition of a LNL model is so simple it almost does not require a definition.

Definition 6 A *linear-non-linear (LNL) category* consists of a symmetric monoidal closed category S, a cartesian closed category C and a symmetric *monoidal adjunction* between them.

Benton also proved that linear categories and LNL models are equivalent, in the specific sense that given a LNL model one obtains a linear category and conversely, given a linear category one can obtain a LNL model. But there is no uniqueness of the categories obtained. Moreover, the category of LNL models and the category of linear categories are *not* equivalent, see below.

Barber and Plotkin improved on Benton's suggestion, by showing that one does not need to require closedness of the cartesian category. Function spaces in the cartesian category can be induced from the linear ones. They also produced a term calculus, called DILL for Dual Intuitionistic and Linear Logic, that looks much more like intuitionistic linear logic than LNL and satisfies all essential syntactic properties of LNL. To model DILL they considered the notion of a DILL-category, which differs from a LNL-category only on not requiring the cartesian closed structure to be given a priori. You only need to start with a cartesian structure and you then prove it cartesian *closed*.

Definition 7 A *dual intuitionistic linear logic (DILL) category* is a symmetric monoidal adjunction between S a symmetric monoidal closed category and C a cartesian category.

By now we have one definition of a categorical model of a generic logic $\mathcal{L}$ and *six* slightly different definitions of a model of intuitionistic linear logic. How do they compare? Which one is best? Why?

5 Comparing Models

As discussed before we know that Lafont models are very specific instances of *linear categories*. We also know that new-Seely categories are linear categories. Further we know that Hyland and Schalk's symmetric monoidal categories with *a linear exponential comonad* are linear categories. Also Benton, Barber, and Mellies have proved, independently, that given a linear category we obtain a DILL-category and given a DILL-category we obtain a linear category. Are all these notions of model equivalent then?

Maietti et al. [12] set out to prove some kind of categorical equivalence of models and discovered that the situation was not quite as straightforward as expected. To wit: Bierman proved that linear categories are sound and complete with respect to ILL. Barber proved that DILL-categories are sound and complete with respect to DILL. Maietti et al. proved (Theorem 18 in [12]) that the category of *theories* of ILL is equivalent to the category of theories of DILL. So we have two calculi, ILL and DILL, which are sound and complete with respect to their models, whose categories of theories are equivalent. One would expect their categories of models (linear categories and symmetric monoidal adjunctions) to be equivalent too. But when we consider the natural morphisms of linear categories and of symmetric monoidal adjunctions (to construct categories **Lin** and **SMA**), we do not obtain a categorical equivalence. Instead we obtain only a retraction (Theorem 41 in [12]). So we faced with the somewhat paradoxical situation of calculi with equivalent categories of theories, whose classes of model are *not* equivalent.

This led Maietti et al. to conclude that soundness and completeness of a notion of categorical model are not enough to determine the "most apropriate" notion of categorical models for a calculus. More than simply soundness and completeness we should say a class of categories is a model for a type theory when we can prove an *internal language theorem* relating the category of models to the category of theories of the calculus.

What does it mean to prove an internal language theorem for a typed calculus and a class of categorical models? Given a type calculus $\mathcal{L}$, one can consider its category of theories, $\mathsf{Th}(\mathcal{L})$ and its category of sound and complete models, together with model morphisms $\mathcal{M}(\mathcal{L})$.

Definition 8 We say that a typed calculus $\mathcal{L}$ provides an *internal language* for the class of models in $\mathcal{M}(\mathcal{L})$ if we can establish an equivalence of categories between the category of $\mathcal{L}$-theories, $\mathsf{Th}(\mathcal{L})$, and the category of $\mathcal{L}$-models $\mathcal{M}(\mathcal{L})$.

The functors establishing this equivalence, say, $L\colon \mathcal{M}(\mathcal{L}) \to \mathsf{Th}(\mathcal{L})$ and $C\colon \mathsf{Th}(\mathcal{L}) \to \mathcal{M}(\mathcal{L})$ allow us to move freely between categorical syntax and categorical semantics. So L assigns to a model M in $\mathcal{M}(\mathcal{L})$ its specific internal language $L(M)$. The functor C associates to each specific $\mathcal{L}$-theory V its syntactic category $C(V)$ in such way that $M \cong C(L(M))$ and $V \cong L(C(V))$.

We note that this definition of internal language is more restrictive than the one (in Barr and Wells—[1] for example) which requires only one of the equivalences, namely $M \cong C(L(M))$ to hold.

Tightening their definitions as above, Maietti et al. obtain the desired situation that equivalent calculi (DILL and ILL), with equivalent categories of theories ($\mathsf{Th}(\text{DILL}) \cong \mathsf{Th}(\text{ILL})$) have equivalent classes of models. But this necessitates postulating that despite being sound and complete for DILL, symmetric monoidal adjunctions between a cartesian and a symmetric monoidal closed category (the category **SMA**) are *not* the models for DILL. Instead we must take as models for DILL a subcategory of **SMA**, the symmetric monoidal adjunctions generated by *finite tensor products of free coalgebras*. (This idea originally due to Hyland was expanded on and explained by Benton and Maietti et al.)

With hindsight one can see that the reason why the calculus DILL does not provide an internal language for *all* the natural symmetric monoidal adjunctions is that the cartesian category of a given symmetric monoidal adjunction $(\mathsf{S}, \mathsf{C}, F, G \dashv)$ may have objects and morphisms that are not in the domain of functor F and hence no DILL-theory can provide syntax corresponding to these objects and morphisms. They are genuinely not part of the linear–nonlinear picture.

Thus the price to pay for the expected result that equivalent categories of theories imply equivalent categories of models is high: not only we have to keep the more complicated notion of model of linear logic, but we need also to insist that categorical modeling requires soundness, completeness, and an internal language theorem. This last requirement makes sense in general: when using categorical models one would want to move between syntax and semantics as easily and freely as possible. But while the notion of satisfying an "internal language theorem" is somewhat up for grabs in the categorical proof theory literature, it seems that in the case of linear logic we really must pay attention to it. Otherwise we end up with 'paradoxical' results as above. This seems the main "take home message" from the work on comparing categorical models of linear logic.

As a side remark, in Maietti et al's work this conclusion comes about from a syntactic perspective, one that enforces very restrictive notions of morphisms of linear categories.

Definition 9 Given linear categories C and S (with respective comonoads $!$ and $!'$) a *linear functor* between C and D is a strong symmetric monoidal closed functor $K\colon \mathsf{C} \to \mathsf{D}$ such that $K! = !'K$ *as monoidal functors*.

By the equality as monoidal functors we mean that $K\epsilon = \epsilon'_K$, $K\delta = \delta'_K$, and two other diagrams relating the monoidal natural transformations $\mathsf{m}^!$ and $\mathsf{m}^{!'}$ that make "$!, !'$" monoidal comonads. Hyland and Schalk, coming from a perspective less syntactic, geared toward categories of games, came up with a more relaxed notion of linear morphism.

Definition 10 Let C and D be two linear categories. A functor $F : \mathsf{C} \to \mathsf{D}$ is *linearly distributive* if and only if F is monoidal and is equipped with a distributive law $\kappa : !F \to F!$ respecting the comonoid structure.

Maietti et al. show that their more strict notion of linear functor is actually a linearly distributive functor as above.

A slightly different approach to carving out the relevant part of the category of symmetric monoidal adjunctions is taken by Møgelberg et al. [15]. They call *linear adjunctions* the symmetric monoidal adjunctions between an smcc and a cartesian category and say that DILL-models are the full subcategory of the category of linear adjunctions on objects equivalent to the objects induced by linear categories, when performing the product of free coalgebras construction.

A very different semantic approach to put together a linear–nonlinear logic is pursued by Maietti et al. [11], who follow the lead of Plotkin and Wadler, consider a calculus (called ILT for Intuitionistic and Linear Type theory) without the modality '!', but with interaction between linear and a intuitionistic implications. Comparison between that approach and the ones in this note is left to future work.

Models with Products

So far we discussed models of a calculus consisting of only tensor products, linear implications, and modalities '!'. If our calculus also has finite cartesian products (or additive conjunctions) the mathematical structures simplify considerably. The point is that a linear category with products yields a co-Kliesli category with products and this is actually cartesian closed. This means that linear categories with products are Seely categories and it makes it easy to show that models of ILL with products and of DILL with products are equivalent, as expected (proposition 54 in [12]. Somewhat surprisingly, given that DILL was conceived after LNL, Maiettti et al. also show that models of DILL with products are only a subcategory of models of LNL with products.

6 Conclusions

We surveyed several notions of categorical model of intuitionistic linear logic and compared them, as categories. The main comparison was between linear categories (in its various guises) and symmetric monoidal adjunctions as proposed by Benton and Barber and Plotkin. The summary is that while the notion of a symmetric monoidal adjunction (between a symmetric monoidal closed category and a cartesian category) is very elegant and appealing, the category of these **SMA** is too big, has objects and morphisms that do not correspond to objects and morphisms in dual intuitionistic and linear logic. We also tried to explain the point, made by Maietti et al. [12], that categorical modeling should be about soundness, completeness, and (*essentially*) internal language theorems. While we agree that this extra criterion is a sensible one, we feel that the only kind of evidence provided (that without it, we have an example of two calculi ILL and DILL with equivalent theories that do not have equivalent classes of models) is not very strong. A traditional logician may prefer to think that the problem discussed is a problem of intuitionistic linear logic, which is not as well behaved as it should be. This is not our conclusion. The problem with modeling the modality 'of course!' of linear logic is similar to the problem of modeling any other modalities and these are pervasive in logic. More research in other modalities should clarify the criteria for categorical modeling, in general. Also new concrete mathematical models, like games, for other logical systems, should help.

References

1. Barr, M., & Wells, C. (1985). *Toposes, triples and theories*, New York: Springer-Verlag.
2. Benton, N., Bierman, G., de Paiva, V., & Hyland, M. (1992). *Term assignment for intuitionistic linear logic*. Technical report, Cambridge: University of Cambridge.
3. Benton, N., Bierman, G., de Paiva, V., & Hyland, M. (1993). *A term calculus for intuitionistic linear logic*. In First International Conference on Typed Lambda Calculi and Applications (TLCA), volume 664 of LNCS.
4. Bierman, G. (1994). *On intuitionistic linear logic*. Technical report 346, Ph.D. Thesis. Computer Laboratory, University of Cambridge.
5. Bierman, G. (1995). What is a categorical model of intuitionistic linear logic? *In Proceedings of the Second International Conference on Typed Lambda Calculus and Applications., volume 902 of Lecture Notes in Computer Science*. Springer Verlag.
6. Blute, R. & Scott, P. (2004). *Category theory for linear logicians*. Linear Logic in Computer Science, Cambridge University Press, Cambridge.
7. de Paiva, V. (1989). The dialectica categories. *In Categories in Computer Science and Logic, volume 92 of Contemporary Mathematics*, Manchester: American Mathematical Society, pp. 47–62.
8. Girard, J. Y. & Lafont, Y. (1987). Linear logic and lazy computation. In TAPSOFT '87, volume 250 of, Lecture Notes in Computer Science, pp. 52–66.
9. Hyland, M. & Schalk, A. (1999). Abstract games for linear logic. Extended abstract. *In CTCS '99, volume 29 of Electronic Notes in Theoretical Computer Science*.
10. MacLane, S. (1971). *Categories for the working mathematician, volume 5 of Graduate Texts in Mathematics*. New York: Springer-Verlag.
11. Maietti, M. E., de Paiva, V., & Ritter, E. (2000). Categorical models for intuitionistic and linear type theory. In J. Tiuryn (Ed.), *Proceedings of Foundations of Software Science and Computation Structures, LNCS*.
12. Maietti, M. E., Mannegia, P., de Paiva, V., & Ritter, E. (2005). Relating categorical models for intuitionistic linear logic. *Applied Categorical Structures*, *13*, 1–36.
13. Maneggia, P. (2004). *Models of linear polymorphism*. PhD thesis, UK: The University of Birmingham.
14. Mellies, P. A. (2002). *Categorical models of linear logic revisited*. submitted to Theoretical Computer Science.
15. Mogelberg, R. E., Birkedal, L., & Petersen, R. L. (Feb 2005). *Categorical models of PILL*. Technical report TR-2005-58, Denmark: IT University of Copenhagen.
16. Schalk, A. (2004). What is a categorical model of linear logic? University of Manchester. http://www.cs.man.ac.uk/schalk/notes/index.html
17. Seely, R. A. G. (1989). Linear logic, *-autonomous categories and cofree coalgebras. *In Categories in computer science and logic (Boulder, CO., 1987)*, volume 92 of Contemporary Mathematics, Providence: American Mathematical Society, pp. 371–382.

Assertions, Hypotheses, Conjectures, Expectations: Rough-Sets Semantics and Proof Theory

Gianluigi Bellin

Abstract In this chapter *bi-intuitionism* is interpreted as an intensional logic which is about the justification conditions of *assertions* and *hypotheses*, extending C. Dalla Pozza and C. Garola's *pragmatic interpretation* [18] of intuitionism, seen as a logic of *assertions* according to a suggestion by M. Dummett. Revising our previous work on this matter [5], we consider two additional illocutionary forces, (*i*) *conjecturing*, seen as making the hypothesis that a proposition is epistemically necessary, and (*ii*) *expecting*, regarded as asserting that a propostion is epistemically possible; we show that a logic of expectations justifies the double negation law. We formalize our logic in a calculus of sequents and study *bimodal* Kripke semantics of bi-intuitionism based on translations in **S4**. We look at *rough set* semantics following P. Pagliani's analysis of "intrinsic co-Heyting boundaries" [40] (after Lawvere). A Natural Deduction system for co-intuitionistic logic is given where proofs are represented as *upside down* Prawitz trees. We give a computational interpretation of co-intuitionism, based on T. Crolard's notion of *coroutine* [16] as the programming construction corresponding to *subtraction introduction*. Our typed calculus of co-routines is dual to the simply typed lambda calculus and shows features of concurrent and distributed computations.

1 Introduction

This chapter aims at developing an intensional logic of the justification conditions of some illocutionary acts, namely, *asserting*, *making hypotheses*, *conjecturing* and *expressing an epistemic expectation*, where the intended interpretation of the logical

---

G. Bellin (✉)
Dipartimento di Informatica, Università di Verona, Strada Le Grazie, 37134 Verona, VR, Italy
e-mail: gianluigi.bellin@univr.it

L. C. Pereira et al. (eds.), *Advances in Natural Deduction*,
Trends in Logic 39, DOI: 10.1007/978-94-007-7548-0_10,

connectives and of the forms of inference are those of intuitionistic logic[1,2]. Our work belongs to the project of a *Logic for Pragmatics*, initiated by the philosopher Carlo Dalla Pozza and by the physicist Claudio Garola [18] and later continued by Dalla Pozza and Bellin [7] and others, in particular Bellin and Biasi [5]. Characteristic of our approach with respect to similar ones, e.g., S. Artemov's *justification logic*, is the focus on illocutionary forces in the elementary expressions of our language, where propositions in the classical sense are never presented without an illocutionary force and thus an "illocutionary mood" (e.g., assertive or hypothetical) is inherited also by composite expressions of the language. This fact is essential in our case study here, *bi-intuitionistic logic*, where intuitionstic and its dual co-intuitionistic logic are joined together. In natural language the acts of asserting, on one hand, and of making hypotheses and expressing a doubt, on the other, may in some sense be regarded as dual. Thus we have an interpretation of *bi-intuitionism* as an intensional logic of assertions and of hypotheses, where the dual intuitionistic and co-intuitionistic parts are "polarised" and kept separate. In this framework it is perfectly appropriate and unproblematic that the law of non-contradiction and the disjunction property hold for the *assertive* notions of intuitionistic negation, conjunction, and disjunction, while the law of excluded middle and para-consistency hold for the *hypothetical* notions of co-intuitionistic negation, conjunction, and disjunction.

In this chapter we revised and sharped the discussion of the logical properties of assertions and conjectures in Bellin and Biasi [5], by distinguishing between *conjectures* and *hypotheses*. In a nutshell, the justification of an *assertion* requires *epistemic necessity* of the truth of the propositional content *p*, which is given, e.g., by a *proof* of *p*; *making a hypothesis* is justified by the *epistemic possibility* of the truth of the propositional content; similarly, *expressing a doubt* about a statement is justified by the epistemic possibility that the statement may be unjustified. But for the justification of a *conjecture* we need the *possibility of the epistemic necessity* of the truth of its propositional content, not just epistemic possibility. Dually, we are led to the distinction between *assertions* and *epistemic expectations*: for the justification of an expectation, it suffices to have the *necessity of epistemic possibility*, which we regard

[1] We wish to thank Prof Andrew Pitts and Dr Valeria de Paiva for their comments on various aspects of this research and Dr Piero Pagliani for his expert support in the Rough Sets semantics. We are indebted with Prof Tristan Crolard for his intriguing work on bi-intuitionistic and classical logic and with Dr Hugo Herbelin for suggestions about our calculus of coroutines for co-intuitionistic logic. Thanks to Dr Carlo Dalla Pozza, Dr Kurt Ranalter, Dr Corrado Biasi, and Dr Graham White for their cooperation in the "logic for pragmatics" enterprise. I am grateful to Prof Dag Prawitz, my first marvellous supervisor in Stockholm 1978 and to Dr Luiz Carlos Pereira, a fellow student then and a supportive colleague now.

[2] Since April 2011 when significant revisions were made to this chapter, categorical models of bi-intuitionism have been studied based on monoidal categories. In particular, if the term assignment to a Gentzen system for co-intuitionistic logic is used in building a categorical model, then disjunction is best given *multiplicative* rules rather than *additive* ones, as it is done in this chapter. Further work on the mathematical structure and the philosophical interpretation of "polarized bi-intuitionism" is in G. Bellin, M. Carrara, D. Chiffi, and A. Menti, A pragmatic dialogic interpretation of bi-intuitionism, submitted to *Logic and Logical Philosophy*, 2013.

as the assertion that in all situations it will be possible for the propositional content to be true. It turns out that a logic of expectations satisfies the law of double negation, a feature of classical logic.
There is a philosophical question about the nature of the epistemic modal notions used here. Every expression of our logic for pragmatics has an interpretation in classical **S4**, the assertion $\vdash p$ and hypothesis $\mathcal{H}p$ of a proposition p are interpreted as $\Box p$ and $\Diamond p$, respectively; similarly, the conjecture $\mathcal{C}p$ and the expectation $\mathcal{E}p$ become $\Diamond\Box p$ and $\Box\Diamond p$. Thus we have an intensional counterpart of all modalities of **S4**; but we do not regard such correspondence with classical **S4** as a *definition* of the new "illocutionary forces" of conjecture and expectation. Indeed, we intend the pragmatic interpretations of intuitionistic and bi-intuitionistic logic as *bona fide* representations of such logics from the viewpoint of an intuitionistic philosopher; moreover, we intend our "logic for pragmatics" to be compatible with the rich proof-theory of intuitionistic logic, including the Curry–Howard correspondence and categorical interpretations. Thus, we are inclined to regard conjectures and expectations as examples of how a theory of *intuitionistic modalities* can be developed starting from the illocutionary forces of assertions and hypotheses as basic. However, this investigation is left for another occasion.

1.1 Logic for Pragmatics: Dalla Pozza and Garola's Approach

The aim of Dalla Pozza and Garola's "logic for pragmatics" is to capture the logical properties of what are called *illocutionary acts*—asserting, conjecturing, commanding, promising, and so on. Consider assertions. In their framework there is a logic of propositions and a logic of assertions. Propositions can be either true or false, according to classical semantics, assertions are acts that can be *justified* or *unjustified*, *felicitous* or *infelicitous*. They propose a two-layer theory with a distinctive informal interpretation, according to which propositions have *truth conditions*, i.e., a semantics, whereas assertions have *justification conditions*, belonging to pragmatics. As a consequence, we can form logical combinations of propositions, which are given a classical semantics as usual, but we can also form logical combinations of assertions, and interpret these combinations along the familiar lines of Heyting's interpretation of intuitionistic connectives. This is Dalla Pozza and Garola's *pragmatic interpretation of intuitionistic logic*: if α denotes a proposition, the *elementary expression* $\vdash\alpha$ stands for an assertion and $\vdash\alpha$ is justified just in case we have *conclusive evidence* that α is true; in the case of a mathematical statement α, "conclusive evidence" is a proof of α. Moreover, an assertive expression of conditional type $A \supset B$ is justified

by providing a method that transforms a justification of an assertive type A into a justification of an assertive type B.[3,4]
It should be noticed that intuitionistic logic is represented in Dalla Pozza and Garola's framework as a theory of pragmatic validity only if the justification of elementary expressions $\vdash \alpha$ does not depend on the logical structure of the *radical expression* α as a classical proposition—e.g., we shall not allow α to be $p \vee \neg p$. Thus in every investigation of intuitionistic theories within the framework of Dalla Pozza and Garola [18] it is assumed that *elementary expressions have atomic radicals,* i.e., $\alpha = p$. This convention is essential also for the present investigation of our co-intuitionistic and bi-intuitionistic logic.
The novelty of Dalla Pozza and Garola's work is that Heyting's semantics is applied to *illocutionary types* of acts, not to propositions; if the justification of an assertion of atomic type $\vdash \alpha$ is related to the semantics of the propositional content α, a complex type has only a *pragmatic* justification value, not a *semantic* one. To recover propositions and semantic values one considers *semantic projections* given by the Gödel, McKinsey, Tarski and Kripke's translation:

$$(\vdash \alpha)^M = \Box \alpha \qquad\qquad (A \supset B)^M = \Box(A^M \to B^M)$$

This modal formalism can be given the usual interpretation through an *epistemic* view of Kripke **S4** semantics. Thus in a Kripke model $(W, R, \Vdash)$ for **S4** every $w \in W$ is seen as a stage of human knowledge and the accessibility relation expresses ways in which our knowledge may evolve; at each stage atomic propositions are locally true or false according to $\Vdash$; reflexivity of R means that what we know must be true also locally and transitivity of R expresses the fact that human knowledge cannot be forgotten or falsified, and so on.[5]

---

[3] We wish to thank Prof Andrew Pitts and Dr Valeria de Paiva for their advice on various aspects of this research and Dr Piero Pagliani for his expert support in the Rough Sets semantics. We are much indebted to Tristan Crolard for his intriguing work on bi-intuitionistic and classical logic and to Hugo Herbelin for important suggestions about the distributed calculus for co-intuitionistic logic. Thanks to Carlo Dalla Pozza, Kurt Ranalter, Corrado Biasi, and Graham White for their cooperation in the "logic for pragmatics" enterprise and thanks to Ugo Solitro for useful discussions. I am grateful to Dag Prawitz, my first marvellous supervisor in Stockholm 1978 and to Luiz Carlos Pereira, a fellow student and a supportive colleague.

[4] Since April 2011 when revisions were made to this chapter, categorical models of bi-intuitionism have been studied based on monoidal categories. In particular, if the term assignment to a Gentzen system for co-intuitionistic logic is used in building a categorical model, then disjunction is best given *multiplicative* rules rather than *additive* ones, as it is done in this chapter. For the conceptual significance of such a choice, see G. Bellin, M. Carrara, D. Chiffi and A. Menti. A pragmatic dialogic interpretation of bi-intuitionism, submitted to *Logic and Logical Philosophy*, 2013.

[5] The interpretation of intuitionistic logic as a logic of *assertions* appears already in Dummett's work. Martin-Löf regards his intuitionistic theory of types as expressing *judgements* about the truth of propositions; in his system well-formed complex types are *propositions* and the terms inhabiting them are witnesses of their truth, intuitionistically understood. This view is disputed by Dalla Pozza: for him only *atomic types* assert the truth of propositions, but complex types neither *are* propositions nor *assert* propositions. To recover a proposition corresponding to the complex type

The basic approach of Dalla Pozza and Garola seems to stand as a helpful conceptual clarification, following Quine saying that *a change of logic reflects a change of the subject matter of the logic*. The remarkable technical developments of the proof-theory of classical logic in the last decades suggest the possibility of a pragmatic interpretation of classical methods of inference; despite some hints in [5], Sect. 5, and the result in Sect. 2.5, this remains an essentially unfinished business.

Justification and Felicity Conditions

Going back to the basic texts of modern pragmatics, such as Austin [2] and Levinson [31], every speech act has a *propositional content*, an *illocutionary force* (or *pragmatic* mood) and *perlocutionary effects*. Now it seems that the *felicity* or *infelicity* conditions of a speech act essentially depend on the actual circumstances of its performance and on its intended or unintended perlocutionary effects. Thus a formalization of the felicity or infelicity conditions of a statement would be based on a *formal theory of actions* including a representation of the *agent* and the *addressees* of a speech act and also its *preconditions* and *postconditions* (for a first formulation of such a theory, see [60]).

On the contrary, the contribution of the *illocutionary mood* to the pragmatics of speech acts can be characterized by *abstracting away* from the actual agents and addressees and from their specific context, effects, and goals. Thus an *impersonal* illocutionary operator of an intensional logic may suffice to express illocutionary force, if the justification of the illocutionary mood of such type of acts makes reference to a relatively stable and uniform context (e.g., scientific knowledge in a given time, obligations within an established legal system, unambiguous linguistic acts within a linguistic community, and so on).

In this framework, several works have explored the "logic for pragmatics" of obligations [19] and then the logic of assertions, obligations with causal reasoning [6, 7, 48, 49]. In general, the development of such logics requires an identification of the appropriate modal operators or non-classical connectives used in the modal projection and their Kripke semantics; then one proceeds to a more abstract treatment of the proof theory, as in Ranalter's work.

(§) $$\vdash\alpha \supset (\vdash\beta \supset \vdash\alpha)$$

we need a *semantic projection*, i.e., $\Box(\Box\alpha \rightarrow \Box(\Box\beta \rightarrow \Box\alpha))$; but justification of the assertion type

(§§) $$\vdash \Box(\Box\alpha \rightarrow \Box(\Box\beta \rightarrow \Box\alpha))$$

is a semantic argument for a sentence of classical **S4** while (§) is justified by something like a program $\lambda x.\lambda y.x$, where $x :\vdash \alpha$ and $y :\vdash \beta$ are variable ranging over proofs of the truth of α and of β.

2 PART I. Conceptual Analysis: Assertions, Hypotheses, and Conjectures

In extending Dalla Pozza and Garola's framework to a logic of hypothetical and conjectural moods, we encounter a variety of moods with different linguistic and logical properties.[6] It is familiar the distinction in Latin between three kinds of *if* clauses, the first one using the indicative to express the condition as a matter of fact, the second the present subjunctive to express possibility of the condition and, finally, the third one using the past subjunctive for counterfactuals. Also consider the *theory of argumentation*. Here six *proof-standards* have been identified from an analysis of legal practice: *no evidence*, *scintilla of evidence*, *preponderance of evidence*, *clear and convincing evidence*, *beyond reasonable doubt*, and *dialectical validity*, in a linear order of strength [24].[7]

It is essential to remember that "legal reasoning is not primarily deductive, but rather a modelling process of shaping an understanding of the facts, based on evidence, and an interpretation of the legal sources, to construct a theory for some legal conclusion" ([12] cited in [24]). More precisely, in order to decide whether to accept or reject each element of a given set of "claims," one constructs a consistent "theory of the generalizations of the domain and the facts of the particular case," together with "a *proof* justifying the decision of each issue, showing how the decision is supported by the theory" [24].

Thus in Argumentation Theory one starts with an inconsistent knowledge base and a set of claims and proceeds to build a consistent theory from them; later, when deriving the claims from such a theory one uses (some fragment of) classical logic. But in this stage it might be desirable to use a logic that retains essential pragmatic information such as the standards of evidence of the premises, rather than classical logic that omits it. Thus some refinement of our logic may have applications to Argumentation Theory to establish a closer correspondence between "theory searching" and deductive reasoning. Here we use the notion of "standards of proof" in an informal way and regard the possibility of developing a theory of positive evidence for hypotheses in our framework as a suggestion for future work.

---

[6] The conceptual development traced results from cooperation with other researchers, in particular with Corrado Biasi, whose doctoral dissertation at Queen Mary University of London is still unfinished.

[7] In the formal treatment of Carneades model of argumentation, *proof-standards* occur in the definition of what it means for an argument with conclusion c from premises P and exceptions E to be *applicable* in a *Carneades argument evaluation structure* $\mathcal{S} = \langle arguments, assumptions, weights, standard \rangle$. The definition relies on a non-logical real-valued function *weights* ranging over arguments. The notion of applicability is recursive, as it depends on the notion of a proposition p being *acceptable* in an argument evaluation structure $\mathcal{S}$. Here a proposition p is *acceptable* with a *scintilla of evidence* if there is at least one applicable argument for p and p is *acceptable* as *dialectically valid* if there is an applicable argument for p and no applicable argument against p. All other proof standards require comparing the weights of arguments for and against p. See [13, 24].

2.1 First Attempt: Assertive and Hypothetical Types

In Bellin and Biasi [5] we have given a logic of *hypothetical types* parallel to Dalla Pozza and Garola's logic of assertions. We start with *elementary illocutionary acts of hypothesis*, denoted by $\mathcal{H}\alpha$: here α is a proposition which is presented as possibly true; such an act is justified if there are grounds for believing that α may be true in some circumstances. Next we consider connectives building up complex hypothetical types from elementary ones. For instance, through the connective of *subtraction* we build the hypothetical expression *possibly C but not D* (written $C \smallsetminus D$); such an expression is justified if it is justified to believe the truth of the hypothetical expression C while also believing that the hypothesis D may never be true; the disjunction $C \curlyvee D$ of the hypothetical expressions C and D is also a hypothetical expression, and so on. The *modal projection* of hypothetical expressions is also in classical **S4**:

$$(\mathcal{H}\alpha)^M = \Diamond\alpha \qquad (C \smallsetminus D)^M = \Diamond(C^M \wedge \neg D^M)$$

Namely, the modal translations of assertions A^M and hypotheses C^M are both interpreted in models $(W, R, \Vdash)$ where R is transitive and reflexive. This choice is crucial for the approach of [5]: other modal candidates are possible as discussed in [5] and in more detail below.

In natural language, illocutionary acts of hypothesis may be embedded into a context consisting of illocutionary act of assertion, for instance,

> *Arturo is the best pianist of his generation and will not refuse to play in this town, although the audience may be slightly noisy;*

an assertive conjunction of two assertions and a hypothetical statement; conversely, assertions may be embedded in a hypothetical context:

> *We may not hear Arturo playing, because he has very high standards and if the audience is slightly noisy then he may refuse to play.*

containing a hypothetical implication with an assertive antecedent and hypothetical consequent. Taking this idea seriously, one obtains a rather unmanageable family of *mixed connectives* [5]; in this chapter we shall consider only the role of *mixed negations* turning assertive expressions into hypothetical ones and conversely.

Three Methodological Principles

Our logical treatment of *assertions and hypotheses* is based on the notion of a *duality* between these two illocutionary moods: informally it is a familiar idea, since a proof of a proposition may be obtained as a refutation of the conjecture that its dual is true. In a formal treatment, there are many aspects to this duality, which are certainly satisfied by the modal translation in **S4**. In [5], Sect 1.1, three methodological principles are stated for a logic expressing the duality between assertive and hypothetical types:

1. The grounds that justify *asserting* a proposition α certainly suffice also for *conjecturing* it, whatever these grounds may be;
2. in any situation, the grounds that justify the assertion of α are also necessary and sufficient to regard the conjecture that $\neg\alpha$ as unjustified;
3. the justification of non-elementary assertive or hypothetical types, built up from elementary types using pragmatic connectives, depends on the justification of the component types, possibly using intensional operations.

The third principle requires a sort of compositionality of justification: this is certainly satisfied by the intended informal interpretation of the connectives.
As it stands, the second principle is inadequate. On one hand, it is indisputable that the grounds allowing one to regard the assertion of α as justified must override any ground in favor of the conjecture of $\neg\alpha$; on the other hand, it is wrong and contrary to common sense to say that if the *conjecture* of $\neg\alpha$ is unjustified then the *assertion* of α is justified: the grounds we may have to dismiss the conjecture that $\neg\alpha$ may be the case may not be strong enough to justify the assertion that α is true. There are at least two issues here.
First, we must distinguish between the illocutionary force of a mere *hypothesis* and that of a *conjecture*, a distinction we shall develop later in this chapter. Let us split the second principle into two parts, replacing "hypothesis" for "conjecture":

2.i If the assertion of α is justified, then the hypothesis that $\neg\alpha$ is true cannot be justified.
2.ii If the hypothesis that $\neg\alpha$ is true is unjustified, then the assertion of α is justified.

Except for the case of counterfactuals, which are not our concern here, (2.*i*) is still correct; as for (2.*ii*), it becomes plausible if we assume that a hypothesis $\mathcal{H}\neg\alpha$ may be justified by a mere *cognitive possibility* of a situation, no matter how unlikely it may be, in which $\neg\alpha$ is true. The epistemic interpretation of the modal interpretation in **S4** validates this reading of (2.*ii*).
This raises a second issue: in our framework there is no theory of *positive evidence*; nevertheless we must be able to distinguish illocutionary forces whose justification depends on different strengths of evidence. Thus the logic of hypothetical reasoning in [5] reduces to a *refutation calculus*; although pure refutation does correspond to common-sense reasoning—indeed it seems to be very close to the medieval practice of disputation [1][8]—it may not suffice for applications, e.g., to a theory of laws and to legal reasoning.
Finally, the first principle is true for any reading of $\mathcal{H}\alpha$, e.g., as *hypothesis* or *conjecture*. Also it is true in argumentation theory: the assertion $\vdash\alpha$ must be justified by

[8] We are grateful to an anonymous referee to [5] for making the point clear and for indicating the reference. The same referee, acknowledging that our "refutation calculus" is dual to intuitionistic logic, questioned whether a calculus based on a theory of positive evidence could be co-intuitionistic: we come back to this issue below.

"standards of proof" at least as strong as those justifying the hypothesis $\mathcal{H}\alpha$. Notice that this principle shows a basic asymmetry between assertions and hypotheses.

A Logic of Assertions and Hypotheses: The Language $\mathcal{L}^{AH}$

The core fragment of the logic of assertions and hypotheses in [5] is a propositional language built from a countable set of atomic formulas $p, p_1, p_2, \ldots$ and symbols of illocutionary force yielding elementary formulas $\vdash p$ (*certainly* p) and $\mathcal{H}p$ (*perhaps* p). It consists of two dual parts:

- an assertive part $\mathcal{L}^A$ built from elementary assertions $\vdash p$, a sentential constant for *validity* ($\curlyvee$), using *assertive conjunction* ($\cap$) and *assertive implication* ($\supset$) and
- a hypothetical part $\mathcal{L}^H$ built from elementary hypotheses $\mathcal{H}p$ and a constant for *absurdity* ($\curlywedge$), using *hypothetical disjunction* ($\curlyvee$), and *hypothetical subtraction* ($\smallsetminus$).

Thus $\mathcal{L}^A$ and $\mathcal{L}^H$ are negation-free fragments of the language of intuitionistic and co-intuitionistic logic. Let **abs** be an absurd statement in $\mathcal{L}^A$ and **val** is a valid statement in $\mathcal{L}^H$. Then $\sim X =_{def} X \supset \mathbf{abs}$ expresses assertively the existence of a method to turn a justification of X into a justification of an absurdity. Similarly $\frown Y =_{def} \mathbf{val} \smallsetminus Y$ expresses the *doubt* that Y may be true, namely, the hypothesis that a valid statement **val** may be compatible with the negation of Y. Thus we have *four negations*:

1. if X is an assertive expression, then $\sim X$ is the usual *intuitionistic negation*;
2. if Y is a hypothetical expression, $\frown Y$ is *co-intuitionistic supplement*;
3. if X is a *hypothetical*, then the mixed expression $X \supset \mathbf{abs}$ is an assertive type;
4. if Y is assertive, then $\mathbf{val} \smallsetminus Y$ is a hypothetical type.[9]

Our logic is therefore *bi-intuitionistic*, in the sense that it has intuitionistic and co-intuitionistic connectives, but it is *polarized*, as elementary formulas are either intuitionistic ($\vdash p$) or co-intuitionistic ($\mathcal{H}p$), *but not both*, and connectives, with the possble exception of negations, preserve the polarity. Thus we have the following grammar of the language of *polarized bi-intuitionistic* logic for the pragmatics of assertions and hypotheses $\mathcal{L}^{AH}$:

$$\begin{array}{lcl|c|c|c|c} A, B & := & \vdash p & \curlyvee & A \supset B & A \cap B & \sim C \\ C, D & := & \mathcal{H}p & \curlywedge & C \smallsetminus D & C \curlyvee D & \frown A \end{array}$$

[9] As in [5], to these pragmatic negations one should add classical negation in the radical part $\neg\alpha$; but no logical property of the radical part can be used in the treatment of intuitionistic pragmatics. To avoid confusions with the "polarized classical logic" in [5], Sect. 5, in the treatment of dualities we shall assume that the atoms occurring in the radical part are either positive p_i^+ or negative p_i^-, i.e., that there is an involution without fixed point on atoms exchanging p_i^+ and p_i^-.

2.2 Second Attempt: More General Modal Translations

In order to approximate alternative treatments of a logic of *assertions*, *hypotheses* and *conjectures*, we consider more general modal translations in *bimodal* **S4**.

Translations in Bimodal S4

Definition 1 *(i) Let p range over a denumerable set of propositional variables* **Var** = $\{p_1, p_1, \ldots\}$. *The bimodal language* $\mathcal{L}_{\Box,\boxminus}$ *is defined by the following grammar.*

$$\alpha := p \mid \neg\alpha \mid \alpha \wedge \alpha \mid \alpha \vee \alpha \mid \alpha \to \alpha \mid \Box\alpha \mid \boxminus\alpha$$

Define $\Diamond\alpha =_{df} \neg\Box\neg\alpha$ *and* $\ominus\alpha =_{df} \neg\boxminus\neg\alpha$.
(ii) Let $\mathcal{F} = (W, R, S)$ *be a multimodal frame, where W is a set, R and S are preorders on W. Given a valuation function* $V : \mathbf{Var} \to \wp(W)$*, the forcing relations are defined as follows:*

- $w \Vdash \Box\alpha$ *iff* $\forall w'.wRw' \Rightarrow w' \Vdash \alpha$,
- $w \Vdash \boxminus\alpha$ *iff* $\forall w'.wSw' \Rightarrow w' \Vdash \alpha$.

(iii) We say that a formula A in the language $\mathcal{L}_{\Box,\boxminus}$ is valid in bimodal **S4** *if A is valid in all bimodal frames* $\mathcal{F} = (W, R, S)$ *where R and S are preorders.*

Lemma 1 *Let* $\mathcal{F} = (W, R, S)$ *be a multimodal frame, where R and S are preorders.*
(i) The following are valid in $\mathcal{F}$

$$\Box\boxminus\Box\alpha \to \Box\alpha \quad \textit{and} \quad \boxminus\Box\boxminus\alpha \to \boxminus\alpha$$

(ii)(a) The following are equivalent:

1.a: $S \subseteq R$;
2.a: the following scheme is valid in $\mathcal{F}$:
. *(Ax.a)* $\Box\alpha \to \Box\boxminus\Box\alpha$;
3.a: the following rule is valid in $\mathcal{F}$:
. *(R.a)*

$$\frac{\ominus\beta \Rightarrow \ominus\neg\Box\alpha}{\Box\alpha \Rightarrow \Box\neg\ominus\beta}$$

(ii)(b) The following are equivalent

1.b: $R \subseteq S$;
2.b: the following scheme is valid in $\mathcal{F}$:
. *(Ax.b)* $\boxminus\alpha \to \boxminus\Box\boxminus\alpha$
3.b: the following rule is valid in $\mathcal{F}$:
. *(R.b)*

$$\frac{\Box\neg\ominus\beta \Rightarrow \Box\alpha}{\ominus\neg\Box\alpha \Rightarrow \ominus\beta}$$

Proof of $(ii)(a)$. (1.a $\Rightarrow$ 2.a) is obvious. (2.a $\Rightarrow$ 1.a): If S is not a subset of R, then given wSv and not wRv define a model on $\mathcal{F}$ where $w' \Vdash p$ for all w' such that wRw' but $v \nVdash p$; thus $\Box p \rightarrow \Box \boxminus \Box p$is false at w. (2.a $\Rightarrow$ 3.a): If $\diamondminus \beta \Rightarrow \diamondminus \neg \Box \alpha$ is valid in $\mathcal{F}$ then so is $\Box \neg \diamondminus \neg \Box \alpha \Rightarrow \Box \neg \diamondminus \beta$ and the conclusion of $(R.a)$ is valid because of $(Ax.a)$. From the conclusion of $(R.a)$ the premise follows using part (i). (3.a $\Rightarrow$ 2.a): $(Ax.a)$ is obtained by applying $(R.a)$ downwards to $\diamondminus \neg \Box \alpha \Rightarrow \diamondminus \neg \Box \alpha$. The other parts are similar.

2.3 Bimodal Interpretations of $\mathcal{L}^{AH}$

Definition 2 *The interpretation* $(\ )^M$ *of* $\mathcal{L}^{AH}$ *into* $\boxminus$ *is defined inductively thus:*

$$
\begin{array}{rll@{\qquad}rll}
(\curlywedge)^M & =_{df} & \bot & (\curlyvee)^M & =_{df} & \top \\
(\vdash p)^M & =_{df} & \Box p & (\mathcal{H} p)^M & =_{df} & \diamondminus p \\
(A \supset B)^M & =_{df} & \Box(A^M \rightarrow B^M) & (C \smallsetminus D)^M & =_{df} & \diamondminus (C^M \wedge \neg D^M) \\
(A_1 \cap A_2)^M & =_{df} & A_1^M \wedge A_2^M & (C_1 \curlyvee C_2)^M & =_{df} & C_1^M \vee C_2^M \\
(\sim C)^M & =_{df} & \Box \neg C^M & (\frown A)^M & =_{df} & \diamondminus \neg A^M
\end{array}
$$

It is easy to prove that $A^M \iff \Box A^M$ *and* $C^M \iff \diamondminus C^M$ *in the semantics of (bimodal)* **S4**.

(i) The propositional theory **PBL** *(polarized bi-intuitionistic logic) is the set of all formulas* δ *in the language* $\mathcal{L}^{AH}$ *such that* δ^M *is valid in every preordered bimodal frame (i.e, in any frame* (W, R, S) *where* R *and* S *are arbitrary preorders).*

(ii) The propositional theory **APBL** *(asymmetric polarized bi-intuitionistic logic) is the set of all formulas* δ *in the language* $\mathcal{L}^{AH}$ *such that* δ^M *is valid in every preordered bi-modal frame* (W, R, S) *where* $S \subseteq R$.

(iii) The propositional theory **AHL** *(bi-intuitionistic logic of assertions and hypotheses) is the set of all formulas* δ *in the language* $\mathcal{L}^{AH}$ *such that* δ^M *is valid in every preordered bi-modal frame* (W, R, S) *where* $R = S$. *In other words, in the modal translation let* $\diamondminus X =_{df} \neg \Box \neg X$; *then* δ *is in* **AHL** *if and only if* δ^M *is valid in* **S4**.

Remark 1 (i) **PBL** is the most *abstract* theory of *bi-intuitionistic* logic where all formulas are *polarized* as assertive or hypothetical. **PBL** is not a suitable candidate for our logic of assertions and hypotheses, since the pair $(\vdash p)^M, (\frown \vdash p)^M$ is consistent in bi-modal **S4**, contrary to the accepted principle $(2.i)$. We will not speculate about the possibility of interpreting $\frown \vdash p$ as a counterfactual.

(ii) On the contrary, the *asymmetric* logic **APBL** satisfies $(2.i)$, but not $(2.ii)$.[10] Thus **APBL** may be the right context for studying assertive and hypothetical reasoning

---

[10] Condition $S \subseteq R$ guarantees that if $w \Vdash \Box p$ then $w \nVdash \diamondminus \neg p$. To see that $w \Vdash \Box p$ is not a valid consequence of $w \nVdash \diamondminus \neg p$, consider a model $\mathcal{M} = (W, R, S, \Vdash)$ with $W = \{w, w'\}$, R and

where hypothetical statements have different degrees of positive evidence and thus are not representable in a pure refutation calculus.

(iii) Finally, our *canonical* system is the *bi-intuitionistic logic of assertions and hypotheses* **AHL**—poorly named *intuitionistic logic for pragmatics* **ILP** in [5]—satisfying both conditions (2.*i*) and (2.*ii*). It is motivated by the epistemic interpretation of (uni-modal) **S4**, where hypotheses are seen as mere epistemic possibilities and assertions as epistemic necessities.

Dualities

Definition 3 *Let* $(\,)^{\perp}$: **Atoms** $\to$ **Atoms** *be an involution without fixed points on the atomic formulas* p_i. Intuitively, we may think of $p_i^{\perp}$ as $\neg p_i$, but intuitionistic dualites are defined best without any reference to the classical part. *We extend* $(\,)^{\perp}$ *to maps* $F : \mathcal{L}^A \to \mathcal{L}^H$ *and* $G : \mathcal{L}^H \to \mathcal{L}^A$ *letting*

$$
\begin{array}{lll}
(a) & F({}^{\vdash}p) = \mathcal{H}p^{\perp} & G(\mathcal{H}p) = {}^{\vdash}p^{\perp} \\
(b) & F(A \cap B) = F(A) \curlyvee F(B) & G(C \curlyvee D) = G(C) \cap G(D) \\
(c) & F(A \supset B) = F(B) \smallsetminus F(A) & G(C \smallsetminus D) = G(D) \supset G(C)
\end{array}
$$

Lemma 2 *In* **AHL** *let* $F(A) = \frown A$ *and* $G(C) = \sim C$. *Then*

1. *if we interpret* $(p^{\perp})^M$ *as* $\neg p$, *then the modal translations of conditions (a)–(c) are valid equivalences in* **S4**;
2. $GF(A) \equiv A$ *and* $FG(C) \equiv C$;
3. $\dfrac{A \Rightarrow G(C)}{C \Rightarrow F(A)}$ *and* $\dfrac{G(C) \Rightarrow A}{F(A) \Rightarrow C}$ (double lines)

Proof. By definition of the modal translation we have

$$
\begin{array}{rl}
(1)(a): & (\frown {}^{\vdash}p)^M = \Diamond\neg\Box p \equiv \Diamond\neg p = (\mathcal{H}p^{\perp})^M \\
& (\sim \mathcal{H}p)^M = \Box\neg\Diamond p \equiv \Box\neg p = ({}^{\vdash}p^{\perp})^M
\end{array}
$$

$$
\begin{array}{rl}
(1)(b): & (\frown (A \cap B))^M = \Diamond\neg(A^M \wedge B^M) \equiv (\Diamond\neg A^M) \vee (\Diamond\neg B^M) = ((\frown A) \curlyvee (\frown B))^M \\
& (\sim (C \curlyvee D))^M = \Box\neg(C^M \vee D^M) \equiv (\Box\neg C^M) \wedge (\Box\neg D^M) = ((\sim C) \cap (\sim D))^M
\end{array}
$$

$$
\begin{array}{rl}
(1)(c): & (\frown (A \supset B))^M = \Diamond\neg\Box(A^M \to B^M) \equiv \Diamond(\Diamond\neg B^M \wedge \neg\Diamond\neg A^M) = ((\frown B) \smallsetminus (\frown A))^M \\
& (\sim (C \smallsetminus D))^M = \Box\neg\Diamond(C^M \wedge \neg D^M) \equiv \Box(\Box\neg D^M \to \Box\neg C^M) = ((\sim D) \supset (\sim C))^M
\end{array}
$$

(2): The conditions

S reflexive and transitive and such that wRw' but not wSw', and $w \Vdash p$ but $w' \Vdash \neg p$. Notice that $\mathcal{H}\neg p$ is not an expression of the language $\mathcal{L}^{AH}$, but the same remark applies to $({}^{\vdash}p)^M = \Box p$ and $(\frown {}^{\vdash}p)^M = \Diamond \neg\Box p$.

$$\sim\frown A \equiv A \qquad \text{and} \qquad C \equiv \frown\sim C \tag{1}$$

follow from Lemma 1 .(i) and (ii). The conditions in (3) follow from rules (R.a) and (R.b) in Lemma 1 .(ii).

Remark 2 (i) Lemma 2 fails for **PBL** and **APBL**.
(ii) As the only mixed formulas in $\mathcal{L}^{AH}$ are negations, Lemma 2 gives us a (meta-theoretic) "method for eliminating mixed formulas in **AHL**" *modulo the atomic involution* $(\,)^{\perp}$, (interpreted in the modal translation as classical negation $\neg p$). e.g., the mixed expression $A \cap \sim (\mathcal{H}p \curlyvee \mathcal{H}q)$ is equivalent in the **S4** semantics to the purely assertive expression $A \cap ({\vdash}p^{\perp} \cap {\vdash}q^{\perp})$.
(iii) Sometimes we shall write $A^{\perp}$ and $C^{\perp}$ for $F(A)$ and $G(C)$, respectively.

Proposition 1 (restricted substitution) *Let σ be a map*

$${\vdash}p_i \mapsto A_i \qquad \mathcal{H}p_j \mapsto C_j$$

sending a vector $\overline{\eta_a}$ of assertive elementary formulas *to a vector $\overline{A}$ of assertive formulas and a vector $\overline{\eta_h}$ of* hypothetical elementary formulas *to a vector $\overline{C}$ of hypothetical formulas. Then $X(\overline{\eta_a}, \overline{\eta_h})$ is a theorem of* **AHL** [**PBL**, **APBL**] *if and only if $X(\sigma(\overline{\eta_a}), \sigma(\overline{\eta_h}))$ is a theorem of* **AHL** [**PBL**, **APBL**].

On the other hand, the theories **AHL**, **PBL** and **APBL** are not closed under substitution of hypothetical formulas for assertive elementary formulas (and symmetrically). An example is the following:

$\sim\sim\sim{\vdash}p \Rightarrow \sim{\vdash}p$ is valid, but $\sim\sim\sim\mathcal{H}p \Rightarrow \sim\mathcal{H}p$ is not.
Indeed $\Box\Diamond\Box\Diamond\neg p \Rightarrow \Box\Diamond\neg p$ is valid, but $\Box\Diamond\Box\neg p \Rightarrow \Box\neg p$ is invalid in **S4**.

2.4 Sequent Calculi for PBL, APBL, AHL

The logics **PBL**, **APBL**, **AHL** can be formalized in **G3**-style sequent calculi [59], where the rules of Weakening and Contraction are implicit, as in [5]. One then proves that the rules of Weakening and Contraction are admissible preserving the depth of the derivation.

Definition 4 *All the sequents S are of the form*

$$\Theta\,;\,\epsilon \Rightarrow \epsilon'\,;\,\Upsilon \tag{2}$$

where

- *Θ is a sequence of assertive formulas $A_1, \ldots, A_m$;*

- *Υ is a sequence of hypothetical formulas $C_1, \ldots, C_n$;*
- *ϵ is hypothetical and ϵ' is assertive and exactly one of ϵ, ϵ' occurs.*

The bi-intuitionistic logic of assertions and conjectures **AHL** *is formalized in the sequent calculus given by the axioms and rules in the Table 1.*[11] *Let us call this fragment standard* **AH-G3**.

The polarized bi-intuitionistic logic **PBL** *and the asymmetric polarized bi-intuitionistic logic* **APBL** *are formalized by restricting the rules of canonical* **AH-G3** *as indicated below: the restrictions only modify the rules* $\supset$-right, $\sim$-right, $\smallsetminus$-left *and* $\frown$-left. *Let us call the resulting sequent calculi abstract* **PB-G3** *and asymmetric* **APB-G3**, *respectively.*

To see why in the *asymmetric* **APB-G3** and in the *canonical* **AH-G3** systems the formulas in Θ are allowed in the antecedent of the sequent-premise of $\smallsetminus$*-left* and of $\frown$*-left*, notice that by the valid scheme $(Ax.a)$ of Lemma 1 .$(ii)(a)$

$$A \Rightarrow \sim\frown A \quad \text{is valid in the semantics of } \mathbf{APB} \text{ and of } \mathbf{AHL} \tag{3}$$

Thus the unrestricted rule $\frown$*-left* of **AP-G3** and **AH-G3** becomes derivable in **PB-G3** using *cut* with scheme (1) taken as axiom:

$$\text{cut}\;\dfrac{\overset{(1)}{B ; \Rightarrow \sim\frown B} \qquad \sim\text{-L}\,\dfrac{\frown\text{-L}\,\dfrac{\frown\text{-R}\,\dfrac{B ; \Rightarrow A ; \Upsilon}{; \Rightarrow A ; \Upsilon, \frown B}}{; \frown A \Rightarrow ; \Upsilon, \frown B}\,(\P)}{\sim\frown B ; \frown A \Rightarrow ; \Upsilon}}{B ; \frown A \Rightarrow ; \Upsilon}$$

Similarly, using the fact that

$$\frown\sim C \Rightarrow C \quad \text{is valid in the semantics of } \mathbf{AHL} \tag{4}$$

we show that in **AH-G3** Υ is allowed in the succedent of the sequent premise of $\supset$*-right* and of $\sim$*-right.*

Using the methods of [5] one may prove the following result:

Theorem 1 *The sequent calculi* **PB-G3** [**APB-G3**, **AH-G3**] *without the rules of cut are sound and complete with respect to the interpretation of* **PBL** [**APBL**, **AHL**, *respectively*] *in* bimodal **S4**.

---

[11] This calculus is essentially the system *Intuitionistic Logic for Pragmatics* **ILP** presented and studied in [5], Sect. 3, restricted to the language $\mathcal{L}^{AH}$ - namely a sequent calculus with axioms and rules for *assertive validity, implication and conjunction, hypothetical absurdity, subtraction and disjunction* and two *mixed negations*.

Table 1 The sequent calculus **AH-G3**

IDENTITY RULES

logical axiom:
$$A, \Theta \,;\, \Rightarrow A \,;\, \Upsilon$$

logical axiom:
$$\Theta \,;\, C \Rightarrow \,;\, \Upsilon, C$$

cut$_1$:
$$\frac{\Theta \,;\, \Rightarrow A \,;\, \Upsilon \qquad A, \Theta' \,;\, \epsilon \Rightarrow \epsilon' \,;\, \Upsilon'}{\Theta, \Theta' \,;\, \epsilon \Rightarrow \epsilon' \,;\, \Upsilon, \Upsilon'}$$

cut$_2$:
$$\frac{\Theta \,;\, \epsilon \Rightarrow \epsilon' \,;\, \Upsilon, C \qquad \Theta' \,;\, C \Rightarrow \Upsilon'}{\Theta, \Theta' \,;\, \epsilon \Rightarrow \epsilon' \,;\, \Upsilon, \Upsilon'}$$

ASSERTIVE LOGICAL RULES

validity axiom:
$$\Theta \,;\, \Rightarrow \curlyvee \,;\, \Upsilon$$

right $\supset$:
$$\frac{\Theta, A_1 \,;\, \Rightarrow A_2 \,;\, \Upsilon}{\Theta \,;\, \Rightarrow A_1 \supset A_2 \,;\, \Upsilon}$$

left $\supset$:
$$\frac{A_1 \supset A_2, \Theta ; \Rightarrow A_1 \,;\, \Upsilon \qquad A_2, \Theta \,;\, \epsilon \Rightarrow \epsilon' \,;\, \Upsilon}{A_1 \supset A_2, \Theta \,;\, \epsilon \Rightarrow \epsilon' \,;\, \Upsilon}$$

right $\cap$:
$$\frac{\Theta \,;\, \Rightarrow A_1 \,;\, \Upsilon \qquad \Theta \,;\, \Rightarrow A_2 \,;\, \Upsilon}{\Theta \,;\, \Rightarrow A_1 \cap A_2 \,;\, \Upsilon}$$

left $\cap$:
$$\frac{A_0, A_1, \Theta \,;\, \epsilon \Rightarrow \epsilon' \,;\, \Upsilon}{A_0 \cap A_1, \Theta \,;\, \epsilon \Rightarrow \epsilon' \,;\, \Upsilon}$$

CONJECTURAL RULES

absurdity axiom:
$$;\, \curlywedge \Rightarrow \,;\, \Upsilon$$

right $\smallsetminus$:
$$\frac{\Theta \,;\, \epsilon \Rightarrow \epsilon' \,;\, \Upsilon, C_1 \qquad \Theta \,;\, C_2 \Rightarrow \,;\, \Upsilon, C_1 \smallsetminus C_2}{\Theta \,;\, \epsilon \Rightarrow \epsilon' \,;\, \Upsilon, C_1 \smallsetminus C_2}$$

left $\smallsetminus$:
$$\frac{\Theta ;\, C_1 \Rightarrow \,;\, \Upsilon, C_2}{\Theta ;\, C_1 \smallsetminus C_2 \Rightarrow \,;\, \Upsilon}$$

right $\curlyvee$:
$$\frac{\Theta \,;\, \epsilon \Rightarrow \epsilon' \,;\, \upsilon, C_0, C_1}{\Theta \,;\, \epsilon \Rightarrow \epsilon' \,;\, \Upsilon, C_0 \curlyvee C_1}$$

left $\curlyvee$:
$$\frac{\Theta \,;\, C_1 \Rightarrow \,;\, \Upsilon \qquad \Theta \,;\, C_2 \Rightarrow \,;\, \Upsilon}{\Theta \,;\, C_1 \curlyvee C_2 \Rightarrow \,;\, \Upsilon}$$

MIXED-TYPE NEGATIONS:

right $\sim$:
$$\frac{\Theta \,;\, C \Rightarrow \,;\, \Upsilon}{\Theta \,;\, \Rightarrow \sim C \,;\, \Upsilon}$$

left $\sim$:
$$\frac{\sim C, \Theta ;\, \epsilon \Rightarrow \epsilon' \,;\, \Upsilon, C}{\sim C, \Theta \,;\, \epsilon \Rightarrow \epsilon' \,;\, \Upsilon}$$

right $\frown$:
$$\frac{\Theta, A \,;\, \epsilon \Rightarrow \epsilon' \,;\, \Upsilon, \frown A}{\Theta \,;\, \epsilon \Rightarrow \epsilon' \,;\, \Upsilon, \frown A}$$

left $\frown$:
$$\frac{\Theta ;\, \Rightarrow A \,;\, \Upsilon}{\Theta \,;\, \frown A \Rightarrow \,;\, \Upsilon}$$

AH-G3:

$$\supset\text{-R}\ \frac{\Theta, A_1\ ;\ \Rightarrow\ A_2\ ;\ \Upsilon^*}{\Theta\ ;\ \Rightarrow\ A_1 \supset A_2\ ;\ \Upsilon} \qquad \smallsetminus\text{-L}\ \frac{\Theta^{**}\ ;\ C_1\ \Rightarrow\ ;\ C_2, \Upsilon}{\Theta\ ;\ C_1 \smallsetminus C_2 \Rightarrow\ ;\ \Upsilon}$$

$$\sim\text{-R}\ \frac{\Theta\ ;\ C\ \Rightarrow\ ;\ \Upsilon^*}{\Theta\ ;\ \Rightarrow \sim C\ ;\ \Upsilon} \qquad \frown\text{-L}\ \frac{\Theta^{**}\ ;\ \Rightarrow\ A\ ;\ \Upsilon}{\Theta\ ;\ \frown A\ \Rightarrow\ ;\ \Upsilon}$$

Υ^* not allowed in **PB-G3, APB-G3**　　　Θ^{**} not allowed in **PB-G3**

2.5 *First Conclusions: Assertions and Conjectures*

Although our approach to the logic for pragmatics does not provide a theory of positive evidence, the epistemic reading of the modal interpretation in **S4** does suggest a way to characterize different degrees of evidence, through the essential distinction between *hypotheses* and *conjectures*. While "epistemic possibility," namely the mere knowledge of a situation in which α happens to be true, does provide enough evidence to justify the hypothesis of α, conjecturing the truth of α requires knowing conditions in which α would be "epistemically necessary." We write $\mathcal{C}\alpha$ to express the *conjecture* that α is true.

Moreover, consider circumstances in which it is unjustified to conjecture the truth of α. This is certainly the case when no matter how our present knowledge evolves, it always reaches a state in which α fails to be true: we may call this epistemic condition *safe expectation* that $\neg\alpha$ eventually becomes true. We write $\mathcal{E}\alpha$ to express the *safe expectation* of α.

Setting $(\mathcal{C}\alpha)^M = \Diamond\Box\alpha$ and $(\mathcal{E}\alpha)^M = \Box\Diamond\alpha$, we have a modal interpretation in **S4** that fits nicely in the above informal interpretation. In Table 2 we find the map of all distinct modalities in **S4**; arrows indicate valid implications between non-equivalent modalities.

Table 2 The modalities of **S4**

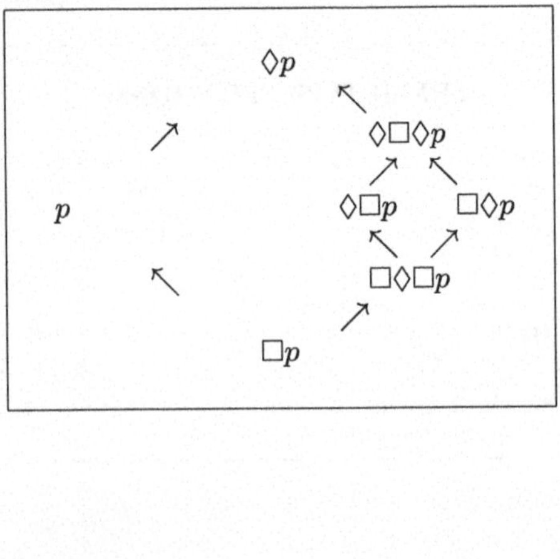

Table 3 presents all pragmatic expressions corresponding to modalities of **S4** and the valid implications between them.

We shall not develop a full theory of assertions, hypotheses, conjectures and expectation with four corresponding types of pragmatic connectives. We are interested in theories obtained by extending the polarized language $\mathcal{L}^{AH}$ of *assertions and hypotheses* with new elementary expressions $\mathcal{C}p$ for *conjectures* and dually, expressions $\mathcal{E}p$ for expectations. Let us write $\mathcal{L}^{AHC}$ [$\mathcal{L}^{AHCE}$] for the extension of $\mathcal{L}^{AH}$ with elementary expressions $\mathcal{C}p$ for conjectures [and $\mathcal{E}p$ for expectations].

Let **AHCE** be the set of all expressions in $\mathcal{L}^{AHCE}$ that are valid in the **S4** modal translation. We conjecture in order to axiomatize **AHCE** in a cut-free sequent calculus it suffices to extend **AH-G3** with the following rules:

Duality Between Safe Expectations and Conjectures

Clearly the **S4** translations of conjectures and of assertions are not dual from the viewpoint of modal logic, but the modal translations of *conjectures* and *safe expectations* certainly are; if in Definition 3 and in Lemma 2 we replace the illocutionary operators "$\mathcal{C}$" (conjectures) and "$\mathcal{E}$" (safe expectations) for the operators "$\mathcal{H}$" (hypotheses) and "$\vdash$" (assertions), respectively, then clearly the conditions of duality are expressible through negations within a logic **AHCEL** of *assertions, hypotheses, conjectures and safe expectations*. For instance, the base case becomes:

The Logic of Safe Expectations is Classical

Let $\mathcal{L}^{E}$ be the language defined by the grammar

$$E, F \quad := \quad \mathcal{E}p \mid \curlyvee \mid E \supset F \mid E \cap F$$

and let bf EL be the set of all formulas δ in the language $\mathcal{L}^{E}$ such that the modal translation δ^{M} is valid in **S4**.

Table 3 Assertions, conjectures, expectations and hypotheses

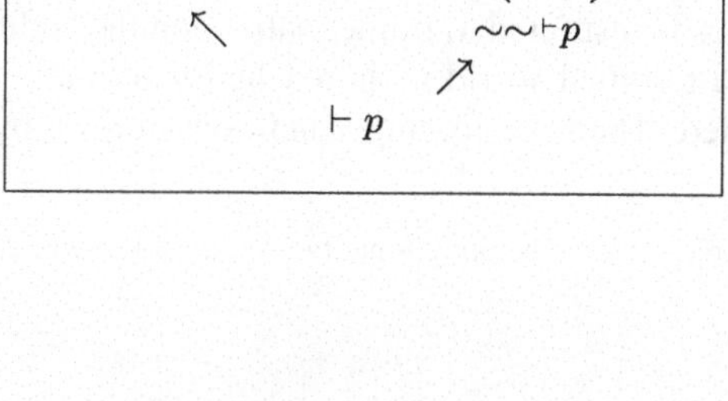

$$\mathcal{C}\text{-R}\ \frac{\Theta\ ;\ \Rightarrow {}^{\vdash}p\ ;\ \Upsilon}{\Theta\ ;\ \Rightarrow\ ;\ \mathcal{C}p,\ \Upsilon} \qquad \mathcal{C}\text{-L}\ \frac{\Theta,{}^{\vdash}p\ ;\ \Rightarrow\ ;\ \Upsilon}{\Theta\ ;\ \mathcal{C}p\ \Rightarrow\ ;\ \Upsilon}$$

$$\mathcal{E}\text{-R}\ \frac{\Theta\ ;\ \Rightarrow\ ;\ \mathcal{H}p,\ \Upsilon}{\Theta\ ;\ \Rightarrow\ \mathcal{E}p\ ;\ \Upsilon} \qquad \mathcal{E}\text{-L}\ \frac{\Theta\ ;\ \mathcal{H}p\ \Rightarrow\ ;\ \Upsilon}{\Theta,\mathcal{E}p\ ;\ \Rightarrow\ ;\ \Upsilon}$$

(a) setting		$F(\mathcal{E}p) = \mathcal{C}p^{\perp}$	and	$G(\mathcal{C}p) = \mathcal{E}p^{\perp}$
	we have	$\frown \mathcal{E}p \equiv \mathcal{C}p^{\perp}$	and	$\sim \mathcal{C}p \equiv \mathcal{E}p^{\perp}$
	since	$\Diamond\neg\Box\Diamond p \equiv \Diamond\Box\neg p$	and	$\Box\neg\Diamond\Box p \equiv \Box\Diamond\neg p.$

Proposition 2 *The theory* **EL** (logic of safe expectations) *is closed under the double negation rule, i.e.,* $\sim\sim E \Rightarrow E$ *is a valid axiom of* **EL**.

The proof shows by induction on the logical complexity that the double negation rule for molecular formulas can be reduced to applications of the double negation rule for elementary formulas (essentially, as in [46]). The base case is then given by the following equivalence:

$$(\sim\sim \mathcal{E}p)^M \quad = \quad \Box\Diamond\Box\Diamond p \quad \equiv \quad \Box\Diamond p \quad = \quad (\mathcal{E}p)^M.$$

On the other hand, if we extend $\mathcal{L}^E$ with intuitionistic disjunction ($\cup$), then $E\cup \sim E$ is not a theorem of the logic of safe expectations extended in this way. Indeed

$$(\mathcal{E}p\cup \sim \mathcal{E}p)^M = \Box\Diamond p \vee \Box\Diamond\Box\neg p$$

is not valid in **S4**.

Historical Note

In Appendix B of [46] Prawitz considers an extension of the language of intuitionistic logic with an involutory negation $\neg$ and then extends intuitionistic natural deduction $\mathbf{NJ}^{\supset\cap}$ with rules $\neg\supset$-I, $\neg\supset$-E, $\neg\cap$-I and $\neg\cap$-E; these new rules are presented as an axiomatization of Nelson's logic of *constructible falsity* [38].[12] Thomason [58] provides a Kripke semantics for Nelson's logic of constructible falsity, where $w \Vdash \neg p$ if and only if $w' \Vdash \neg p$ for all w' with wRw'; this implies that the evaluation function must be *partial*. Miglioli et al. [37] introduce an operator **T** which represents classical truth within the context of Nelson's logic of constructive negation: in particular we have A is classically valid if and only if $\sim\sim A$ is intuitionistically valid (by Gödel's translation) if and only if **T**A is valid in the constructive extended system. In [37] a Kripke semantics for the constructive logic with **T** operator is presented, where Thomason's semantics is restricted to frames satisfying the additional condition that from each world w a terminal world w' is reached where all atoms and negations of atoms are evaluated. Then the forcing conditions for **T**p by Miglioli et al. are

[12] The negation "$\neg$" corresponds to the orthogonality $(\ )^{\perp}$, as in Remark 2.

expressible as $w \Vdash \mathbf{T}p$ if and only if $w \Vdash \Box\Diamond p$ and $w \Vdash \neg\mathbf{T}p$ if and only if in all w' with wRw' we have that p is either not evaluated or false in w'. Comparing the operator $\mathbf{T}$ to our operator $\mathcal{E}$ of *safe expectation*, when applied to atomic formulas, we can say that their properties are similar, but in the context of a *polarized* bi-intuitionistic system they can be expressed in a simpler way. We cannot discuss the intriguing work by Miglioli and his co-workers in more detail here; a recent discussion of their approach is in Pagliani's book [39].

3 PART II. Rough-Set Semantics

Proofs and Refutations

The idea that a characterization of constructive logic must include a definition not only of what proofs of a formula A are but also of *refutations* of A goes back at least to Nelson [38] and comes up in various contexts related to game semantics and in particular the construction of Chu spaces. Thus we may say that *a proof* of $A \supset B$ is a method transforming a proof of A into a proof of B and that a *refutation* of $A \supset B$ is a pair consisting of a proof of A together with a refutation of B; in some contexts instead of proofs and refutations we may speak of *evidence for* and *against* A. To study bi-intuitionistic logic and its dualities one may say that a proof of $C \setminus D$ is a pair consisting of a proof of C and of a refutation of D and that a refutation of $C \setminus D$ is a method transforming a proof of C into a proof of D. But we will not go very far if the spaces of proofs and of refutations of A coincide with the spaces of refutations and of proofs of $A^{\perp}$, respectively. This is certainly not the case if we consider the semantics of *assertions, hypotheses, and conjectures* rather than that of assertions and hypotheses, as discussed informally in Sect. 2.1. Moreover, it turns out that *Rough-Set semantics* applied to our canonical polarized system **AHCB** does provide new insight and also a bridge to geometric models [57].

3.1 Rough Sets

As pointed out in [5], any topological space provides a mathematical model of bi-intuitionistic logic, thus also of our canonical system **AHL**, if we interpret the assertive expressions by *open sets* and the hypothetical ones by *closed sets*. A more interesting suggestion comes from the interpretation in terms of Rough Sets, following Piero Pagliani's work (in particular, see [40, 41] and Lech Polkowski's book [45], Chap. 12).

Definition 5 *Given an indiscernibility space (U, E), where U is a finite set and $E \subseteq U \times U$ an equivalence relation, identifying objects that may be indiscernible* *from some point of view, let* $\mathbf{AS}(U)$ *be the atomic Boolean algebras having the set of equivalence classes* U/E *as atoms; then* $(U, \mathbf{AS}(U))$ *is a topological space, called the Approximation Space of* (U, E)*, which induces an interior operator and a closure operator* $\mathcal{I}, \mathcal{C} : \wp(U) \to \mathbf{AS}(U)$. *If two subsets* $G', G'' \subseteq U$ *have the same interior and the same closure, then they are roughly equal, i.e., indistinguishable either by the coarsest classification given by* $\mathcal{C}$*, or by the finest classification* $\mathcal{I}$*; thus each subset* G *is a representative of a class of subsets identified by the pair* $(\mathcal{I}(G), \mathcal{C}(G))$*; only a clopen* G *for which* $\mathcal{I}(G) = \mathcal{C}(G)$ *is fully characterized in* (U, E).

For our purpose it is more convenient the *disjoint representation* $(\mathcal{I}(G), -\mathcal{C}(G))$ using the *complement of the closure* of G, the set of object different from G even for the coarser classification, instead of $\mathcal{C}(G)$. Thus we may regard the two clopen sets ($\mathcal{I}(G)$ and $-\mathcal{C}(G)$) as representing the space of proofs and of refutations of an intuitionisitc formula.

Following Pagliani, we can define the following data and operations on pairs

1. $1 = (U, \emptyset)$, $\quad 0 = (\emptyset, U)$;
2. $(X^+, X^-) \wedge (Y^+, Y^-) = (X^+ \cap Y^+, X^- \cup Y^-)$ (*conjunction*);
3. $(X^+, X^-) \vee (Y^+, Y^-) = (X^+ \cup Y^+, X^- \cap Y^-)$ (*disjunction*);
4. $(X^+, X^-) \to (Y^+, Y^-) = (-X^+ \cup Y^+, X^+ \cap Y^-)$; (*Nelson's implication*)
5. $\ulcorner (X^+, X^-) = (-X^+, X^+)$ (*weak negation* or *supplement*);
6. $(X^+, X^-)^\perp = (X^-, X^+)$ (*orthogonality*);
7. $(X^+, X^-) \Rightarrow (Y^+, Y^-) = ((-X^+ \cup Y^+) \cap (-Y^- \cup X^-), -X^- \cap Y^-)$ (*Heyting's implication*);
8. $\urcorner (X^+, X^-) = (X^+, X^-) \Rightarrow (\emptyset, U) = (X^-, -X^-)$ (*intuitionistic negation*);
9. $(X^+, X^-) \setminus (Y^+, Y^-) = (X^+ \cap -Y^+), (-X^+ \cup Y^+) \cap (-Y^- \cup X^-)$ (*co-intuitionistic subtraction*).

(see Pagliani [41], Polkowski [45], p. 363—with an equivalent definition of Heyting implication).[13]

Of course one will not obtain a complete semantics for intuitionistic logic starting from a finite base of clopen sets. Thus we need to look at general topological spaces. Since the language of our logic of assertions, hypotheses, and conjectures **AHCL** is *polarized*, in order to turn Pagliani's operations into a topological model of **AHCL**

[13] Notational decisions are nightmarish if we try to match the uses in the literature of Rough Sets, in Rauszer's bi-intuitionistic logic and our own.
In our polarized bi-intuitionistic logic [5] we used $\sim A$ for intuitionistic negation and $\frown C$ for co-intuitionistic supplement, leaving $\neg\alpha$ for classical negation, as required in Dalla Pozza and Garola's framework and following the meaning originally given to the symbol "$\neg$" by Frege.
C.Rauszer uses $\urcorner A$ for intuitionistic negation and $\ulcorner A$ for co-intuitionistic supplement; but in the later literature on bi-intuitionistic logic $\sim A$ is used for co-intuitionistic supplement.
In the literature on Rough Sets, weak-negation is sometimes written $\neg C$; intuitionistic negation is written in various ways (Pagliani uses $\div A$, in Polkowski's book there is $\dagger A$), while the symbol $\sim A$ is used exactly in the sense of *orthogonality* $A^\perp$.
However, it is unnecessary to make notations uniform across three areas, where similar connectives have different meanings: e.g., in Rough Sets negations are defined in a more general algebraic setting than Heyting algebras.
Hence it seems reasonable for us to retain the notation of [5] for our polarized logic, while using "$\urcorner$," "$\ulcorner$," and "$()^\perp$" for intuitionistic negation, co-intuitionistic supplement and orthogonality in Rough Sets.

we need to make sure that the interpretation of an assertive expression is an *open set* and a hypothetical expression is assigned a *closed set*; this is not always the case for Pagliani's operations, in particular implications and negations, which have to be modified as follows.

Definition 6 *Let (U, O) be a topological space, where O is the collection of open sets on U, and $\mathcal{I}(X)$ and $\mathcal{C}(X)$ are the interior and the closure of X,*[14] *respectively. We write (A_o^+, A_c^-) and (C_c^+, C_o^-) for pairs of disjoint sets of the types (open, closed) and (closed, open), respectively. We define the rough set interpretation $()^R$ of the language of assertions, hypotheses, and conjectures $\mathcal{L}^{AC}$ (in the disjoint representation) as follows.*

Fix an assignment $R : (\vdash p_i)^R = (A_{i_o}^+, A_{i_c}^-)$ and $(\mathcal{H}\, p_i)^R = (C_{i_c}^+, C_{i_o}^-)$ to the elementary expressions of $\mathcal{L}^{AH}$. Then

1 $\curlyvee^R = (U, \emptyset)$ *and* $\curlywedge^M = (\emptyset, U)$ *(clopen, clopen);*
2 $(A \cap B)^R = (A_o^+, A_c^-) \wedge (B_o^+, B_c^-) = (A_o^+ \cap B_o^+, A_c^- \cup B_c^-)$;
3 $(C \curlyvee D)^R = (C_c^+, C_o^-) \vee (D_c^+, D_O^-) = (C_c^+ \cup D_c^+, C_o^- \cap D_o^-)$;
4 $(A_o^+, A_c^-) \rightarrow (B_o^+, B_c^-) = \big(\mathcal{I}(-A_o^+ \cup B_o^+), \mathcal{C}(A_o^+ \cap B_c^-)\big)$[15];
5 $(\frown C)^R = \ulcorner (C_c^+, C_o^-) = \big(\mathcal{C}(-C_c^+), \mathcal{I}(C_c^+)\big)$ *and*
. $(\frown A)^R = \ulcorner (A_o^+, A_c^-) = (-A_o^+, A_o^+)$;
6 $(A_o^+, A_c^-)^\perp = (A_c^-, A_o^+)$ *and* $(C_c^+, C_o^-)^\perp = (C_o^-, C_c^+)$[16];
7 $(A \supset B)^R = (A_o^+, A_c^-) \Rightarrow (B_o^+, B_c^-) =$
. $= \big(\mathcal{I}(-A_o^+ \cup B_o^+) \cap \mathcal{I}(-B_c^- \cup A_c^-), \mathcal{C}(-A_c^- \cap B_c^-)\big)$;
8 $(\sim A)^R = \urcorner (A_o^+, A_c^-) = \big(\mathcal{I}(A_c^-), \mathcal{C}(-A_c^-)\big)$ *and*
. $(\sim C)^R = \urcorner (C_c^+, C_o^-) = (C_o^-, -C_o^-)$;
9 $(C \smallsetminus D)^R = (C_c^+, C_o^-) \smallsetminus (D_c^+, D_c^-) =$
. $= \big(\mathcal{C}(C_c^+ \cap -D_c^+), \mathcal{I}(-C_c^+ \cup D_c^-) \cap \mathcal{I}(-D_o^- \cup C_o^-)\big)$.

Let $\mathcal{L}^{AHC}$ a language of assertions, hypotheses, and conjectures built from a set of propositional atoms $p_0, p_1, \ldots$ and let $(\,)^\perp$ be an involution without fixed points on the atoms. A rough set interpretation *$\mathcal{M} = (U, O, R)$ of the language $\mathcal{L}^{AHC}$ (with an involution $(\,)^\perp$ on the atoms) is a topological space (U, O) together with an assignment R to the elementary expressions of disjoint pairs of the following forms:*

$$(\vdash p)^R = (A_o^+, A_c^-);$$
$$(\mathcal{H}p)^R = (C_c^+, C_o^-);$$
$$(\mathcal{C}p_i)^R = \big(\mathcal{C}(X^+), \mathcal{I}(X^-)\big), \qquad \text{where } (\vdash p_i)^R = (X^+, X^-).$$

[14] The notation $\mathcal{C}X$ is overloaded, for the illocutionary operator of conjecture in the syntax of the language of pragmatics and for the closure operator in a topological space. No confusion is possible, given the difference in context.

[15] There is no connective to represent Nelson's implication as distinct from intuitionistic implication in $\mathcal{L}^{AH}$.

[16] There is no specific connective for orthogonality in $\mathcal{L}^{AH}$.

Lemma 3 *Let $\mathcal{M} = (U, O, R)$ be an interpretation of $\mathcal{L}^{AHC}$, with an involution $(\,)^{\perp}$ on the atoms. Then $\mathcal{M}$ is a model of* **AHCL** *if and only if the assignment R to elementary expressions of $\mathcal{L}^{AHC}$ satisfies the following duality conditions:*

$$\begin{array}{ll} ({\vdash} p^{\perp})^R = (C_o^-, C_c^+) = (C_c^+, C_o^-)^{\perp} & \text{where } (\mathcal{H} p)^R = (C_c^+, C_o^-) \\ (\mathcal{H} p^{\perp})^R = (A_c^-, A_o^+) = (A_o^+, A_c^-)^{\perp} & \text{where } ({\vdash} p)^R = (A_o^+, A_c^-) \end{array}$$

and moreover for every (A_o^+, A_c^-) and (C_c^+, C_o^-) in R we have

$$A_c^- = -A_o^+ \quad \textit{and} \quad C_o^- = -C_c^+. \tag{5}$$

Proof. Concerning conditions $\sim\frown A \equiv A$ and $\frown\sim C \equiv C$ of Lemma 2, notice that

$$\neg\ulcorner\ (A_o^+, A_c^-) = \neg\,(-A_o^+, A_o^+) = (A_o^+, -A_o^+) = (A_o^+, A_c^-)$$

if and only if $A_c^- = -A_o^+$ and similarly

$$\ulcorner\neg\ (C_c^+, C_o^-) = \ulcorner\,(C_o^-, -C_o^-) = (-C_o^-, C_o^-) = (C_c^+, C_o^-)$$

where the last equality holds if and only if $C_c^+ = -C_o^-$. Moreover, the conditions $(b) - (c)$ in the definition of duality between $\mathcal{L}^A$ and $\mathcal{L}^H$ (Definition 3) are clearly satisfied by the standard Rough Set definition. As for condition (a), given the involution $(\,)^{\perp}$ on the atoms, we have

$$(\frown {\vdash} p)^R = \ulcorner\,(A_o^+, A_c^-) = (-A_o^+, A_o^+) = (A_c^-, A_o^+) = (\mathcal{H} p^{\perp})^R$$

where the third equality holds by condition (5) and the fourth by the condition of duality in a model. Similarly,

$$(\sim \mathcal{H} p)^R = \neg\,(C_c^+, C_o^-) = (C_o^-, -C_o^-) = (C_o^-, C_c^+) = ({\vdash} p^{\perp})^R$$

as required.

Remark 3 In a model $\mathcal{M} = (U, O, R)$ for **AHCL** *intuitionistic negation* and *Nelson's negation* coincide:

$$\begin{aligned} (A_o^+, A_c^-) \supset (B_o^+, B_c^-) &= (\mathcal{I}(-A_o^+ \cup B_o^+) \cap \mathcal{I}(-B_c^- \cup A_c^-), \mathcal{C}(-A_c^- \cap B_c^-)) \\ &= (\mathcal{I}(A_c^- \cup B_o^+), \mathcal{C}(A_o^+ \cap B_c^-)) \end{aligned}$$

Thus to exploit Rough-Set semantics in full, we may want to consider notions of duality where condition (5) does not hold.

Fig. 1 "Kripke model"

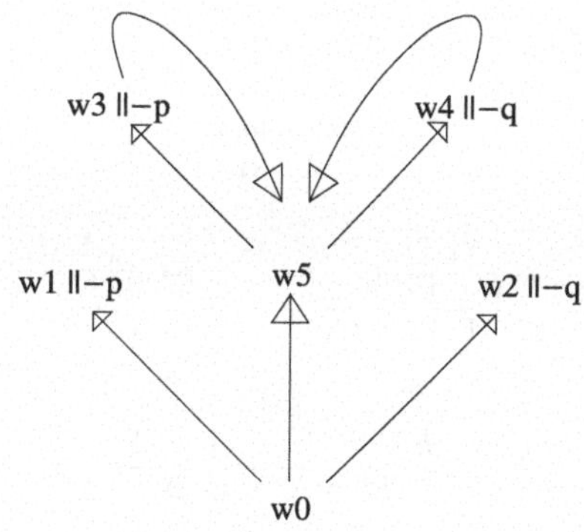

3.2 *Algebra of Regions*

A main reason of interest in bi-intuitionistic logic are its topos-theoretic models studied by Lawvere [32] Reyes and Zolfaghari [52], recently reconsidered by Stell and Worboys [57] in their "algebra of regions." It is clearly impossible here to compare Reyes and Zolfaghari's modal logic to our polarized bi-intuitionistic systems, but we must say something about Stell and Worboys' geometric examples.

The first one is Reyes and Zolfaghari's motivating example [52]: it provides a model of bi-intuitionistic logic based on the subgraphs of arbitrary *undirected* graphs. It ought to be possible to define graphic models of **AHL** and **PBL**, but we shall not attempt this here. On the other hand, "two stages sets" in the second example are just a geometric representation of the basic notion of "rough equality": in an approximation space each subset G of the universe is identified only by the pair $(\mathcal{I}(G), \mathcal{C}(G))$—or with $(\mathcal{I}(G), -\mathcal{C}(G))$ in the disjoint representation—where the interior and closure operator result from two stages of process of classification.

Now it is evident that condition (5) on models of **AHL** restricts the interpretation to sets G that are fully characterized in (U, E), i.e., such that $\mathcal{I}(G) = \mathcal{C}(G)$. We illustrate more interesting semantics applications with an example. Consider the Kripke model K for **S4** obtained from the reflexive and transitive closure of the graph in Fig. 1.

Writing α_K for the set of possible worlds satisfying α, we have $(\vdash p)_K = \{w_1\}$, $(\mathcal{C}p)_K = \{w_0, w_1\}$ and $(\mathcal{H}p)_K = \{w_0, w_1, w_3, w_4, w_5\} = K \setminus \{w_2\}$. We are satisfied with the Rough Set interpretation of *assertions* in the disjoint representation as $(\vdash p)^R = (\{w_1\}, K \setminus \{w_1\})$: after all, the grounds for an assertion ought to be a "stable" state of knowledge; by duality the representation of *hypotheses* as $(\mathcal{H}p)^R = (K \setminus \{w_2\}, \{w_2\})$ is appropriate. On the other hand, the state of knowledge justifying *conjectures* is "unstable"; thus there seems to be a meaningful "two-stage set" representation of *conjectures* of the form $(\mathcal{C}p)^R = (\{w_1\}, K \setminus \{w_0, w_1\})$, of type (*open, open*). We notice that such an interpretation is possible for the logic **AHCL** of assertions, conjectures, and hypotheses, as it does not interfere with the basic symmetry between assertions and hypotheses. It remains an open problem whether these very

Table 4 Natural Deduction $\mathbf{NJ}^{\smallsetminus\curlyvee}$

assumption

$$A \vdash A$$

$$\curlyvee_0\text{-I}\quad \frac{E \vdash \Gamma, C_0}{E \vdash \Gamma, C_0 \curlyvee C_1} \qquad \curlyvee_1\text{-I}\quad \frac{E \vdash \Gamma, C_0}{E \vdash \Gamma, C_0 \curlyvee C_1} \qquad \curlyvee\text{-E}\quad \frac{E \vdash \Gamma, C_0 \curlyvee C_1 \qquad C_0 \vdash \Gamma_0 \qquad C_1 \vdash \Gamma_1}{E \vdash \Gamma, \Gamma_0, \Gamma_1}$$

$$\smallsetminus\text{-I}\quad \frac{E \vdash \Gamma, C \qquad D \vdash \Delta}{E \vdash \Gamma, C \smallsetminus D, \Delta} \qquad \smallsetminus\text{-E}\quad \frac{E \vdash \Gamma, C \smallsetminus D \qquad C \vdash (D)^k, \Delta}{E \vdash \Gamma, \Delta}$$

$(D)^k$ is a multiset with k occurrences of D.

conjectural remarks can be developed into an interesting rough-set semantics of a logic of assertions, hypotheses, conjectures, and expectations.

4 PART III. Proof Theory

We shall start with the definition of a sequent-style *single-assumption multiple-conclusions natural deduction* system for the subtractive-disjunctive fragment $\mathbf{NJ}^{\smallsetminus\curlyvee}$ of co-intuitionistic logic. We have sequents of the form

$$A \vdash C_1, \ldots, C_m$$

where A indicates the only open assumption in a derivation with the multiset $C_1, \ldots,$ C_m of open conclusions. The rules of inference are in Table 4.

Definition 7 *We say that $C_1, \ldots, C_m$ is derivable from A if there is a natural deduction derivation of the sequent $A \vdash \Gamma$ where all formulas in the multiset Γ are among $C_1, \ldots, C_m$.*

Remark 4 (i) Looking at the deduction rules in Table 4, notice that $\smallsetminus$-*introductions*, $\curlyvee$-*eliminations* and $\smallsetminus$-*eliminations* discharge the open assumption(s) of the sequent-premise(s) to the right, but a $\smallsetminus$-*elimination* discharges also a *multiset of open conclusions*. As a consequence, $\smallsetminus$-*eliminations* are the only inferences that cannot be permuted freely with other inferences. From another point of view, here we have a limit to the "parallelization of the syntax," a *box* in the sense of Girard. To remove such a box, a device is needed to discharge open conclusions preserving as much as possible the geometry of proofs. In this section we recover *Prawitz trees* as an appropriate representation of proofs in $\mathbf{NJ}^{\smallsetminus\curlyvee}$.

(ii) As in Prawitz's natural deduction *weakening* is not explicitly represented in proof-trees and *contraction* appears only in the discharging of conclusions in a $\smallsetminus$-E inference.

Definition 8 (i) (active and passive formula-occurrences) *In assumptions and in rules of inference the indicated formula-occurrences in the succedent of a sequent are* active *and all occurrences in the multisets* Γ, Γ_i, Δ *are* passive. *Also the discharged assumptions in a* $\smallsetminus$*-I,* $\curlyvee$*-E and* $\smallsetminus$*-E are* active, *all other assumptions are* passive. *An active formula in the sequent-conclusion of an inference is also called the* conclusion *of the inference.*

(ii) (segments) *If* Υ *occurs in the premise and in the conclusion of an inference then an occurrence* $D_i \in \Upsilon$ *in the premise is the* immediate ancestor *of the occurrence* D_i *in the conclusion. Then as in Prawitz [46] we define a* segment *as a sequence* D_1, ..., D_m *of occurrences of the same formula where* D_1 *and* D_m *are active, and* D_i *is the immediate ancestor of* D_{i+1}, *for* $i < m$.

(iii) *Thus we may speak of a* segment *as the conclusion or the premise of some inference.*

(iv) *A* maximal segment *is the conclusion of an introduction rule which is premise of an elimination. A derivation is* normal *if it does not have maximal segments.*

4.1 Structure of Normal Proofs

The structure of normal deductions in co-intuitionistic logic $\mathbf{NJ}^{\smallsetminus,\curlyvee}$ mirrors that of normal deductions in intuitionistic logic $\mathbf{NJ}^{\supset,\cap}$.

Definition 9 (i) *A* Prawitz path *in a normal deduction is a sequence* $C_1, \ldots, C_i, \ldots, C_n$ *of segments such that*

- C_1 *is an assumption, either open or discharged by a* $\smallsetminus$*-introduction;*
- *for* j *with* $1 \leq j < i$, $C_j = C \smallsetminus D$ *is a premise of a* $\curlyvee$*- or* $\smallsetminus$*-elimination and* $C_{j+1} = C$ *is an assumption discharged by the inference;*
- *for* j *with* $i \leq j < n$, C_j *is a premise of a* $\curlyvee$*- or* $\smallsetminus$*-introduction with conclusion* C_{j+1};
- C_n *is a conclusion of the derivation, either open or discharged by a* $\smallsetminus$*-E.*

(ii) *The collection of all Prawitz paths in a derivation is a graph, called the* tree of Prawitz paths τ. *If we collapse segments to their formulas, the resulting tree yields a graphical representation of proofs which we shall call* Prawitz tree *for* $\mathbf{NJ}^{\smallsetminus\curlyvee}$. *Such trees are similar to those in Prawitz-style Natural Deduction derivation for* $\mathbf{NJ}^{\supset\cap}$, *but in* $\mathbf{NJ}^{\smallsetminus\curlyvee}$ *the logical flow goes from the root to the leaves, rather than from the leaves to the root as in* $\mathbf{NJ}^{\supset\cap}$.

(iii) *The definition of the* depth *of a path* π *in a tree* τ *is familiar: the depth of* π *is 0 if its first formula* C_1 *is open; the depth of* π *is* $n+1$ *if* C_1 *is discharged by a* $\smallsetminus$*-introduction with conclusion in a path of depth* n*.*

From this analysis we derive as usual the *subformula property* for normal deductions:

Proposition 3 *Every formula occurring in a normal deduction of* $A \vdash C_1, \ldots, C_m$ *is a subformula either of* A *or of* C_i *for some* i*.*

Example We constuct a derivation d in $\mathbf{NJ}^{\smallsetminus\curlyvee}$ of

$$C \smallsetminus A \vdash ((C \smallsetminus (B \curlyvee D)) \smallsetminus A, ((B \smallsetminus A) \curlyvee (D \smallsetminus A))^2$$

It may be helpful to think of the dual derivation in $\mathbf{NJ}^{\supset\cap}$ of

$$(A \supset B) \cap (A \supset D), A \supset ((B \cap D) \supset C) \vdash A \supset C.$$

We write $\mathbf{F}$ for $(B \smallsetminus A) \curlyvee (D \smallsetminus A)$ and $\mathbf{G}$ for $C \smallsetminus (B \curlyvee D)$.

$$\dfrac{\dfrac{\dfrac{C \vdash C \qquad B \curlyvee D \vdash B \curlyvee D}{C \vdash C \smallsetminus (B \curlyvee D), B \curlyvee D}\smallsetminus\text{-I} \qquad B \vdash B \qquad D \vdash D}{C \vdash C \smallsetminus (B \curlyvee D), B, D}\curlyvee\text{-E} \qquad A \vdash A}{C \vdash \mathbf{G} \smallsetminus A, A, B, D}\smallsetminus\text{-I}$$

$$\dfrac{C \smallsetminus A \vdash C \smallsetminus A \qquad \dfrac{\dfrac{\dfrac{\dfrac{\vdots \qquad A \vdash A}{C \vdash \mathbf{G} \smallsetminus A, (A)^2, B \smallsetminus A, D}\smallsetminus\text{-I} \qquad A \vdash A}{C \vdash \mathbf{G} \smallsetminus A, (A)^3, B \smallsetminus A, D \smallsetminus A}\smallsetminus\text{-I} \qquad A \vdash A}{C \vdash \mathbf{G} \smallsetminus A, (A)^3, (B \smallsetminus A) \curlyvee (D \smallsetminus A), D \smallsetminus A}\curlyvee_0\text{-I}}{C \vdash (A)^3, \mathbf{G} \smallsetminus A, (\mathbf{F})^2}\curlyvee_1\text{-I}}{C \smallsetminus A \vdash \mathbf{G} \smallsetminus A, (\mathbf{F})^2}\smallsetminus\text{-E}$$

In Fig. 2 we find the tree-structure of "Prawitz' paths" of the derivation d.

4.2 Sequents with Tail Formula

A very perspicuous representation of derivations in co-intuitionistic logic is through *sequent calculus with tail formula* q-$\mathbf{LJ}^{\smallsetminus\curlyvee}$, the exact dual of the well-known sequent calculus with head formula t-$\mathbf{LJ}^{\supset\cap}$.[17] Here sequents have the form

[17] The "q" in q-$\mathbf{LJ}^{\smallsetminus\curlyvee}$ stands for *queue*, tail, as the "t" in t-$\mathbf{LJ}^{\supset\cap}$ stands for *tête*, head.

Fig. 2 A Prawitz tree

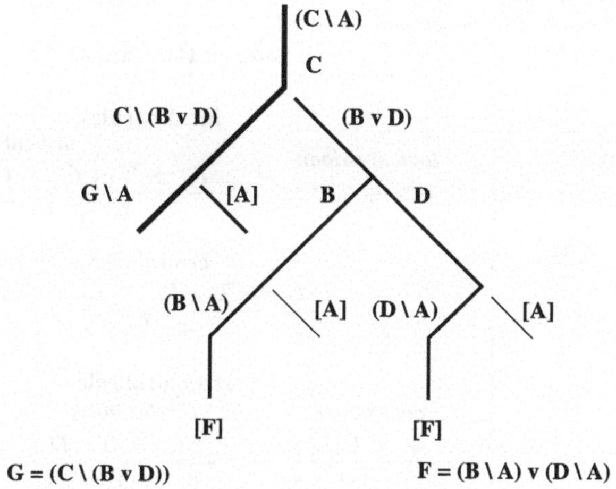

$$E \Rightarrow \Upsilon ; C$$

with one formula in the antecedent, a multiset in the *ordinary area*, and at most one formula in the *linear area* (*stoup*) of the succedent. The principal formulas of the *right*-rules are in the stoup and *left*-rules require empty stoup in the sequent-premies.[18] The rules of q-$\mathbf{LJ}^{\smallsetminus\curlyvee}$ are given in Table 5.

The following fact is the dual of a well-known correspondence between Natural Deduction derivations in $\mathbf{NJ}^{\supset\cap}$ and Sequent Calculus derivations in t-$\mathbf{LJ}^{\supset\cap}$. For sequent calculi with head formulas or tail formulas see, for instance, [20].

Proposition 4 *There is a bijection between trees of Prawitz paths of normal derivations in* $\mathbf{NJ}^{\smallsetminus\curlyvee}$ *and cut-free derivations in q-*$\mathbf{LJ}^{\smallsetminus\curlyvee}$ *(modulo the order of structural inferences).*

Proof. Given a Prawitz tree τ, by induction on τ we construct a q-$\mathbf{LJ}^{\smallsetminus\curlyvee}$ derivation with the property that the formula in the stoup (*tail formula*), if any, is the conclusion of a path of depth 0 (*main path*) of τ. If τ begins with an elimination rule, the result is immediate by the inductive hypothesis applied to the immediate subtree(s) from the top, since we may assume that the corresponding cut-free derivations have conclusions with empty stoup. If τ begins with an introduction rule, then there is only one main path and we remove the last inference of it: if the conclusion was a formula $C \smallsetminus D$, the inductive hypothesis yields two q-$\mathbf{LJ}^{\smallsetminus\curlyvee}$ derivations; in one the endsequent must have C in the stoup, since C belongs to the main path; in the other

[18] It ought to be clear that the use of *focalization* in the sequent calculus q-$\mathbf{LJ}^{\smallsetminus\curlyvee}$ and in the dual t-$\mathbf{LJ}^{\supset\cap}$ (see Table 8, Appendix III), is unrelated to the use of the "stoup" in our sequent calculi **AH-G3**, **PB-G3**, and **APB-G3** for bi-intuitionistic logic, where it is used simply to highlight the restrictions of intuitionistic systems.

Table 5 The sequent calculus q-$\mathbf{LJ}^{\smallsetminus\curlyvee}$

Sequent Calculus q-$\mathbf{LJ}^{\smallsetminus\curlyvee}$

identity rules

logical axiom:

$$C \Rightarrow \; ; C$$

tail-cut:

$$\frac{E \Rightarrow \Upsilon_1 \; ; C \qquad C \Rightarrow \Upsilon_2 \; ; D}{E \Rightarrow \Upsilon_1, \Upsilon_2 \; ; D}$$

central-cut:

$$\frac{E \Rightarrow \Upsilon_1, C \; ; \qquad C \Rightarrow \Upsilon_2 ;}{E \Rightarrow \Upsilon_1, \Upsilon_2 \; ;}$$

structural rules

contraction:

$$\frac{E \Rightarrow \Upsilon, C, C \; ; D}{E \Rightarrow \Upsilon, C \; ; D}$$

weakening:

$$\frac{E \Rightarrow \Upsilon \; ; D}{E \Rightarrow \Upsilon, C \; ; D}$$

dereliction:

$$\frac{E \Rightarrow \Upsilon \; ; D}{E \Rightarrow \Upsilon, D \; ;}$$

logical rules

$\smallsetminus$ *right:*

$$\frac{E \Rightarrow \Upsilon_1 \; ; C \qquad D \Rightarrow \Upsilon_2 \; ;}{E \Rightarrow \Upsilon_1, \upsilon_2 \; ; C \smallsetminus D}$$

$\smallsetminus$ *left:*

$$\frac{C \Rightarrow \Upsilon, D \; ;}{C \smallsetminus D \Rightarrow \Upsilon \; ;}$$

right $\curlyvee_i$:

$$\frac{E \Rightarrow \Upsilon \; ; C_i}{E \Rightarrow \Upsilon \; ; C_0 \curlyvee C_1}$$

left $\curlyvee$:

$$\frac{C_0 \Rightarrow \Upsilon ; \qquad C_1 \Rightarrow \Upsilon ;}{C_0 \curlyvee C_1 \Rightarrow \Upsilon ;}$$

the endsequent has D in the antecedent and we may assume that it has no formula in the stoup, by applying dereliction if necessary. Therefore we may apply $\smallsetminus$-R to obtain the desired derivation. The other cases are obvious.
The fact that two derivations d' and d'' corresponding to the same tree τ can only differ for the order of structural inferences is due to the fact that in q-$\mathbf{LJ}^{\smallsetminus\curlyvee}$ logical inferences cannot be permuted with each other. Indeed, the principal formulas of all inferences occur either in the antecedent or in the stoup, and the rule of dereliction is irreversible.

Example (cont.) The following sequent derivation d_q corresponds to the natural deduction derivation d:

$$
\dfrac{
\dfrac{C \Rightarrow ;\, C \qquad
\dfrac{
\dfrac{
\dfrac{
\dfrac{B \Rightarrow ;\, B \qquad \dfrac{A \Rightarrow ;\, A}{A \Rightarrow A\,;}\,\textit{der}}{B \Rightarrow A\,;\, B \smallsetminus A}\,\smallsetminus\text{-R}
}{B \Rightarrow A\,;\, \mathbf{F}}\,\curlyvee_0\text{-R}
}{B \Rightarrow A, \mathbf{F}\,;}\,\textit{der}
\qquad
\dfrac{
\dfrac{
\dfrac{D \Rightarrow ;\, D \qquad \dfrac{A \Rightarrow ;\, A}{A \Rightarrow A\,;}\,\textit{der}}{D \Rightarrow A\,;\, D \smallsetminus A}\,\smallsetminus\text{-R}
}{D \Rightarrow A\,;\, \mathbf{F}}\,\curlyvee_1\text{-R}
}{D \Rightarrow A, \mathbf{F}\,;}\,\textit{der}
}{B \curlyvee D \Rightarrow A, A, \mathbf{F}, \mathbf{F}\,;}
}{C \Rightarrow A, A, \mathbf{F}, \mathbf{F}\,;\, C \smallsetminus (B \curlyvee D)}\,\smallsetminus\text{-R}
\qquad
\dfrac{A \Rightarrow ;\, A}{A \Rightarrow A\,;}\,\textit{der}
}{
\dfrac{
\dfrac{C \Rightarrow A, A, A, \mathbf{F}, \mathbf{F}\,;\, \mathbf{G} \smallsetminus A}{C \Rightarrow A, \mathbf{F}, \mathbf{G} \smallsetminus A\,;}\,\textit{contr, der}
}{C \smallsetminus A \Rightarrow \mathbf{F}, \mathbf{G} \smallsetminus A\,;}\,\smallsetminus\text{-L}
}
$$

5 PART IV.Term Assignment for Co-intuitionistic Logic

In a tantalising pair of papers [42, 44] Michel Parigot introduced *Free Deduction*, a formalism consisting of *elimination rules only*, with the property that both Natural Deduction and the Sequent Calculus could be represented in it simply by restricting the order of deduction, e.g., by permutations of inferences. Free Deduction was conceived to study the computational properties of classical logic, but it can be adapted to intuitionistic and co-intuitionistic logic through the analog of Gentzen's restrictions on sequents.

For instance, although they do not appear in this form in [42], the rules for *multiplicative implication and subtraction* can be formulated as follows:

multiplicative implication

$$
\dfrac{\Gamma, \mathbf{A} \to \mathbf{B} \vdash \Delta \qquad \Pi, \mathbf{A} \vdash \mathbf{B}, (\Sigma^{\P})}{\Gamma, \Pi \vdash \Delta, \Sigma} \to \textit{elim left}
$$

$$
\dfrac{\Gamma \vdash \Delta, \mathbf{A} \to \mathbf{B} \qquad \Pi \vdash \Sigma, \mathbf{A} \qquad \Pi', \mathbf{B} \vdash \Sigma'}{\Gamma, \Pi, \Pi' \vdash \Delta, \Sigma, \Sigma'} \to \textit{elim right}
$$

multiplicative subtraction

$$
\dfrac{\Gamma, \mathbf{A} \smallsetminus \mathbf{B} \vdash \Delta \qquad \Pi \vdash \Sigma, \mathbf{A} \qquad \Pi', \mathbf{B} \vdash \Sigma'}{\Gamma, \Pi, \Pi' \vdash \Delta, \Sigma, \Sigma'} \smallsetminus \textit{elim left}
$$

$$
\dfrac{\Gamma \vdash \Delta, \mathbf{A} \smallsetminus \mathbf{B} \qquad (\Pi^{\P}), \mathbf{A} \vdash \mathbf{B}, \Sigma}{\Gamma, \Pi \vdash \Delta, \Sigma} \smallsetminus \textit{elim right}
$$

The intuitionistic restriction ($\Sigma^{\P}$), namely that Σ is empty, applies to the secondary premise of the $\to$*-left elimination* rule, and the dual restriction holds for $\smallsetminus$*-right elimination.* The sequent calculus rules are obtained by *killing the main premise* (i.e., keeping it only as an axiom). Here are the rules for subtraction:
Natural Deduction, on the other hand, is given by *keeping all inputs on the left.* Namely: for left elimination rules, *we kill the main premise*; for right elimination

subtraction rules, as in the sequent calculus

$$\frac{\mathbf{A} \setminus \mathbf{B} \vdash A \setminus B \qquad \Pi \vdash \Sigma, \mathbf{A} \qquad \Pi', \mathbf{B} \vdash \Sigma'}{\Pi, \Pi' \vdash A \setminus B, \Sigma, \Sigma'} \setminus\text{-R}$$

$$\frac{A \setminus B \vdash \mathbf{A} \setminus \mathbf{B} \qquad [\Pi^{\P}], \mathbf{A} \vdash \Sigma, \mathbf{B}}{A \setminus B, \Pi \vdash \Sigma} \setminus\text{-L}$$

rules, *we kill the secondary premises which have only a left active formula.* Thus no premise is killed in *subtraction elimination right*.

subtraction rules, as in natural deduction

$$\frac{\mathbf{A} \setminus \mathbf{B} \vdash A \setminus B \qquad \Pi \vdash \mathbf{A}, \Sigma \qquad \Pi', \mathbf{B} \vdash \Sigma'}{\Pi, \Pi' \vdash A \setminus B, \Sigma'} \setminus \textit{intro}$$

$$\frac{\Gamma \vdash \mathbf{A} \setminus \mathbf{B}, \Delta \qquad [\Pi^{\P}], \mathbf{A} \vdash \mathbf{B}, \Sigma}{\Gamma, \Pi \vdash \Delta, \Sigma} \setminus \textit{elim}$$

Since Free Deduction yields a *multiple conclusion* natural deduction system in a very straightforward way, one would expect that a term assignment to Free Deduction might be *distributed* to all formula in the succedent of sequents. On the contrary in 1992 Michel Parigot introduced the λ-μ calculus as "an algorithmic interpretation of classical Natural Deduction," which is based on a notion of "central control." In the last part of this chapter we propose a distributed term assignment to co-intuitionistic logic.

5.1 Term Assignment to the Subtraction Rules in the λ-μ Calculus

Recently the proof theory of bi-intuitionistic (*subtractive*) logic has been studied by Crolard [15, 16]: in [16] a Natural Deduction system is presented with a *calculus of coroutines* as term assignment.[19] Crolard works in the framework of Parigot's $\lambda\mu$-calculus: sequents may be written in the form[20] $\Gamma \vdash t : A \mid \Delta$, with contexts $\Gamma = x_1 : C_1, \ldots, x_m : C_m$ and $\Delta = \alpha_1 : D_1, \ldots, \alpha_n : D_n$, where the x_i are *variables* and the α_j are *μ-variables*. In addition to the rules of the simply typed lambda calculus, there are *naming rules*

$$\frac{\Gamma \vdash t : A \mid \alpha : A, \Delta}{\Gamma \vdash [\alpha]t : \bot \mid \alpha : A, \Delta}[\alpha] \qquad \frac{\Gamma \vdash t : \bot \mid \alpha : A, \Delta}{\Gamma \vdash \mu\alpha.t : A \mid \Delta}\mu$$

[19] This part is joint work with Corrado Biasi and incorporates important contributions from his still unfinished doctoral dissertation at Queen Mary, University of London.

[20] Parigot and Crolard actually write sequents in the form $t : \Gamma \vdash \Delta; A$, where the term t is given the type of the formula A in the stoup, if such a formula exists. If the stoup is empty, the notation allows one to think of t as being assigned to the entire sequent or to a formula $\bot$ implicitly present in the stoup.

It is well-known that the $\lambda\mu$-calculus provides a computational interpretation of classical logic and a typing system for functional programs with continuations (see, e.g., [17, 54]).

Crolard extends the $\lambda\mu$ calculus with introduction and elimination rules for subtraction[21]:

$$\frac{\Gamma \vdash t : A \mid \Delta}{\Gamma \vdash \texttt{make-coroutine}(t, \beta) : A \setminus B \mid \beta : B, \Delta} \setminus I$$

$$\frac{\Gamma \vdash t : A \setminus B \mid \Delta \qquad \Gamma, x : A \vdash u : B \mid \Delta}{\Gamma \vdash \texttt{resume}\ t\ \texttt{with}\ x \mapsto u : C \mid \Delta} \setminus E$$

The reduction of a redex of the form $\texttt{resume}(\texttt{make-coroutine}(t, \beta))\ \texttt{with}\ x \mapsto u : C$ yields $\mu\gamma.[\beta]u[t/x]$, where the μ-variables are typed as $\beta : B$ and $\gamma : C$. Namely

$$\frac{\dfrac{\Gamma \vdash t : A \mid \Delta}{\Gamma \vdash \texttt{make-coroutine}(t, \beta) : A \setminus B \mid \beta : B, \Delta} \setminus\text{-I} \qquad \Gamma, x : A \vdash u : B \mid \Delta}{\Gamma \vdash \texttt{resume}\ (\texttt{make-coroutine}(t, \beta))\ \texttt{with}\ x \mapsto u\ : C \mid \beta : B, \Delta} \setminus\text{-E}$$

reduces to

$$\frac{\dfrac{\dfrac{\Gamma \vdash t : A \mid \Delta \qquad \Gamma, x : A \vdash u : B \mid \Delta}{\Gamma \vdash u[t/x] : B \mid \Delta}\ \textit{substitution}}{\Gamma \vdash [\beta]u[t/x] : \bot \mid \beta : B, \gamma : C, \Delta'}\ [\beta]}{\Gamma \vdash \mu\gamma.[\beta]u[t/x] : C \mid \beta : B, \Delta'}\ \mu$$

Working with the full power of classical logic, if a constructive system of bi-intuitionistic logic is required, then the implication right and subtraction left rules must be restricted by considering *relevant dependencies*.[22] Crolard is able to show that the term assignment for such a restricted logic is a calculus of *safe coroutines*, namely terms in which no coroutine can access the local environment of another coroutine.

---

[21] In Crolard [16] the introduction rule corresponds to the more general form $\setminus$-I given above, and more general *continuation contexts* occur in place of β; the above formulation suffices for our purpose here.

[22] For instance, in the derivation of the right premise of a subtraction elimination ($\setminus E$), there should be no relevant dependency between the formula B and the assumptions in Γ, but only between B and A.

5.2 A Distributed Term Assignment for the Subtractive Fragment

When we consider a term assignment for the Natural Deduction system $\mathbf{NJ}^{\setminus\curlyvee}$ of dual intuitionistic logic only, we are led to ask what Crolard's calculus becomes when separated from its $\lambda\mu$ context. Indeed the naming rules of the $\lambda\mu$ calculus allow us to represent the action of an operating system jumping from one thread of computation to another: when a name β for a coroutine has been created by `make-coroutine`, it can be later accessed by the system and the coroutine executed.

On the contrary in our proposal different terms are simultaneously assigned to the multiple conclusions of a sequent in a sequent-style Natural Deduction, (or in the Sequent Calculus with tail formula). There is no mechanism to simulate the passage of control from one "thread" to another. A process is stopped by the operator assigned to subtraction elimination (called here `postpone` rather than Crolard's `resume`) and becomes active only in the normalization process. Thus in the presence of different processes running in parallel, one wonders whether our system can still be regarded as a calculus of *coroutines*: it is perhaps closer to an abstract representation of a *multiprocessing system*.

Before giving formal definitions, let us survey the most distinctive features of our calculus for the terms assignment to the *subtractive* fragment only. Most characteristic is the treatment of *variables*: there is no operator for *explicitly binding* variables or *delimiting the scope* of an implicitly binding operation. We may say that a *computational context* is characterized by exactly one free variable and that a free variable a becomes bound when its computational context $\mathcal{S}_a$ is plunged into the computational context $\mathcal{S}_b$ associated with another variable b. In this case, the variable a is replaced everywhere by $\mathtt{a}(t)$ for some term t containing b; here the function $\mathtt{a}$ is vaguely reminiscent of a *Herbrand function*. In the normalization process the term $\mathtt{a}(t)$ may later be replaced by another term u throughout the new computational context; thus we assume that a mechanism is in place for *broadcasting* substitutions throughout an environment.

We have the following operators:

- the term $\mathtt{mkc}(t, \mathtt{y})$, which is assigned to the conclusion of a $\setminus$-*introduction*, connects two disjoint computational contexts, say, $\mathcal{S}_x$ and $\mathcal{S}_y$. Every term in $\mathcal{S}_x$ contains exactly one free variable x, and we assume that the term t represents a *thread* starting from x.[23] The computational context $\mathcal{S}_y$ contains the free variable y and all threads starting from y. When the term $\mathtt{mkc}(t, \mathtt{y})$ is introduced, the substitution $y := \mathtt{y}(t)$ must be performed throughout the environment $\mathcal{S}_y$. Thus the term $\mathtt{mkc}(t, \mathtt{y})$ represents a *jump* extending the thread t to all threads in $\mathcal{S}_y\{y := \mathtt{y}(t)\}$; the substitution of $\mathtt{y}(t)$ for y throughout $\mathcal{S}_y$ has the effect that the extended computational context contains only the free variable x. Here we retain

[23] Here we use the term "thread" in the sense of Prawitz [46], p. 25; namely, a *thread* is a branch in the proof-tree from the a leaf to the root. The equivalent notion here is that of a branch in *Prawitz' tree* τ from the root to the leaf. No claim is made here about the computer science usage of the term "thread."

Crolard's name `make-coroutine` for historic reasons; a more precise but more redundant description would be the following:

$$\texttt{mkc}(t, \texttt{y}) \quad \text{stands for} \quad \texttt{extend thread}\ t\ \texttt{from}\ \texttt{y}(t).$$

- The term $\texttt{postp}(\texttt{z} \mapsto \ell, t)$, which is assigned to the conclusion of a $\searrow$-*elimination*, takes a computational context $\mathcal{S}_z$ containing the only free variable z, and plunges it into another context $\mathcal{S}_x$ where the only free variable is x; this is done by selecting the list ℓ of threads starting from z and the term t with free variable x, replacing z with $\texttt{z}(t)$ throughout $\mathcal{S}_z$ and *freezing* $\ell\{z := \texttt{z}(t)\}$ until through normalization the term t is transformed to a term of the form `extend thread`. A fuller description is therefore the following:

$$\texttt{postp}(\texttt{z} \mapsto \ell, w) \text{ stands for } \texttt{postpone subthreads}\ \ell\{z := \texttt{z}(w)\} \text{ until } w.$$

Let **M** be $\texttt{mkc}(t, \texttt{y})$ and let $\mathbf{P}(v)$ be $\texttt{postp}(\texttt{z} \mapsto \ell, v)$. Then

$$\mathbf{P}(\mathbf{M}) = \texttt{postp}(\texttt{z} \mapsto \ell, \texttt{mkc}(t, \texttt{y}))$$

is a *redex*. The main idea of a reduction is to replace the jump from t to $\texttt{y}(t)$ with each one of the subthreads in ℓ. But such an operation has important side effects. A redex $\mathbf{P}(\mathbf{M})$ occurs in a *computational context* $\mathcal{S}_x$ of the form

$$\mathcal{S}_x : \quad \texttt{postp}(\texttt{z} \mapsto \ell, \texttt{mkc}(t, \texttt{y})), \quad \overline{\kappa}, \quad \overline{\zeta}_{\texttt{y}}, \quad \overline{\xi}_{\texttt{z}}$$

where $\overline{\zeta}_{\texttt{y}}$ is a sequence of terms containing $\texttt{y}(t)$, $\overline{\xi}_{\texttt{z}}$ a sequence of terms containing $\texttt{z}(\texttt{mkc}(t, \texttt{y}))$ and $\overline{\kappa}$ a sequence containing neither $\texttt{y}(t)$ nor $\texttt{z}(\texttt{mkc}(t, \texttt{y}))$. Thus the *side effects* consist in the replacement of $\texttt{z}(\texttt{mkc}(t, \texttt{y}))$ with t in $\overline{\xi}_{\texttt{z}}$ and in each subthread s_k of ℓ; let $\ell' = s'_1, \ldots, s'_n$ be the resulting sequence. Finally, we replace $\texttt{y}(t)$ in $\overline{\zeta}_{\texttt{y}}$ with each one of the subthreads s'_k, thus *expanding* the sequence $\overline{\zeta}_{\texttt{y}}$ in a sense to be made precise below. To indicate such a rewriting process we shall use the notation

$$\mathcal{S}' = \mathcal{S}_x - \mathbf{P}(\mathbf{M}) \quad \{z := t\} \quad \{y := \ell\{z := t\}\}$$

where $z = \texttt{z}(\texttt{mkc}(t, \texttt{y}))$ and $y = \texttt{y}(t)$.
In an enterprise where notation is in danger of growing out of control, readability is essential. The notations $\texttt{mkc}(t, \texttt{y})$ and $\texttt{postp}(\texttt{z} \mapsto \ell, w)$ are already effective abbreviations, as from them we can recover the terms $\texttt{y}(t)$ and $\texttt{z}(w)$ present in the context. Further simplification is given by Corrado Biasi's elegant notations:

$$t \to \texttt{y} \ \text{ for } \ \texttt{mkc}(t, \texttt{y}) \quad \text{and} \quad \xleftarrow{\texttt{z} \mapsto \ell} t \ \text{ for } \ \texttt{postp}(\texttt{z} \mapsto \ell, t).$$

If we consider the typed version of the above rewriting we have the following reduction. Let us write[24]

$$\begin{array}{rcl} \mathcal{S}_x : \Delta & \text{for} & \pi_0 : \bullet \mid \overline{\kappa} : \Delta \\ \mathcal{S}_y : \Upsilon & \text{for} & \pi_1 : \bullet \mid \overline{\zeta} : \Upsilon \\ \mathcal{S}_z : \Xi & \text{for} & \pi_2 : \bullet \mid \overline{\xi} : \Xi \\ & \text{and also} & \\ \mathcal{S}_x : \Delta, \mathcal{S}_y : \Upsilon, \mathcal{S}_z : \Xi & \text{for} & \pi_0, \pi_1, \pi_2 : \bullet \mid \overline{\kappa} : \Delta, \overline{\zeta} : \Upsilon, \overline{\xi} : \Xi. \\ & \text{Next} & \\ \text{let} \quad \mathcal{S}_y : \Upsilon & \text{be} & \mathcal{S}_y\{y := \mathtt{y}(t)\} : \Upsilon, \\ \text{let} \quad \mathcal{S}_z : \Xi & \text{be} & \mathcal{S}_z\{z := \mathtt{z}((\mathtt{mkc}(t, \mathtt{y}))\} : \Xi. \end{array}$$

Then we have:

$$\dfrac{\dfrac{x : E \vdash \mathcal{S}_x : \Delta, t : C \qquad y : D \vdash \mathcal{S}_y : \Upsilon}{x : E \vdash \mathcal{S}_x : \Delta, \mathcal{S}_y : \Upsilon,\ \mathtt{mkc}(t, \mathtt{y}) : C \setminus D}\setminus\text{-I} \qquad z : C \vdash \mathcal{S}_z : \Xi, \ell : D}{x : E \vdash \mathtt{postp}(\mathtt{z} \mapsto \ell, \mathtt{mkc}(t, \mathtt{y})) : \bullet, \mathcal{S}_x : \Delta, \mathcal{S}_y : \Upsilon, \mathcal{S}_z : \Xi}\setminus\text{-E}$$

reduces to

$$\dfrac{\dfrac{x : E \vdash \mathcal{S}_x : \Delta, t : C \qquad z : C \vdash \mathcal{S}_z : \Xi, \ell : D}{x : E \vdash \mathcal{S}_x : \Delta, \mathcal{S}_t : \Xi, \ell\{z := t\} : D}\mathit{subst} \qquad y : D \vdash \mathcal{S}_y : \Upsilon}{x : E \vdash \mathcal{S}_x : \Delta, \mathcal{S}_{\ell\{z:=t\}} : \Upsilon, \mathcal{S}_t : \Xi}\mathit{subst}$$

where $\mathcal{S}_t = \mathcal{S}_z\{z := t\}$, $\mathcal{S}_{\ell\{z:=t\}} = \mathcal{S}_y\{y := \ell\{z := t\}\}$

6 A Distributed Term Assignment for Co-intuitionistic Logic $\mathbf{NJ}^{\setminus\Upsilon}$

We present the grammar and the basic definitions of our distributed calculus for the fragment of co-intuitionistic logic with subtraction and disjunction.

Definition 10 *We are given a countable set of* free variables *(denoted by x, y, z …), and a countable set of* unary functions *(denoted by* $\mathtt{x}, \mathtt{y}, \mathtt{z}, \ldots$*).*

(i) *Terms and lists of terms are defined by the following grammar:*

$$\begin{array}{l} t := x \mid \mathtt{x}(t) \mid \mathtt{inl}(t) \mid \mathtt{inr}(t) \mid \mathtt{casel}(t) \mid \mathtt{caser}(t) \mid \mathtt{mkc}(t, \mathtt{x}) \\ \ell := () \mid t \cdot \ell \end{array}$$

(ii) *Let* $t_1, t_2, \ldots$ *an enumeration in a given order of all the* terms *freely generated by the above grammar starting with a special symbol* $*$ *and no variables (a selected variable a would also do the job). Thus we have a fixed* bijection $t_i \mapsto x_i$ *between terms and free variables.*

[24] The expression $\bullet$ is not a formula, but a non-logical expression, which cannot be part of other formulas; its meaning could be though of as an absurdity.

(iii) *Moreover, if* t *is a term and* ℓ *is a list such that for each term* $u \in \ell$, y *occurs in* u, *then* $\mathtt{postp}(\mathtt{y} \mapsto \ell\{y := \mathtt{y}(t)\}, t)$ *is a p-term.*

We use the abbreviations $(t \to \mathtt{y})$ *for* $\mathtt{mkc}(t, \mathtt{y})$ *and* $\xleftarrow{\mathtt{z}\mapsto\ell} w$ *for* $\mathtt{postp}(\mathtt{z} \mapsto \ell, w)$.

Thus a p-term cannot be a subterm of other terms. In the official definition above lists appear only as arguments of postp,[25] It is notationally convenient to extend the above definition so that our operators apply to *lists* in addition to terms:

Definition 11 *Let* $\mathtt{op}(\;)$ *be one of* $\mathtt{x}(\;)$, $\mathtt{inl}(\;)$, $\mathtt{inr}(\;)$, $\mathtt{casel}(\;)$, $\mathtt{caser}(\;)$, $\mathtt{mkc}((\;), \mathtt{x})$, $\mathtt{postp}(\mathtt{x} \mapsto \ell, (\;))$.
Then the term expansion $\mathtt{op}(\ell)$ *is the list of terms defined inductively thus:*

$$\mathtt{op}(()) = () \qquad\qquad \mathtt{op}\,(t \cdot \ell) = \mathtt{op}\,(t) \cdot \mathtt{op}\,(\ell)$$

Remark 5 By term expansion, a term consisting of an operator applied to a list of terms is turned into a list of terms; thus terms may always be transformed into an expanded form where operators are applied only to terms, *except for expressions* ℓ occurring in terms of the form $\mathtt{postp}(\mathtt{y} \mapsto \ell, u)$.

Definition 12 (i) *The* free variables $FV(\ell)$ *in a list of terms* ℓ *are defined as follows:*

$$\begin{aligned}
FV(()) &= \emptyset\\
FV(t \cdot \ell) &= FV(t) \cup FV(\ell)\\
FV(x) &= \{x\}\\
FV(\mathtt{x}(t)) &= FV(t)\\
FV(\mathtt{inl}(t)) = FV(\mathtt{inr}(t)) &= FV(t)\\
FV(\mathtt{casel}(t)) = FV(\mathtt{caser}(t)) &= FV(t)\\
FV(\mathtt{mkc}\,(t, \mathtt{x}) &= FV(t)\\
FV(\mathtt{postp}(\mathtt{x} \mapsto \ell, t) &= FV(\ell) \cup FV(t).
\end{aligned}$$

(ii) *A* computational context $\mathcal{S}_x$ *is a set of terms and p-terms containing the free variable x and no other free variable.*

Definition 13 *Substitution of a term t for a free variable x in a term u is defined as follows:*

[25] In our definition we use lists of terms where multisets are intended. A multiset can be represented as a list $\ell = (t_1, \ldots, t_n)$ with the action of the group of permutations $\sigma : n \to n$ given by $\ell_\sigma = (t_{\sigma(1)}, \ldots, t_{\sigma(n)})$.

$$x\{x := t\} = t, \quad y\{x := t\} = y \text{ if } x \neq y;$$
$$\mathtt{y}(u)\{x := t\} = \mathtt{y}(u\{x := t\});$$
$$\mathtt{inl}(r)\{x := t\} = \mathtt{inl}(r\{x := t\}), \quad \mathtt{inr}(r)\{x := t\} = \mathtt{inr}(r\{x := t\});$$
$$\mathtt{casel}(r)\{x := t\} = \mathtt{casel}(r\{x := t\}), \quad \mathtt{caser}(r)\{x := t\} = \mathtt{caser}(r\{x := t\});$$
$$\mathtt{mkc}(r, \mathtt{y})\{x := t\} = \mathtt{mkc}(r\{x := t\}, \mathtt{y}),$$
$$\mathtt{postp}(\mathtt{y} \mapsto (\ell), s)\{x := t\} = \mathtt{postp}(\mathtt{y} \mapsto (\ell\{x := t\}), s\{x := t\}).$$

We define substitution of a list of terms ℓ for a variable x in a list of terms κ:

$$()\{x := \ell\} = () \qquad (t \cdot \kappa)\{x := \ell\} = t\{x := \ell\} \cdot \kappa\{x := \ell\}$$
$$t\{x := ()\} = () \qquad t\{x := u \cdot \ell\} = t\{x := u\} \cdot t\{x := \ell\}$$

If $\overline{\zeta}$ is a vector of lists $\ell_1, \ldots, \ell_m$, then $\overline{\zeta}\{x := \ell\} = \ell_1\{x := \ell\}, \ldots, \ell_m\{x := \ell\}$.

Definition 14 *β-reduction of a* redex $\mathcal{R}ed$ *in a* computational context $\mathcal{S}_x$ *is defined as follows.*
(i) *If $\mathcal{R}ed$ is a term u of the following form, then the reduction is local and consists of the rewriting $u \rightsquigarrow_\beta u'$ in $\mathcal{S}_x$ as follows:*

$$\mathtt{casel}\ (\mathtt{inl}(t)) \rightsquigarrow_\beta t; \qquad \mathtt{caser}\ (\mathtt{inr}(t)) \rightsquigarrow_\beta t.$$
$$\mathtt{casel}\ (\mathtt{inr}(t)) \rightsquigarrow_\beta (); \qquad \mathtt{caser}\ (\mathtt{inl}(t)) \rightsquigarrow_\beta ();$$

(ii) *If $\mathcal{R}ed$ has the form* $\xleftarrow{\mathtt{z}\mapsto\ell} (t \rightarrow \mathtt{y})$, *i.e.,* $\mathtt{postp}(\mathtt{z} \mapsto \ell, \mathtt{mkc}(t, \mathtt{y}))$, *then $\mathcal{S}_x$ has the form*

$$\mathcal{S}_x = \mathcal{R}ed, \ \overline{\kappa}, \ \overline{\zeta}_{\mathtt{y}}, \ \overline{\xi}_{\mathtt{z}}$$

where $\mathtt{y}(t)$ occurs in $\overline{\zeta}_{\mathtt{y}}$ and $\mathtt{z}((t \rightarrow \mathtt{y}))$ occurs in $\overline{\xi}_{\mathtt{z}}$ and neither $\mathtt{y}(t)$ nor $\mathtt{z}((t \rightarrow \mathtt{y}))$ occurs in $\overline{\kappa}$. Writing $y = \mathtt{y}(t)$ and $z = \mathtt{z}((t \rightarrow \mathtt{y}))$, a reduction *of $\mathcal{R}ed$ transforms the computational context as follows:*

$$\mathcal{S}_x \ \rightsquigarrow \ \overline{\kappa}, \ \overline{\zeta}\{y := \ell\{z := t\}\}, \ \overline{\xi}\{z := t\}.$$

Thus for $\overline{\zeta} = u_1, \ldots, u_k$, for $\overline{\xi} = r_1, \ldots, r_m$ and for $\ell = s_1, \ldots, s_n$ we have:

$$\begin{aligned}
\overline{\xi}\{z := t\} &= r_1\{z := t\}, \ldots, r_m\{z := t\};\\
\overline{\zeta}\{y := \ell\{z := t\}\} &= u_1\{y := s_1\{z := t\}\}, \ldots, u_1\{y := s_n\{z := t\}\}, \ldots\\
&\ldots u_k\{y := s_1\{z := t\}\}, \ldots, u_k\{y := s_n\{z := t\}\};\\
&= \overline{\zeta}\{y := s_1\{z := t\}\}, \ldots, \overline{\zeta}\{y := s_n\{z := t\}\}.
\end{aligned}$$

Given the correspondence between Prawitz style Natural Deduction derivations in $\mathbf{NJ}^{\supset\cap}$ and sequent derivations in t-$\mathbf{LJ}^{\supset\cap}$, and the dual correspondence between Prawitz trees for **co** – $\mathbf{NJ}^{\smallsetminus\curlyvee}$ and sequent derivations in q-$\mathbf{LJ}^{\smallsetminus\curlyvee}$, we find it convenient to define the term assignment directly to sequent calculus in q-$\mathbf{LJ}^{\smallsetminus\curlyvee}$, given in Appendix III, Table 7.

Definition 15 (term assignment) *The assignment of terms of the distributed calculus to sequent calculus derivation in q-*$\mathbf{LJ}^{\searrow\curlyvee}$ *is given in Appendix III, Table 7. In Table 8 we give the familiar assignment of* λ*-terms to sequent calculus with head formulas t-*$\mathbf{LJ}^{\supset\cap}$.

Remark on free variables and α conversion. Since in our calculus the binding of a free variable x is expressed through its substitution with a term $\mathtt{x}(t)$, the so-called "capture of free variables" takes a different form. Suppose a free variable y has been replaced by $\mathtt{y}(t)$ in the construction of a term $M = \mathtt{mkc}(t, \mathtt{y})$ or $P(t) = \mathtt{postp}(\mathtt{y}, \ell\{y := \mathtt{y}(t)\}, t)$: all other occurrences of y in the previous context have been replaced with $\mathtt{y}(t)$ in the current context, represented, say, by a vector $\overline{\ell}$, and we may say that M or $P(t)$ is a *binder* of $\mathtt{y}(t)$ in $\overline{\ell}$.

In the process of normalization such a "bound" term $\mathtt{y}(t)$ may be replaced by another term u. It would be natural to think of such a replacement as a two-step process, first recovering the free variable y and then applying a substitution $\{y := u\}$ to the current computational context. However, it may also happen that in the process of normalization different occurrences of the term $\mathtt{y}(t)$ evolve to $\mathtt{y}(t')$ and to $\mathtt{y}(t'')$ so that distinct variables y' and y'' are needed for distinct substitutions. For this reason we have established a bijection between freely generated terms and free variables.

This may not solve all problems: indeed in the *untyped* formulation of our calculus it might happen that the same free variable y has been replaced with $\mathtt{y}(t)$ in the construction of *two distinct terms* of $\overline{\ell}$: our syntax may not have tools to disambiguate the "scope" of the bindings and some further restrictions may be needed to block such pathologies. However, if the calculus is used for assigning term to derivations in $\mathbf{NJ}^{\searrow\curlyvee}$, then to avoid "capture of free variables" it is enough to set the following condition.

Convention. We assume that

- Derivations have the *pure parameter property*, i.e., that in a derivation free variables assigned to distinct open assumptions are distinct;

Since to distinct free variables x, y there correspond distinct unary functions $\mathtt{x}, \mathtt{y}$, then it is clear that in the term assignment to derivations with the *pure parameter property* the above indicated ambiguity cannot occur. Moreover, a derivation resulting by normalization from a derivation with the pure parameter property can be transformed again into a derivation with the pure parameter property. Indeed, the set of terms assigned to a $\mathbf{NJ}^{\searrow\curlyvee}$ derivation encode a tree-structure, and it is easy to see that if different occurrences of the term $\mathtt{y}(t)$ evolve to $\mathtt{y}(t')$ and to $\mathtt{y}(t'')$ *in a tree*, then the terms t' and t'' are distinct as they encode distinct threads. Thus once again applying the bijection between terms and free variables can be used to produce a derivation with the pure parameter property.

Example (i) Assigning terms to the derivation d_q in Sect. 4.2 we obtain the following assignment to the endsequent:

$$z : C \setminus A \Rightarrow \xleftarrow{\mathsf{c}\mapsto(a',a'',a''')} z : \bullet \mid (t', t'') : \mathbf{F}, ((\mathsf{c}(z) \to \mathsf{e}) \to \mathsf{a}''') : \mathbf{G} \setminus A$$

where we have

$$a' = \mathsf{a}_1(\mathtt{casel}(\mathsf{e}(\mathsf{c}(z)))), a'' = \mathsf{a}_2(\mathtt{caser}(\mathsf{e}(\mathsf{c}(z)))), a''' = \mathsf{a}_3((\mathsf{c}(z) \to \mathsf{e})) : A;$$
$$t' = \mathtt{inl}((\mathtt{casel}(\mathsf{e}(\mathsf{c}(z))) \to \mathsf{a}_1)), t'' = \mathtt{inr}((\mathtt{caser}(\mathsf{e}(\mathsf{c}(z))) \to \mathsf{a}_2)) : \mathbf{F},$$
$$\mathbf{F} = (B \setminus A) \curlyvee (D \setminus A), \mathbf{G} = (C \setminus (B \curlyvee D)).$$

(ii) Applying cut-elimination to the derivation

$$\dfrac{\dfrac{a : A \Rightarrow\ ;\ a : A}{a : A \Rightarrow :\ \mathtt{inl}(a) : A \curlyvee B} \qquad \begin{array}{c}\vdots\\ \\ \end{array}\ \dfrac{a' : A \Rightarrow\ ;\ a' : A \qquad \dfrac{b : B \Rightarrow\ ;\ b : B \qquad c : C \Rightarrow\ c : C\ ;}{b : B \Rightarrow\ \mathsf{c}(b) : C\ ;\ (b \to \mathsf{c}) : B \setminus C}}{e : A \curlyvee B \Rightarrow\ \mathtt{casel}(e) : A, \mathsf{c}(\mathtt{caser}(e)) : C, (\mathtt{caser}(e) \to \mathsf{c}) : B \setminus C\ ;}}{a : A \Rightarrow\ t_1 : A, t_2 : C, t_3 : B \setminus C\ ;}$$

we obtain the following rewritings: $t_1 = \mathtt{casel}(\mathtt{inl}(a)) \rightsquigarrow a;$

$$t_2 = \mathsf{c}(\mathtt{caser}(\mathtt{inl}(a))) \rightsquigarrow (), \qquad t_3 = (\mathtt{caser}(\mathtt{inl}(a)) \to \mathsf{c}) \rightsquigarrow ()$$

and the term assignment

$$a : A \Rightarrow\ a : A, () : C, () : B \setminus C\ ; .$$

6.1 *Duality Between the Distributed Calculus and the Simply Typed λ Calculus*

Consider the term assignment in Appendix III, Tables 7 and 8. In this setting the following facts are clear:

- given a sequent S in q-$\mathbf{LJ}^{\setminus\curlyvee}$, there is a dual sequent $S^{\perp}$ in h-$\mathbf{LJ}^{\supset\cap}$, and conversely;
- given a derivation d of S in q-$\mathbf{LJ}^{\setminus\curlyvee}$, there is a dual derivation $d^{\perp}$ of $S^{\perp}$ in h-$\mathbf{LJ}^{\supset\cap}$, and conversely.

Therefore any cut-elimination procedure in h-$\mathbf{LJ}^{\supset\cap}$ induces a cut-elimination procedure for q-$\mathbf{LJ}^{\setminus\curlyvee}$; clearly the steps of such reduction procedure for q-$\mathbf{LJ}^{\setminus\curlyvee}$ must be seen as "macro" instructions for several steps of rewriting, which may nevertheless be seen as a unit. Thus we have the following fact:

Theorem 2 *There is a correspondence between reduction sequences starting from a derivation d of S in q-*$\mathbf{LJ}^{\setminus\curlyvee}$ *and reduction sequences from a derivation* $d^{\perp}$ *of* $S^{\perp}$*, and conversely.*

In the present setting this result seems obvious and its proof straightforward. Going through the details of the construction, as done in [8], does give an insight into the structure of terminating computations in our distributed calculus. Assigning terms to derivations in q-$\mathbf{LJ}^{\setminus\curlyvee}$ as in in Appendix III, Table 7 makes the structure of the calculus clearer and provides a bridge to the representation of computations in the graphical notation of *Prawitz trees* as in Appendix II.

7 Conclusions

In this chapter we have given an account of research in the logic for pragmatics of *assertions* and *conjectures*, following the paper Bellin and Biasi [5] and also of work in the proof-theory of co-intuitionistic logic aiming at defining natural deduction system and a distributed term-assignment for it.

A conceptual clarification of the distinction between *hypotheses* and *conjectures* with respect to their interpretation in epistemic **S4**, where hypotheses are justified by mere *epistemic possibility* of the truth of their propositional content and conjectures require *possible necessity*, has shown connections with other areas of logic and semantics. On one hand, within our framework we can make distinctions which may be relevant to work on *standards of evidence* in the theory of argumentation [13, 24]. On the other hand, the semantics of *rough sets* and the notion of an *approximation space* provide another semantics to a theory of assertions, hypotheses, conjectures, and expectations, in addition to Kripke models through the translation in epistemic **S4** and in *bimodal* **S4**, as in [5]. Rough sets point at promising connections with research by Pagliani [40, 41].

Abstract relations between functional programming and concurrent programming have been studied extensively, e.g., through translations of the λ calculus into R. Milner's π-calculus. Abstract forms of the continuation-passing style, e.g., as in Thielecke's work, have been typed in classical logic, suggesting an interpretation of these relations as a logical duality between classical and intuitionistic logic. In this way, the $\lambda\mu$ calculus is naturally invoked here. In [8] and this chapter we propose the duality between intuitionistic and co-intuitionistic logic as the most basic type theoretic setting for studying the relations between distributed and functional programming calculi. Our calculus distributed displays exactly the programming features that are required in order to implement such a logical duality. In this way this chapter and other still unpublished work by Corrado Biasi give a type-theoretic framework for studying the relations between *safe* and *unsafe coroutines* in the sense of Crolard: typically, safe coroutines are those which can be represented *as constructs of a distributed calculus* without making essential use of the $\lambda\mu$ calculus *and* can be typed in co-intuitionistic logic. Thus the term assignment to proofs in co-intuitionistic logic can be seen as a contribution to a challenging problem, namely providing a logical foundation to distributed calculi by means of a typing system, in the Curry-Howard approach. Clearly solving such a problem has a clear interest in computer science, if only to ensure properties of such systems such as termination and confluence.

All the paths followed in this research are open and point at possible directions of work, as already suggested. Other projects could explore the proof-net representation of co-intuitionistic logic and the construction of a term model for co-Cartesian Closed Categories. The proof theory of classical logic is the framework of Crolard's investigations [15, 16] and the concern of Bellin, Hyland, Robinson, and Urban [9]: it is expected that eventually research in bi-intuitionistic logic may improve our understanding of classical logic. But this is now a good point to take a rest.

Appendix I. Polarized Rauszer's Logic

The main stream of bi-intuitionistic logic follows the tradition of Cecylia Rauszer, who created the theory of *bi-Heyting algebras* [50, 51], and defined its Kripke semantics, later studied with categorical methods by Lawvere [32], Makkai, Reyes and Zolfaghari [33, 52]; more recently, proof theoretic treatments of *subtractive* or *bi-intuitionistic logic* have been given by Rajeev Gore [25], Tristan Crolard [14, 15] and others.

In Rauszer's possible-world semantics the forcing conditions for *implication* refer to *up-sets* of possible worlds with respect to the accessibility relation, while the forcing conditions for subtraction refer to *down-sets*. Namely, $(A \supset B)^M = \Box(A^M \rightarrow B^M)$ is true in a world w if for all w' such that wRw' $A^M \rightarrow B^M$ is true in w'; on the other hand, $(C \smallsetminus D)^M = \diamondminus (C^M \wedge \neg D^M)$ is true in a world w if for some w' such that $w'Rw$ we have $C^M \wedge \neg D^M$ is true in w'; in other words, modal translations are interpreted in models $\mathcal{M} = (W, R, S, \Vdash)$ where R and S are pre-orders such that $S = R^{-1}$. This suggests a *temporal dimension* in the bi-modal translation: the forcing condition for the operator $\Box$ may be seen as referring to "future knowledge" and those for $\diamondminus$ to "past knowledge."

We see at once that Rauzer's bi-intuitionistic logic is as inadequate for a representation of assertions and hypotheses as **PBL**: letting $\vdash p = \Box p$ and $\mathcal{H}\neg p = \diamondminus \neg p$, it is consistent to assert p (with respect to "the future") and also to conjecture $\neg p$ (in the past). Although the issue is beyond the range of the present chapter, it may be interesting to catch a glimpse of what *polarized* Rauszer's logic looks like in our framework.

Tense-Sensitive Polarization

Is there a *pragmatic interpretation* of dual intuitionistic logic which retains such a temporal element and is thus closer to Rauszer's tradition? The question does make sense. Clearly, the justification conditions for assertions and conjectures concerning the future and the past are different in several important ways: for instance, direct observations of some future events will be possible, but never of past events. Thus it would be plausible to introduce *tense-sensitive* illocutionary operators, giving *assertive* force to statements about the *future* ($\vdash^{\bullet}\alpha$) and about the *past* ($^{\bullet}\!\vdash\alpha$) and, similarly, *conjectural* force to statements about the *future* ($\mathcal{H}^{\bullet}\alpha$) and about the *past* ($^{\bullet}\mathcal{H}\alpha$). Moreover, we would have *strong negation* about the future ($\sim^{\bullet}$) and about the past ($^{\bullet}\!\sim$) and *weak negation* about the future ($\frown^{\bullet}$) and about the past ($^{\bullet}\!\frown$). More generally, all pragmatic formulas would become tense-sensitive and could polarized in four ways:

$$(A^{\bullet})\ \textit{future} - \textit{assertive}, \qquad ({}^{\bullet}A)\ \textit{past} - \textit{assertive},$$
$$(C^{\bullet})\ \textit{future} - \textit{hypothetical} \ \text{ and } \ ({}^{\bullet}C)\ \textit{past} - \textit{hypothetical}:$$

We define a language $\mathcal{L}_t^{PB}$ according to the grammar in Table 6.

This would lead to the development of a *tense-sensitive polarized bi-intuitionistic logic* (**PBL**$_t$). The "semantic reflection" of $\mathcal{L}_t^{PB}$ is in *temporal* **S4**, where formulas of $\mathcal{L}_{\Box,\boxminus}$ are interpreted in bimodal frames $\mathcal{F} = (W, R, S)$ with R a preorder and $S = R^{-1}$.

The following fact is standard (see, e.g., Ryan and Shobbens [53]):

Proposition 5 *Given a bimodal frame $\mathcal{F}$, the following are equivalent:*

1. $S = R^{-1}$;
2. $\alpha \rightarrow \Box \diamondminus \alpha$ *and* $\diamondminus \Box\alpha \rightarrow \alpha$ *are valid in every Kripke model over* $\mathcal{F}$;
3. *the following rule is valid and semantically invertible in* $\mathcal{F}$

$$\frac{\diamondminus\alpha \rightarrow \beta}{\alpha \rightarrow \Box\beta}$$

Therefore the following are equivalent:

1. *the modal interpretation* $(\ )^M$ *of the language* $\mathcal{L}_t^{PB}$ *is in temporal S4;*
2. *for any formula* $\delta^{\bullet}$ *and* ${}^{\bullet}\delta$, *the sequents* $\delta^{\bullet} \Rightarrow \sim^{\bullet}\ {}^{\bullet}\!\sim \delta^{\bullet}$ *and* ${}^{\bullet}\!\frown\frown^{\bullet}\ {}^{\bullet}\delta \Rightarrow {}^{\bullet}\delta$ *are valid axioms of* **PBL**$_\mathbf{t}$.

We leave the task of finding a suitable formalization of **PBL**$_t$ as an open problem.

Appendix II. Example of Computation

In this section we consider an example of computation that is dual to a familiar reduction sequence for Church's numerals.

Table 6 Tense-sensitive polarized bi-intuitionistic language

$A^{\bullet} :=$	$\vdash^{\bullet} p$	$\mid A^{\bullet} \supset B^{\bullet}$	$\mid A^{\bullet} \cap B^{\bullet}$	$\mid \sim^{\bullet} X$
$C^{\bullet} :=$	$\mathcal{H}^{\bullet} p$	$\mid C^{\bullet} \smallsetminus D^{\bullet}$	$\mid C^{\bullet} \curlyvee D^{\bullet}$	$\mid \frown^{\bullet} X$
${}^{\bullet}A :=$	${}^{\bullet}\!\vdash p$	$\mid {}^{\bullet}A \supset {}^{\bullet}B$	$\mid {}^{\bullet}A \cap {}^{\bullet}B$	$\mid {}^{\bullet}\!\sim X$
${}^{\bullet}C :=$	${}^{\bullet}\!\frown p$	$\mid {}^{\bullet}C \smallsetminus {}^{\bullet}D$	$\mid {}^{\bullet}C \curlyvee {}^{\bullet}D$	$\mid {}^{\bullet}\!\frown X$

Two Times Zero

We consider the dual of a computation of the term representing $\mathbf{2} \times \mathbf{0}$:

$$(\lambda m.\lambda n.\lambda f.m(nf))(\lambda g.\lambda x.g(gx))(\lambda h.\lambda z.z) : (A \supset A) \supset (A \supset A)$$

We follow a *call by value* strategy:

$$\begin{array}{lll} \lambda f.(\lambda g.\lambda x.g(gx))((\lambda h.\lambda z.z)f) & \rightsquigarrow \lambda f.(\lambda g.\lambda x.g(gx))(\lambda z.z) & (i) \\ & \rightsquigarrow \lambda f.\lambda x.(\lambda z.z)((\lambda z'.z')x))) & (ii) \\ & \rightsquigarrow \lambda f.\lambda x.((\lambda z.z)x)) & (iii) \\ & \rightsquigarrow \lambda f.\lambda x.x & (iv) \end{array}$$

Labelled Prawitz' Trees

As trees in Prawitz style Natural Deduction $\mathbf{NJ}^{\supset}$ can be decorated with λ terms, so we can assign terms of our dual calculus to Prawitz trees of subformulas for $\mathbf{co-NJ}^{\setminus}$ derivations. For convenience, we still draw trees with the root at the bottom, keeping in mind that here derivations are built from bottom up. We shall use Biasi's notation $(t \rightarrow \mathtt{a})$ for $\mathtt{mkc}(t, a)$ and $\xrightarrow{e \mapsto \ell} t$ for $\mathbf{postp}(e \mapsto \ell, t)$.

$$\mathcal{S}_0 : \xleftarrow{\mathtt{y}\mapsto(b)} g, \quad \xleftarrow[\mathcal{R}ed_1]{\mathtt{g}\mapsto((y\to\mathtt{a}),(a\to\mathtt{b}))} (e \to \mathtt{j}), \quad \xleftarrow{\mathtt{x}\mapsto(x)} h, \quad \xleftarrow[\mathcal{R}ed_0]{\mathtt{h}\mapsto()} (j \to \mathtt{f}), \xleftarrow{\mathtt{e}\mapsto(f)} n$$

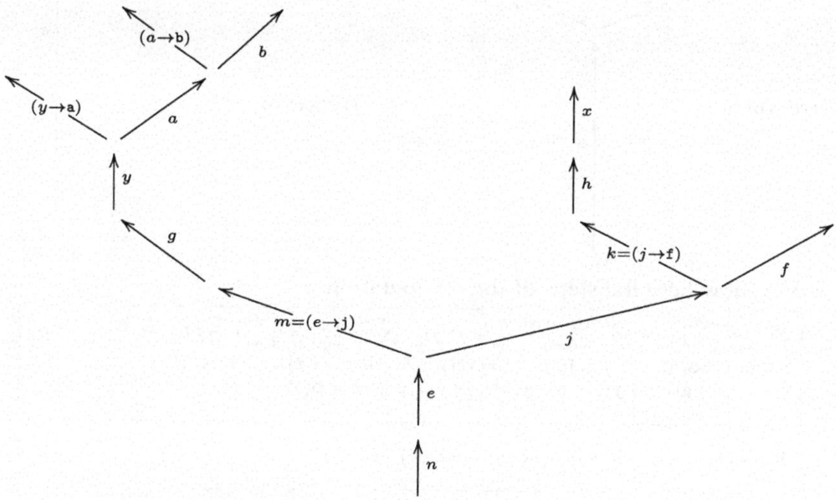

reduces to

$$\mathcal{S}_1 \xleftarrow[\mathcal{R}ed_1]{\mathtt{g}\mapsto((y\to\mathtt{a})^{c}(a\to\mathtt{b})} (e \to \mathtt{j})^{c}$$

$$\xleftarrow{\mathtt{y}\mapsto(b)} g^{c} \quad \xleftarrow{\mathtt{x}\mapsto(x)} j^{c} \quad \xleftarrow{\mathtt{e}\mapsto()} n$$

$$\mathcal{S}_2 : \xleftarrow[\mathcal{R}ed_2]{\mathtt{x}'\mapsto(x')} (y \to \mathtt{a}) \qquad \xleftarrow{\mathtt{y}\mapsto(b)} e$$

$$\xleftarrow[\mathcal{R}ed_3]{\mathtt{x}''\mapsto(x'')} (a \to \mathtt{b}) \qquad \xleftarrow{\mathtt{e}\mapsto()} n$$

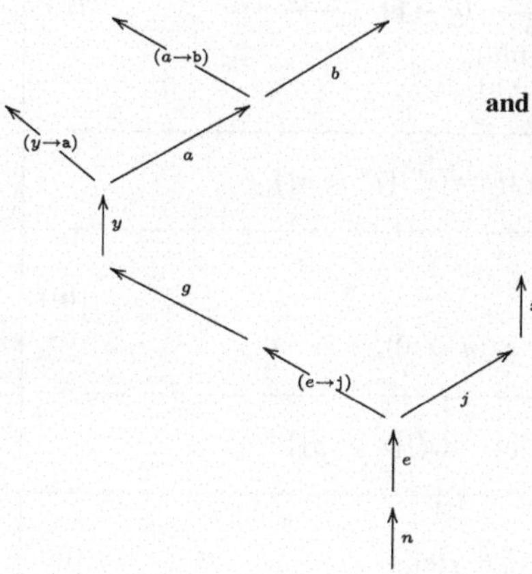

and to

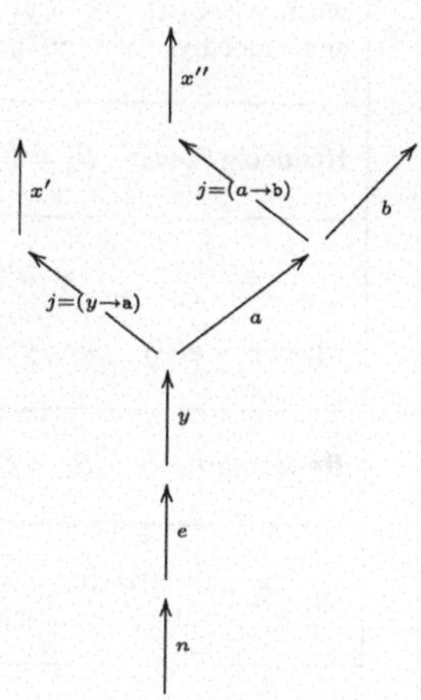

$$\mathcal{S}_3 : \xleftarrow[\mathcal{R}ed_2]{\mathtt{x}'\mapsto(x')} (y \to \mathtt{a}) \quad \xleftarrow{\mathtt{y}\mapsto(a)} e \quad \xleftarrow{\mathtt{e}\mapsto()} n \qquad \mathcal{S}_4 : \xleftarrow{\mathtt{y}\mapsto(y)} e \quad \xleftarrow{\mathtt{e}\mapsto()} n$$

reduces to

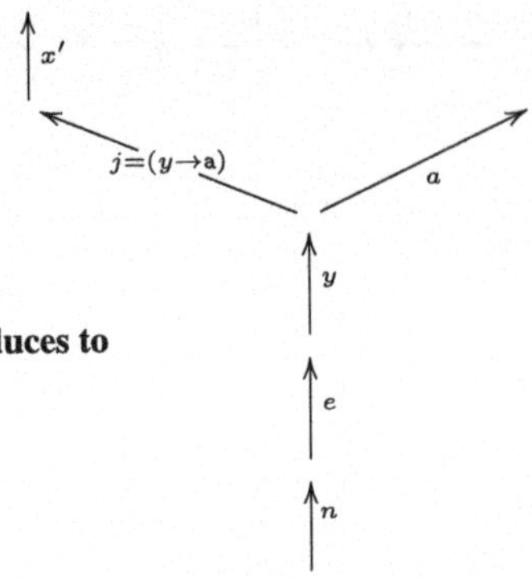

reduces to

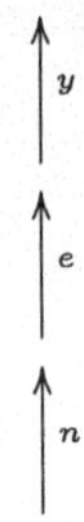

We show here the steps of the computation:

$$\mathcal{S}_0 : \xleftarrow{\mathsf{y}\mapsto(b)} g,\quad \xleftarrow[\mathcal{R}ed_1]{\mathsf{g}\mapsto((y\to\mathsf{a}),(a\to\mathsf{b}))} (e\to \mathsf{j}),\quad \xleftarrow{\mathsf{x}\mapsto(x)} h,\quad \xleftarrow[\mathcal{R}ed_0]{\mathsf{h}\mapsto()} (j\to \mathsf{f}),\quad \xleftarrow{\mathsf{e}\mapsto(f)} n \qquad (o)$$

where $e=\mathbf{e}(n),\quad j=\mathbf{j}(e),\quad f=\mathbf{f}(j),\quad h=\mathbf{h}((j\to\mathsf{f})),\quad x=\mathbf{x}(h)$
$g=\mathbf{g}((e\to\mathsf{j})),\quad y=\mathbf{y}(g),\quad a=\mathbf{a}(y),\quad b=\mathbf{b}(a)$.

Reducing $\mathcal{R}ed_0$: $\mathcal{S}_1=\mathcal{S}_0-\mathcal{R}ed_0\quad \{h:=j\}\,\{f:=()\}$

$$\mathcal{S}_1 : \xleftarrow{\mathsf{y}\mapsto(b)} g,\quad \xleftarrow[\mathcal{R}ed_1]{\mathsf{g}\mapsto((y\to\mathsf{a}),(a\to\mathsf{b}))} (e\to \mathsf{j}),\quad \xleftarrow{\mathsf{x}\mapsto(x)} j,\quad \xleftarrow{\mathsf{e}\mapsto()} n \qquad (i)$$

where $e=\mathbf{e}(n),\quad j=\mathbf{j}(e),\quad x=\mathbf{x}(j)$,
$g=\mathbf{g}((e\to\mathsf{j})),\quad y=\mathbf{y}(g),\quad a=\mathbf{a}(y)\quad b=\mathbf{b}(a)$.

Reducing $\mathcal{R}ed_1$: $\mathcal{S}_2=\mathcal{S}_1-\mathcal{R}ed_1\quad \{g:=e\}\,\{j:=((y\to\mathsf{a}),(a\to\mathsf{b}))\{g:=e\}\}$.

$$\mathcal{S}_2 : \xleftarrow{\mathsf{y}\mapsto(b)} e,\quad \xleftarrow[\mathcal{R}ed_2]{\mathsf{x}\mapsto(x')} (y\to\mathsf{a}),\quad \xleftarrow[\mathcal{R}ed_3]{\mathsf{x}\mapsto(x'')} (a\to\mathsf{b}),\quad \xleftarrow{\mathsf{e}\mapsto()} n \qquad (ii)$$

where $e=\mathbf{e}(n),\quad y=\mathbf{y}(e),\quad a=\mathbf{a}(y),\quad b=\mathbf{b}(a)$,
and, crucially, $x'=\mathbf{x}((y\to\mathsf{a})),\quad x''=\mathbf{x}((a\to\mathsf{b}))$.

Reducing $\mathcal{R}ed_3$: $\mathcal{S}_3=\mathcal{S}_2-\mathcal{R}ed_3\quad \{x'':=a\}\,\{b:=(x'')\{x'':=a\}\}$.

$$\mathcal{S}_3 : \xleftarrow{\mathsf{y}\mapsto(a)} e,\quad \xleftarrow[\mathcal{R}ed_2]{\mathsf{x}\mapsto(x')} (y\to\mathsf{a}),\quad \xleftarrow{\mathsf{e}\mapsto()} n \qquad (iii)$$

where $e=\mathbf{e}(n),\quad y=\mathbf{y}(e),\quad a=\mathbf{a}(y),\quad x'=\mathbf{x}((y\to\mathsf{a}))$.

Reducing $\mathcal{R}ed_2$: $\mathcal{S}_4=\mathcal{S}_3-\mathcal{R}ed_2\quad \{x':=y\}\,\{a:=(x')\{x':=y\}\}$.

$$\mathcal{S}_4 : \xleftarrow{\mathsf{y}\mapsto(y)} e,\quad \xleftarrow{\mathsf{e}\mapsto()} n, \qquad \text{where } e=\mathbf{e}(n),\quad y=\mathbf{y}(e). \qquad (iv)$$

Appendix III. Term Assignment to q-$\mathbf{LJ}^{\smallsetminus\curlyvee}$

Table 7 The sequent calculus q-$\mathbf{LJ}^{\smallsetminus\curlyvee}$

Sequent Calculus q-$\mathbf{LJ}^{\smallsetminus\curlyvee}$

identity rules

logical axiom:

$$x : C \;\Rightarrow\; \pi : \bullet \,|\, ;\, x : C$$

tail-cut:

$$\dfrac{x : E \;\Rightarrow\; \pi_1 : \bullet \,|\, \overline{\ell} : \Upsilon_1 \,;\, t : C \qquad y : C \;\Rightarrow\; \pi_2 : \bullet \,|\, \overline{\kappa} : \Upsilon_2 \,;\, u : D}{x : E \;\Rightarrow\; \pi_1, \pi_2\{y := t\} : \bullet \,|\, \overline{\ell} : \Upsilon_1, \overline{\kappa}\{y := t\} : \Upsilon_2 \,;\, u\{y := t\} : D}$$

central-cut:

$$\dfrac{x : E \;\Rightarrow\; \pi_1 : \bullet \,|\, \overline{\ell} : \Upsilon_1, \ell : C \,; \qquad y : C \;\Rightarrow\; \pi_2 : \bullet \,|\, \overline{\kappa} : \Upsilon_2 ;}{x : E \;\Rightarrow\; \pi_1, \pi_2\{y := \ell\} : \bullet \,|\, \overline{\ell} : \Upsilon_1, \overline{\kappa}\{y := \ell\} : \Upsilon_2 \,;}$$

structural rules

contraction:

$$\dfrac{x : E \;\Rightarrow\; \pi : \bullet \,|\, \overline{\ell} : \Upsilon, \ell : C, \kappa : C \,; u : D}{x : E \;\Rightarrow\; \pi : \bullet \,|\, \overline{\ell} : \Upsilon, \ell * \kappa : C \,; u : D}$$

weakening:

$$\dfrac{x : E \;\Rightarrow\; \pi : \bullet \,|\, \ell : \Upsilon \,;\, u : D}{x : E \;\Rightarrow\; \pi : \bullet \,|\, \ell : \Upsilon, () : C \,;\, u : D}$$

dereliction:

$$\dfrac{x : E \;\Rightarrow\; \pi : \bullet \,|\, \ell : \Upsilon \,;\, u : D}{x : E \;\Rightarrow\; \pi : \bullet \,|\, \ell : \Upsilon, (u) : D \,;}$$

logical rules

$\smallsetminus$ *right:*

$$\dfrac{x : E \;\Rightarrow\; \pi_1 : \bullet \,|\, \overline{\ell} : \Upsilon_1 \,; t : C \qquad y : D \;\Rightarrow\; \pi_2 : \bullet \,|\, \overline{\kappa} : \Upsilon_2 \,;}{E \;\Rightarrow\; \pi_1, \pi_2\{y := \mathtt{y}(x)\} : \bullet \,|\, \overline{\ell} : \Upsilon_1, \overline{\kappa}\{y := \mathtt{y}(x)\} : \Upsilon_2 \,;\, \mathtt{mkc}(t, \mathtt{y}) : C \smallsetminus D}$$

$\smallsetminus$ *left:*

$$\dfrac{x : C \;\Rightarrow\; \pi_1 : \bullet \,|\, \overline{\ell} : \Upsilon, \ell : D \,;}{y : C \smallsetminus D \;\Rightarrow\; \pi_1 : \bullet, \mathtt{postp}(\mathtt{x} \mapsto \ell\{x := \mathtt{x}(y)\}, y) : \bullet \,|\, \overline{\ell}\{x := \mathtt{x}(y)\} : \Upsilon \,;}$$

$\curlyvee_0$ *right:*

$$\dfrac{x : E \Rightarrow \pi : \bullet \,|\, \overline{\ell} : \Upsilon \,;\, t : C_0}{x : E \Rightarrow \pi : \bullet \,|\, \overline{\ell} : \Upsilon \,;\, \mathtt{inl}(t) : C_0 \curlyvee C_1}$$

$\curlyvee_1$ *right:*

$$\dfrac{x : E \;\Rightarrow\; \pi : \bullet \,|\, \overline{\ell} : \Upsilon \,;\, t : C_1}{x : E \;\Rightarrow\; \pi : \bullet \,|\, \overline{\ell} : \Upsilon \,;\, \mathtt{inr}(t) : C_0 \curlyvee C_1}$$

$\curlyvee$ *left:*

$$\dfrac{x : C_0 \;\Rightarrow\; \pi_0 : \bullet \,|\, \overline{\ell}_0 : \Upsilon_0 \,; \qquad y : C_1 \;\Rightarrow\; \pi_1 : \bullet \,|\, \overline{\ell}_1 : \Upsilon_1}{z : C_0 \curlyvee C_1 \;\Rightarrow\; \pi_0', \pi_1' \;: \bullet \,|\, \overline{\ell}_0' : \Upsilon_0, \overline{\ell}_1' : \Upsilon_1}$$

where $\pi_0' = \pi_0\{x := \mathtt{casel}(z)\}, \pi_1' = \pi_1\{y := \mathtt{caser}(z)\}$

$\overline{\ell}_0'\{x := \mathtt{casel}(z)\}, \overline{\ell}_1' = \overline{\ell}_1\{\{y := \mathtt{caser}(z)\}.$

Table 8 The sequent calculus t-$\mathbf{LJ}^{\supset\cap}$

Sequent Calculus t-$\mathbf{LJ}^{\supset\cap}$

identity rules

logical axiom:

$$x : A\,;\ \Rightarrow\ x : A$$

head-cut:

$$\frac{x : C\,;\ \overline{x} : \Theta_1 \Rightarrow t : A \qquad y : A\,;\ \overline{y} : \Theta_2 \Rightarrow u : B}{x : C\,;\ \overline{x} : \Theta_1, \overline{y} : \Theta_2 \Rightarrow u\{y := t\} : B}$$

central-cut:

$$\frac{;\ \overline{x} : \Theta_1 \Rightarrow t : A \qquad ;\ y : A, \overline{y} : \Theta_2 \Rightarrow u : B}{;\ \overline{x} : \Theta_1, \overline{y} : \Theta_2 \Rightarrow u\{y := t\} : B}$$

structural rules

contraction:

$$\frac{y : C\,;\ \overline{x} : \Theta, x_1 : A, x_2 : A \Rightarrow t : B}{y : C\,;\ \overline{x} : \Theta, x : A \Rightarrow t\{x_1 := x, x_2 := x\} : B}$$

weakening:

$$\frac{y : C\,;\ \overline{x} : \Theta, \Rightarrow t : B}{y : C\,;\ \overline{x} : \Theta, x : A \Rightarrow t : B}$$

dereliction:

$$\frac{x : A\,;\ \overline{x} : \Theta \Rightarrow t : B}{;\ x : A, \overline{x} : \Theta \Rightarrow t : B}$$

logical rules

$\supset$ *right*:

$$\frac{;\ \overline{x} : \Theta, x : A \Rightarrow t : C}{;\ \overline{x} : \Theta \Rightarrow \lambda x.t : A \supset B}$$

$\supset$ *left:*

$$\frac{;\ \overline{x} : \Theta_0 \Rightarrow t : A \qquad y : B\,;\ \overline{y} : \Theta_1 \Rightarrow u : C}{f : A \supset B\,;\ \overline{x} : \Theta_0, \overline{y} : \Theta_1 \Rightarrow u\{y := f(t)\} : C}$$

$\cap$ *right*:

$$\frac{;\ \overline{x} : \Theta_0 \Rightarrow t_0 : A_0 \qquad ;\ \overline{y} : \Theta_1 \Rightarrow t_1 : A_1}{;\ \overline{x} : \Theta_0, \overline{y} : \Theta_1 \Rightarrow \langle t_0, t_1 \rangle : A_0 \cap A_1}$$

$\cap_0$ *left*:

$$\frac{x : A_0\,;\ \overline{x} : \Theta \Rightarrow t : C}{z : A_0 \cap A_1\,;\ \overline{x} : \Theta \Rightarrow t[\pi_0(z)/x] : C}$$

$\cap_1$ *left*:

$$\frac{x : A_1\,;\ \overline{x} : \Theta \Rightarrow t : C}{z : A_0 \cap A_1\,;\ \overline{x} : \Theta \Rightarrow t\{x := \pi_1(z)\} : C}$$

References

1. Angelelli, I. (1970). The techniques of disputation in the history of logic. *The Journal of Philosophy*, *67*(20), 800–815.
2. Austin, J. L. (1970). *Philosophical papers* (2nd ed.). USA: Oxford University Press.
3. Bellin, G., & Scott, P. J. (1994). On the Pi-calculus and linear logic. *Theoretical Computer Science*, *135*, 11–65.
4. Bellin, G. (2003). Chu's Construction: A Proof-theoretic Approach. In Ruy J.G.B. de Queiroz (Ed.), Logic for Concurrency and Synchronisation (Vol. 18, pp. 93–114). Dordrecht: Kluwer Trends in Logic.
5. Bellin, G., & Biasi, C. (2004). Towards a logic for pragmatics. Assertions and conjectures. Journal of Logic and Computation, 14(4), 473–506. Special Issue with the Proceedings of the Workshop on Intuitionistic Modal Logic and Application (IMLA-FLOC 2002), V. de Paiva, R. Goré & M. Mendler (Eds.).
6. Bellin, G., & Ranalter, K. (2003). A Kripke-style semantics for the intuitionistic logic of pragmatics ILP. Journal of Logic and Computation, 13(5), 755–775. Special Issue with the Proceedings of the Dagstuhl Seminar on Semantic Foundations of Proof-search, Schloss Dagstuhl.
7. Bellin, G., & Dalla Pozza, C. (2002). A pragmatic interpretation of substructural logics. Reflection on the Foundations of Mathematics (Stanford, CA, 1998), Essays in honor of Solomon Feferman, W. Sieg, R. Sommer and C. Talcott (Eds.), Association for Symbolic Logic, Urbana, IL, Lecture Notes in Logic, (Vol. 15, pp. 139–163).
8. Bellin, G. (2005). A Term Assignment for Dual Intuitionistic Logic. Conference paper presented at the LICS'05-IMLA'05 Workshop, Chicago, IL, June 30.
9. Bellin, G., Hyland, M., Robinson, E., & Urban, C. (2006). Categorical proof theory of classical propositional calculus. *Theoretical Computer Science*, *364*(2), 146–165.
10. Biasi, C. (2003). *Verso una logica degli operatori prammatici asserzioni e congetture*. Tesi di Laurea: Facoltà di Scienze, Università di Verona.
11. Biasi, C., & Aschieri, F. (2008). A term assignment for polarized bi-intuitionistic logic and its strong normalization. Fundamenta Informaticae, 84(2), 185–205. Special issue on Logic for Pragmatics.
12. Bing, J. (1982). Uncertainty, decisions and information systems. In C. Ciampi (Ed.), *Artificial Intelligence and Legal Information Systems*. Amsterdam: North-Holland.
13. Brewka, G. & Gordon, T. (2010). Carneades and Abstract Dialectical Frameworks: A Reconstruction. In: P. Baroni, M. Giacomin & G. Simari (Ed.), Computational Models of Argument, Proceedings of COMMA 2010, IOS Press.
14. Crolard, T. (1996). *Extension de l'isomorphisme de Curry-Howard au traitement des exceptions (application d'une ètude de la dualité en logique intuitionniste)* (p. 7). Thèse de Doctorat: Université de Paris.
15. Crolard, T. (2001). Subtractive logic. *Theoretical Computer Science*, *254*(1–2), 151–185.
16. Crolard, T. (2004). A formulae-as-types interpretation of subtractive logic. Journal of Logic and Computation, 14(4), 529–570. Special Issue with the Proceedings of the Workshop on Intuitionistic Modal Logic and Application (IMLA-FLOC 2002), V. de Paiva, R. Goré and M. Mendler (Eds.).
17. Curien, P. L. (2002). Abstract Machines, Control, and Sequents. In G. Barthe, P. Dybjer, L. Pinto, J. Saraiva (Eds.), Applied Semantics, International Summer School, APPSEM 2000, Caminha, Portugal, September 9–15, Advanced Lectures. Lecture Notes in Computer Science, Vol. 2395, Springer.
18. Dalla Pozza, C., & Garola, C. (1995). A pragmatic interpretation of intuitionistic propositional logic. *Erkenntnis*, *43*(1995), 81–109.
19. Dalla Pozza, C. (1997). Una logica prammatica per la concezione "espressiva" delle norme. In A. Martino (Ed.), *Logica delle Norme*. Pisa: Servizio Editoriale Universitario.
20. Danos, V., Joinet, J.-B., & Schellinx, H. (1995). LKQ and LKT: Sequent calculi for second order logic based upon dual linear decompositions of classical implication. In Proceedings of

the workshop on Advances in linear logic (pp. 211–224). New York: Cambridge University Press.
21. Dummett, M. (1991). *The Logical Basis of Metaphysics*. Cambridge: Cambridge University Press.
22. Girard, J.-Y. (1993). On the unity of logic. *Annals of Pure and Applied Logic*, *59*, 201–217.
23. Gödel, K. (1933). Eine Interpretation des Intuitionistischen Aussagenkalküls. Ergebnisse eines Mathematischen Kolloquiums IV, pp. 39–40.
24. Gordon, T. F., & Walton, D. (2009). Proof burdens and standards In: I. Rahwan & G. Simari (Eds.), Argumentations in Artificial Intelligence (pp. 239–258). New York: Springer.
25. Goré, R. (2000). Dual intuitionistic logic revisited. In TABLEAUX00: Automated reasoning with analytic tableaux and related methods, LNAI (Vol. 1847, pp. 252–267). Berlin: Springer.
26. Hyland, J. M., & De Paiva, V. C. V. (1993). Full intuitionistic linear logic. *Annals of Pure and Applied Logic*, *64*, 273–291.
27. Kripke, S. A. (1963). Semantical analysis of modal logic I: Normal modal propostional calculi. Zeitschrift für Mathematische Logik und Grundlagen der Mathematik, 9.
28. Kripke, S. A. (1965). Semantical analysis of intuitionistic logic I. Formal Systems and Recursive Functions. In J. N. Crossley & M. A. E. Dummett (Eds.), Studies in logic and the foundations of mathematics (pp. 92–130). Amsterdam: North-Holland.
29. Krivine, J. L. (1990). *Lambda-calcul, types et modèles*. Paris: Masson.
30. Lambek, J., & Scott, P. J. (1986). *Introduction to higher order categorical logic*. Cambridge: Cambridge University Press.
31. Levinson, S. C. (1983). *Pragmatics*. Cambridge: Cambridge University Press.
32. Lawvere, F. W. (1991). Intrinsic co-Heyting boundaries and the Leibniz rule in certain toposes. In A. Carboni, M. C. Pedicchio, & G. Rosolini (Eds.), *Category Theory (Como 1990), Lecture Notes in Mantehmatics* (Vol. 1488, pp. 279–297). Verlag: Springer.
33. Makkai, M., & Reyes, G. E. (1995). Completeness results for intuitionistic and modal logic in a categorical setting. *Annals od Pure and Applied Logic*, *72*, 25–101.
34. Martin-Löf, P. (1985). On the meaning and justification of logical laws. In Bernardi & Pagli (Eds.) Atti degli Incontri di Logica Matematica, vol. II, Universita‘ di Siena.
35. Martin-Löf, P. (1987). Truth of a proposition, evidence of a judgement, validity of a proof. *Synthese*, *73*, 407–420.
36. McKinsey, J. C. C., & Tarski, A. (1948). Some theorems about the sentential calculi of Lewis and Heyting. *Journal of Symbolic Logic*, *13*, 1–15.
37. Miglioli, P., Moscato, U., Ornaghi, M., & Usberti, U. (1989). A constructivism based on classical truth. *Notre Dame Journal of Formal Logic*, *30*, 67–90.
38. Nelson, D. (1949). Constructible falsity. *The Journal of Symbolic Logic*, *14*, 16–26.
39. Pagliani, P., & Chakraborty, M. (2008). *A geometry of approximation rough set theory: Logic algebra and topology of conceptual patterns*. Verlag: Springer.
40. Pagliani, P. (2009). Intrinsic co-Heyting boundaries and information incompleteness in Rough Set Analysis. In L. Polkowski & A. Skowron (Eds.), RSCTC 1998. Lecture Notes in Computer Science (Vol. 1424, pp. 123–130). Heidelberg: Springer.
41. Pagliani, P. (1997). Information gaps as communication needs: A new semantic foundation for some non-classical logics. *Journal of Logic, Language and Information*, *6*(1), 63–99.
42. Parigot, M. (1991). Free Deduction: An Analysis of "Computations" in Classical Logic. In A. Voronkov (Ed.), Logic Programming, Proceedings of the First Russian Conference on Logic Programming, Irkutsk, Russia, September 14–18, 1990–Second Russian Conference on Logic Programming, St. Petersburg, Russia, September 11–16, Springer Lecture Notes in Computer Science, ISBN 3-540-55460-2, London: Springer, pp. 361-380.
43. Parigot, M. (1992). Lambda-My-Calculus: An Algorithmic Interpretation of Classical Natural Deduction. In A. Voronkov (Ed.), Logic Programming and Automated Reasoning, Proceedings of the International Conference LPAR'92, St. Petersburg, Russia. Springer Lecture Notes in Computer Science 624, 1992, ISBN 3-540-55727-X, pp. 190-201.

44. Parigot, M.,(1993). Strong Normalization for Second Order Classical Natural Deduction. In LICS,. (1993). *Proceedings* (pp. 39–46). Canada: Eighth Annual IEEE Symposium on Logic in Computer Science Montreal.
45. Polkowski, L. (2002). *Rough Sets*. Physical-Verlag: Mathematical Foundations. Advances in Soft Computing. Heidelberg.
46. Prawitz, D. (1965). *Natural deduction. A proof-theoretic study*. Stockholm: Almquist and Wikksell.
47. Prawitz, D. (1987). Dummett on a theory of meaning and its impact on logic. In B. Taylor (Ed.), *Michael Dummett: Contributions to philosophy* (pp. 117–165). The Hague: Nijhoff.
48. Ranalter, K. (2008). A semantic analysis of a logic for pragmatics with assertions, obligations, and causal implication. *Fundamenta Informaticae*, *84*(3–4), 443–470.
49. Ranalter, K. (2008). Reasoning about assertions, obligations and causality: on a categorical semantics for a logics for pragmatics. PhD Thesis, Queen Mary, University of London and University of Verona.
50. Rauszer, C. (1974). Semi-Boolean algebras and their applications to intuitionistic logic with dual operations. *Fundamenta Mathematicae*, *83*, 219–249.
51. Rauszer, C. (1977). Applications of Kripke models to Heyting-Brouwer logic. *Studia Logica*, *36*, 61–71.
52. Reyes, G., & Zolfaghari, H. (1996). Bi-Heyting algebras, toposes and modalities. *Journal of Philosophical Logic*, *25*, 25–43.
53. Ryan, M., & Schobbens, P.-Y. (1997). Counterfactuals and updates as inverse modalities. *Journal of Logic, Language and Information*, *6*(2), 123–146.
54. Selinger, P. (2001). Control categories and duality: On the categorical semantics of the lambda-mu calculus. *Mathematical Structures in Computer Science*, *11*, 207–260.
55. Shapiro, S. (1985). Epistemic and Intuitionistic Arithmetic. In S. Shapiro (Ed.), *Intensional mathematics* (pp. 11–46). Amsterdam: North-Holland.
56. Shapiro, S. (1997). *Philosophy of mathematics : structure and ontology*. New York: Oxford University Press.
57. Stell, J., & Worboys, M. (1997). The Algebraic Structure of Sets of Regions. In S. Hirtle & A. Frank (Eds.), *Spatial Information Theory, COSIT'97 Proceedings Lecture Notes in Computer Science* (Vol. 1329, pp. 163–174). Verlag: Springer.
58. Thomason, R. (1969). A semantical study of constructible falsity. *Zeitschrift fur mathematische Logik und Grundlagen der Mathematik*, *15*, 247–257.
59. Troelstra, A. S. & Schwichtenberg, H. (1996). Basic proof theory. Cambridge tracts in theoretical computer science (Vol. 43). Cambridge: Cambridge University Press.
60. White, G. (2008). Davidson and Reiter on actions. *Fundamenta Informaticae*, *84*(2), 259–289.

Decomposition of Reduction

Ernst Zimmermann

Abstract The method of decomposition of reduction is presented for proofs of confluence and termination in rewriting and natural deduction. It is shown that any subclass of reductions, which is locally confluent and terminating and commutes with the full reduction in a specific way, transfers confluence and termination to the full reduction. The method is introduced on a pure rewriting level and subsequently applied to natural deduction, i.e. its calculi for Intuitionistic Logic and Intuitionistic Linear Logic, where respective subclasses of reductions are shown to exist. A key observation is that local confluence holds for reductions in natural deduction.

1 Natural Deduction and Rewriting

From time to time it may be useful in science to keep different theories apart from each other as far as possible, such that their genuine basic concepts can be settled and their own merits shown. But in other times it may be of interest to relate different theories because they can be shown to shed new light onto each other. This chapter would like to give evidence that the two theories of Rewriting on the one side and Natural Deduction on the other side, could benefit very much from each other if applied to each other more intensively as previously done. Of course there are various points of touch of Rewriting and Natural Deduction, especially if the formulas-as-types interpretation of Natural Deduction is envisaged. Most proofs of strong normalisation for Natural Deduction Intuitionistic Logic employe type theoretical formalisms, which in turn use Rewriting for the description of reductions of terms; see Prawitz, Gandy, Girard, Schwichtenberg, Joachimski and Matthes [3, 5–9]. And concepts like confluence and termination being of special interest for reductions

E. Zimmermann (✉)
DFG and ANR Project Hypotheses, Wilhelm-Schickard-Institut, Universität Tübingen, Sand 13, 72076 Tübingen, Germany
e-mail: ernst.z@gmx.de

L. C. Pereira et al. (eds.), *Advances in Natural Deduction*, Trends in Logic 39, DOI: 10.1007/978-94-007-7548-0_11,

in calculi of Natural Deduction constitute the heart of Rewriting. Nevertheless, the chapter would like to show that there is much more to be done. Rewriting for instance does not know only one concept of confluence, besides full confluence there are the concepts of local confluence and of semi confluence. Moreover, there is the theorem of decidability of local confluence for term rewriting systems, due to Knuth and Bendix. So, it may be regarded as a legitimate question to ask if these concepts give some insight into reduction classes of Natural Deduction or not. And, indeed, it will be shown in this chapter that reductions in Natural Deduction are locally confluent, i.e. it is not only the case that local confluence of reductions in Natural Deduction is decidable, but better, reductions in Natural Deduction *are* locally confluent. This surprising fact can be considered as a result of more interest for Natural Deduction than for example a theorem on confluence. First local confluence can be locally checked in contrast to full confluence; and second full confluence is a consequence of local confluence given termination, shown by Newman.

But even the other way round, the view on Rewriting from the point of Natural Deduction shows remarkable things. It is explored in this chapter that reductions of Natural Deduction allow something of a general interest to Rewriting. There are indeed nice subclasses of reductions in Natural Deduction, with very nice properties. To be more precise. The chapter details that reductions in Natural Deduction can be decomposed into important subclasses. Such a subclass of reduction, called its decomposition, is locally confluent and terminating; so, the decomposition is fully confluent and terminating. Furthermore, since this subclass of reduction does interact or commute in a specific way with the full reduction, the properties of confluence and termination transfer to the full reduction class. It is this method, the decomposition of reductions, which primarily constitute a new stock of results in Rewriting and secondarily may be applied to various term rewriting systems, here to Natural Deduction.

The chapter is organised as follows. In the first part a confluence and a termination theorem are proved for Abstract Rewriting Systems by use of decomposition of reductions. In the second part two calculi of Natural Deduction are defined, Intuitionistic Logic and Intuitionistic Linear Logic and the lemma on local confluence of reductions in Natural Deduction is proved. The final two parts consist of confluence and termination theorems for each of the defined calculi by applying the method of decomposition of reductions.

The investigations in the sequel do not contain quantifiers. But the author believes that the universal quantifier and the existential quantifier can be added to the calculi of Natural Deduction, Intuitionistic Logic and Intuitionistic Linear Logic, while retaining the stated results on confluence and termination, let alone the respective conversions and permutations are included. Of course, there is one thing which is a true prerequisite: the sensitive treatment of variables, i.e. the separation of free and bound variables, the separation of proper and non-proper variables. But this is cumbersome and well-known stuff to the logician.

2 Abstract Reduction Systems

The section establishes basic concepts of Abstract Reduction Systems, together with diagram proofs for confluence and termination theorems. A new proof of confluence and a new proof of termination of a rewriting relation $\succ$ constitute the heart of the section resulting from decomposing $\succ$ in subrelations.

Definition 1 An Abstract Reduction System is a pair $\mathcal{A} = \langle A, \succ \rangle$, where $A \neq \emptyset$ and $\succ \subseteq A \times A$ a relation of reduction. Products of $\succ$:
$\alpha \succ^0 \beta$ iff $\alpha = \beta$;
$\alpha \succ^1 \beta$ iff $\alpha \succ \beta$;
$\alpha \succ^{k+1} \beta$ iff $\exists\gamma$ s.t. $\alpha \succ^k \gamma$ and $\gamma \succ \beta$.
$\alpha \succeq \beta$ iff $\alpha \succ \beta$ or $\alpha = \beta$.
$\succ^* = \bigcup_k \succ^k$, the transitive, reflexive closure of $\succ$.
$\alpha \prec \beta$ iff $\beta \succ \alpha$.
Similarly, other reduction relations $\dot{\succ} \subseteq A \times A$ are defined.

$\succ$ is *confluent* iff $\forall\alpha, \beta, \gamma(\alpha \succ^* \beta \wedge \alpha \succ^* \gamma \rightarrow \exists\delta(\beta \succ^* \delta \wedge \gamma \succ^* \delta))$.

$\succ$ is *locally confluent* iff $\forall\alpha, \beta, \gamma(\alpha \succ \beta \wedge \alpha \succ \gamma \rightarrow \exists\delta(\beta \succ^* \delta \wedge \gamma \succ^* \delta))$.

$\succ$ is *strictly locally confluent* iff
$\forall\alpha, \beta, \gamma(\alpha \succ \beta \wedge \alpha \succ \gamma \rightarrow \exists\delta(\beta \succ \delta \wedge \gamma \succ^* \delta \vee \beta \succ^* \delta \wedge \gamma \succ \delta))$.

$\succ$ is *normal* iff $\forall\alpha\exists\beta, k\forall\gamma(\alpha \succ^k \beta \wedge \neg\beta \succ \gamma)$.

$\succ$ is *terminating* iff $\neg\exists\alpha_0, \alpha_1, \alpha_2, \ldots(\alpha_0 \succ \alpha_1 \wedge \alpha_1 \succ \alpha_2 \wedge \ldots)$

$\succ^*$ *induces well-founded* iff $\forall\alpha(\forall\beta(\alpha \succ^* \beta \rightarrow P(\beta)) \rightarrow P(\alpha)) \rightarrow \forall\gamma P(\gamma)$.

$\dot{\succ}$ *commutes with* $\succ$ iff $\forall\alpha, \beta(\alpha \succ \beta \rightarrow \exists\gamma, \delta, m, n(\alpha \dot{\succ}^m \gamma \wedge \gamma \dot{\succ} \delta \wedge \beta \dot{\succ}^n \delta))$.

The here defined concept of commutation is not equivalent to other concepts of commutation in rewriting. Except strict local confluence the other concepts are well-known.

The listed properties of reduction have a well-known graphical content shown below. The graphs follow the usual convention that unbroken lines indicate universal quantifiers and broken lines existential ones.

$\bullet \longrightarrow \bullet \quad = \quad \alpha \succ \beta \qquad \bullet \rightsquigarrow \bullet \quad = \quad \exists k : \alpha \succ^k \beta$

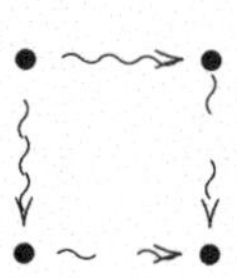

Confluence

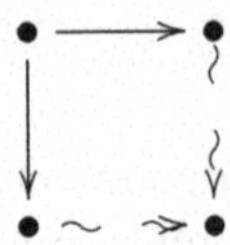

Local Confluence

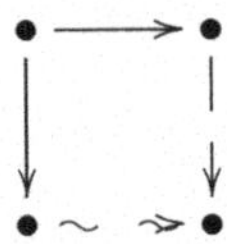

Strict Local Confluence

Theorem 2 *Well-founded induction holds on* $\succ^*$ *iff* $\succ$ *is terminating.*

Proof well-known, see for example Baader and Nipkow [1]. The only-if-direction. If well-founded induction of $\succ^*$ does not hold, there is by first-order reasoning an infinite descending chain, i.e. $\succ$ does not terminate. The if-direction. Define $P(\alpha)$ to be the statement that termination holds for α. Now, assume that for every point β *s.t.* $\alpha \succ^* \beta$, $P(\beta)$; then $P(\alpha)$; so, by well-founded induction $P(\alpha)$ for all α.

Theorem 3 $\succ$ *is confluent if locally confluent and terminating.*

This is Newman's Lemma and its proof by well-founded induction from Huet.

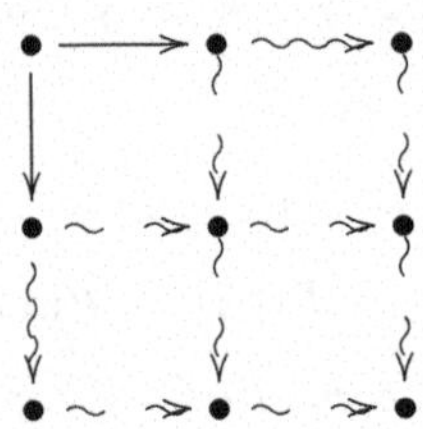

Lemma 4 *If* $\dot{\succ}$ *is confluent and commutes with* $\succ$ *it holds:*
$\forall\alpha\beta k(\alpha \succ^k \beta \rightarrow \exists\gamma mn(\alpha\dot{\succ}^m\gamma \wedge \beta\dot{\succ}^n\gamma)$.

Proof by induction on k, with induction step below.

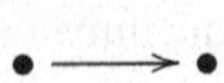 $=$ $\alpha \succ \beta$  $=$ $\exists k : \alpha \succ^k \beta$

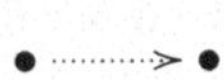 $=$ $\exists\beta, k : \alpha\dot{\succ}^k\beta$

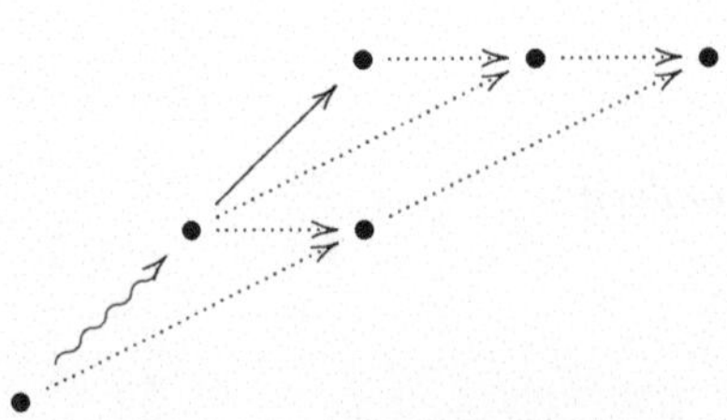

Theorem 5 *Rewriting Confluence by Decomposition*
$\succ$ is confluent if $\dot{\succ}$ exists: $\dot{\succ}\ \subseteq \succ$, confluent and commuting with $\succ$.

Proof by the lemma above. This proof of confluence of $\succ$ is key to this chapter.

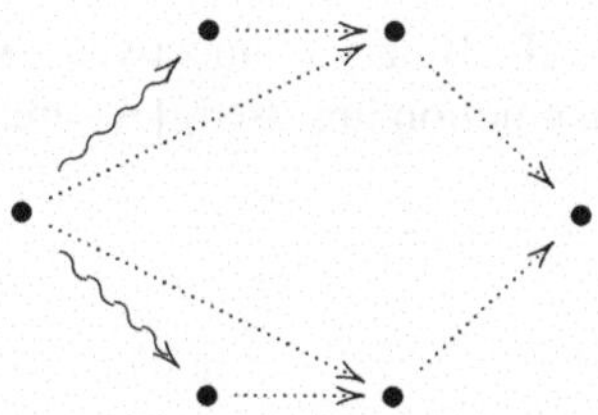

Theorem 6 *Rewriting Termination by Decomposition*
$\succ$ terminates if $\dot{\succ}$ exists: confluent, terminating and commuting with $\succ$.

Proof by contrary. This proof of termination is key to this chapter.

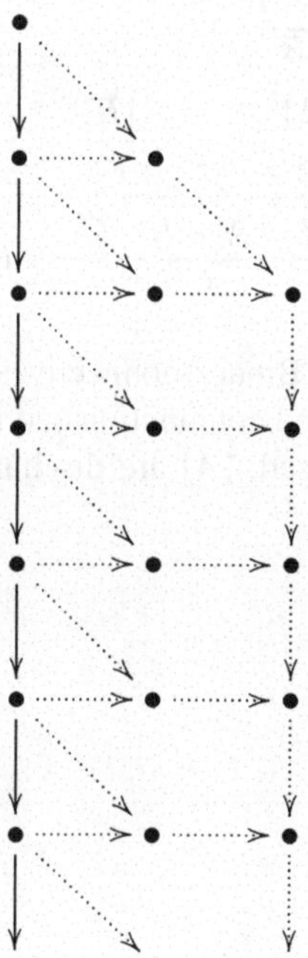

3 Basics of Natural Deduction

This section contains the definition of two calculi of Natural Deduction, which are in the focus of this chapter: intuitionistic logic and intuitionistic linear logic, see [4, 10]. And it contains the lemma on strict local confluence.

Definition 7 Intuitionistic Logic IL. Binary connectives $\to$, $\wedge$, $\vee$ for intuitionistic implication, conjunction, and disjunction are used. $[A]$ are discharged assumption sets.

$A \quad BR$

$$\frac{\begin{array}{c}[A]^v\\ \vdots\\ B\end{array}}{A \to B}\to Iv \qquad \frac{\begin{array}{cc}\vdots & \vdots\\ A \to B & A\end{array}}{B}\to E$$

$$\frac{\begin{array}{cc}\vdots & \vdots\\ A & B\end{array}}{A \wedge B}\wedge I \qquad \frac{\begin{array}{c}\vdots\\ A \wedge B\end{array}}{A}\wedge El \qquad \frac{\begin{array}{c}\vdots\\ A \wedge B\end{array}}{B}\wedge Er$$

$$\frac{\begin{array}{c}\vdots\\ A\end{array}}{A \vee B}\vee Il \qquad \frac{\begin{array}{c}\vdots\\ B\end{array}}{A \vee B}\vee Ir \qquad \frac{\begin{array}{ccc}[A]^v & & [B]^w\\ \vdots & \vdots & \vdots\\ C & A \vee B & C\end{array}}{C}\vee Evw$$

Definition 8 Intuitionistic Linear Logic ILL. Binary connectives $\to$, $\bullet$, $\cap$, $\cup$ for intuitionistic linear multiplicative implication and conjunction and intuitionistic linear additive conjunction and disjunction are used. $[A]$ are discharged assumption singletons, Γ are multisets

$A \quad BR$

$$\frac{\begin{array}{c}[A]^v\\ \vdots\\ B\end{array}}{A \to B}\to Iv \qquad \frac{\begin{array}{cc}\vdots & \vdots\\ A & A \to B\end{array}}{B}\to E$$

$$\frac{\begin{array}{cc}\vdots & \vdots\\ A & B\end{array}}{A \bullet B}\bullet I \qquad \frac{\begin{array}{cc} & [A]^v[B]^w\\ \vdots & \vdots\\ A \bullet B & C\end{array}}{C}\bullet Evw$$

$$
\frac{\begin{matrix}\Gamma & [\Gamma]^v\\ \vdots & \vdots\\ A & B\end{matrix}}{A\cap B}\cap Iv
\qquad
\frac{\begin{matrix}[\Gamma]^v & \Gamma\\ \vdots & \vdots\\ A & B\end{matrix}}{A\cap B}\cap Iv
\qquad
\frac{\begin{matrix}\vdots\\ A\cap B\end{matrix}}{A}\cap El
\qquad
\frac{\begin{matrix}\vdots\\ A\cap B\end{matrix}}{B}\cap Er
$$

$$
\frac{\begin{matrix}\vdots\\ A\end{matrix}}{A\cup B}\cup Il
\qquad
\frac{\begin{matrix}\vdots\\ B\end{matrix}}{A\cup B}\cup Ir
\qquad
\frac{\begin{matrix}[A\Delta]^v & & [B]^w\Delta\\ \vdots & \vdots & \vdots\\ C & A\cup B & C\end{matrix}}{C}\cup Evw
\qquad
\frac{\begin{matrix}[A]^v\Delta & & [B\Delta]^w\\ \vdots & \vdots & \vdots\\ C & A\cup B & C\end{matrix}}{C}\cup Evw
$$

In the sequel the variable labels of rules in deductions are treated more or less relaxed, so often variables in one and the same rule clash to one variable.

The *degree* of a formula is a natural number, defined inductively: the degree of formula AbC with binary connective b is larger than the degrees of formulas A and C. A *premise* of a rule is a formula immediately above a line of a rule, a *conclusion* is a formula immediately below a line of a rule. A *major premise* is a premise of an introduction rule or that premise of an elimination rule which contains the logical connective of the rule. A *minor premise* is a premise being not major. A *cut segment* is a sequence of occurrences $C_1, \ldots, C_k$ of one and the some formula C, where C_{i+1} is a conclusion with C_i as premise, C_1 is a conclusion of an introduction rule and C_k is the major premise of an elimination rule; the *length* of a cut segment is the number of occurrences in the sequence; the degree of a cut segment is the degree of the formula of the sequence. A *cut* is a cut segment of length 1. The *context* of a rule is the class of assumptions of premises occurring pairwise in rules $\cap I$ and the class of assumptions of minor premises occurring pairwise in $\cup E$.

Lemma 9 *Strict Local Confluence in ND*
$\succ$ *is strictly locally confluent in IL and ILL:*
$\forall \alpha, \beta, \gamma(\alpha \succ \beta \wedge \alpha \succ \gamma \rightarrow \exists \delta, m(\beta \succ \delta \wedge \gamma \succ^m \delta \vee \beta \succ^m \delta \wedge \gamma \succ \delta)).$

Proof By investigating all possible pairs of reductions, conversions and permutations, operators of constructive propositional natural deduction can be shown to be strictly locally confluent: from intuitionistic $\rightarrow, \wedge, \vee$ to linear $\rightarrow, \bullet, \cap, \cup$.
Before starting the proof by cases of all possible pairs of reductions in a certain logic let us observe the following general facts about conversions and permutations, which are defined in the sequel:
Conversions and permutations are uniquely defined, i.e. given cut segment A in a deduction, there is exactly one conversion or permutation to be applied on A.
Conversions and permutations sometimes remove or multiply whole subdeductions and other cut segments therein.
Permutations may lengthen cut segments B while applied on cut segment A for shortening.

Now a proof of the lemma for intuitionistic logic $\rightarrow, \wedge, \vee$ is given, which proceeds by investigating all possible pairs of reductions.

Assume $\mathcal{D}$ with two cut segments $A \to B$ and E and $\mathcal{D} \succ \mathcal{D}'$ as below and $\mathcal{D} \succ \mathcal{D}''$ for the reduction of E and $k = |[A]|$. There are 3 cases.
E is in $\mathcal{D}^1$: then $\exists \mathcal{D}'''$ s.t. $\mathcal{D}'' \succ \mathcal{D}'''$ and $\mathcal{D}' \succ \mathcal{D}'''$.
E is in $\mathcal{D}^2$: then $\exists \mathcal{D}''', k$ s.t. $\mathcal{D}'' \succ \mathcal{D}'''$ and $\mathcal{D}' \succ^k \mathcal{D}'''$.
E is in $\mathcal{D}^3$: these subcases are covered by the other subcases.
So, in all cases strict local confluence holds.

$$\mathcal{D} = \begin{matrix} \dfrac{\dfrac{\begin{matrix}[A]^u\\ \mathcal{D}^1\\ B\end{matrix}}{A \to B}{\to}Iu \quad \begin{matrix}\mathcal{D}^2\\ A\end{matrix}}{B}{\to}E \\ \mathcal{D}^3 \end{matrix} \succ \begin{matrix}\mathcal{D}^2\\ [A]\\ \mathcal{D}^1\\ B\\ \mathcal{D}^3\end{matrix} = \mathcal{D}'$$

Assume $\mathcal{D}$ with two cut segments $A \wedge B$ and E and $\mathcal{D} \succ \mathcal{D}'$ as below and $\mathcal{D} \succ \mathcal{D}''$ for the reduction of E. There are 3 cases.
E is in $\mathcal{D}^1$: then $\exists \mathcal{D}'''$ s.t. $\mathcal{D}'' \succ \mathcal{D}'''$ and $\mathcal{D}' = \mathcal{D}'''$.
E is in $\mathcal{D}^2$: then $\exists \mathcal{D}'''$ s.t. $\mathcal{D}'' \succ \mathcal{D}'''$ and $\mathcal{D}' \succ \mathcal{D}'''$.
E is in $\mathcal{D}^3$: these subcases are covered by the other subcases.
So, in all cases strict local confluence holds.

$$\mathcal{D} = \begin{matrix} \dfrac{\dfrac{\begin{matrix}\mathcal{D}^1\\ A\end{matrix} \quad \begin{matrix}\mathcal{D}^2\\ B\end{matrix}}{A \wedge B}{\wedge}I}{B}{\wedge}E \\ \mathcal{D}^3 \end{matrix} \succ \begin{matrix}\mathcal{D}^2\\ B\\ \mathcal{D}^3\end{matrix} = \mathcal{D}'$$

Assume $\mathcal{D}$ with two cut segments $A \vee B$ and E and $\mathcal{D} \succ \mathcal{D}'$ as below by permutation and $\mathcal{D} \succ \mathcal{D}''$ for the reduction of E and $k = |[A]|$. There are 4 cases.
E is in $\mathcal{D}^1$: then $\exists \mathcal{D}'''$ s.t. $\mathcal{D}'' \succ \mathcal{D}'''$ and $\mathcal{D}' \succ \mathcal{D}'''$.
E is in $\mathcal{D}^2$: then $\exists \mathcal{D}'''$ s.t. $\mathcal{D}'' \succ \mathcal{D}'''$ and $\mathcal{D}' = \mathcal{D}'''$.
E is in $\mathcal{D}^3$: then $\exists \mathcal{D}''', k$ s.t. $\mathcal{D}'' \succ \mathcal{D}'''$ and $\mathcal{D}' \succ^k \mathcal{D}'''$.
E is in $\mathcal{D}^4$: these subcases are covered by the other subcases.
So, in all cases strict local confluence holds.

$$\mathcal{D} = \begin{matrix} \dfrac{\begin{matrix}[A]^u\\ \mathcal{D}^1\\ C\end{matrix} \quad \dfrac{\begin{matrix}\mathcal{D}^3\\ A\end{matrix}}{A \vee B}{\vee}I \quad \begin{matrix}[B]^u\\ \mathcal{D}^2\\ C\end{matrix}}{C}{\vee}Eu \\ \mathcal{D}^4 \end{matrix} \succ \begin{matrix}\mathcal{D}^3\\ [A]\\ \mathcal{D}^1\\ C\\ \mathcal{D}^4\end{matrix} = \mathcal{D}'$$

Assume $\mathcal{D}$ with two cut segments C and E and $\mathcal{D} \succ \mathcal{D}'$ as below by permutation and $\mathcal{D} \succ \mathcal{D}''$ for the reduction of E. There are 6 cases.
E is in $\mathcal{D}^1, \mathcal{D}^2, \mathcal{D}^3$: then $\exists \mathcal{D}'''$ s.t. $\mathcal{D}'' \succ \mathcal{D}'''$ and $\mathcal{D}' \succ \mathcal{D}'''$.

E is in $\mathcal{D}^4, \mathcal{D}^5$: then $\exists \mathcal{D}'''$ s.t. $\mathcal{D}'' \succ \mathcal{D}'''$ and $\mathcal{D}' \succ^2 \mathcal{D}'''$.
E is in $\mathcal{D}^6$: these subcases are covered by the other subcases.
So, in all cases strict local confluence holds.

$$\dfrac{\overset{\mathcal{D}^4}{F} \quad \dfrac{\overset{\mathcal{D}^1}{C} \quad \overset{\mathcal{D}^2}{A \vee B} \quad \overset{\mathcal{D}^3}{C}}{C}\vee E \quad \overset{\mathcal{D}^5}{F}}{\underset{\mathcal{D}^6}{F}}\cdot E \ \succ$$

$$\dfrac{\dfrac{\overset{\mathcal{D}^4}{F} \ \overset{\mathcal{D}^1}{C} \ \overset{\mathcal{D}^5}{F}}{F} \quad \overset{\mathcal{D}^2}{A \vee B} \quad \dfrac{\overset{\mathcal{D}^4}{F} \ \overset{\mathcal{D}^3}{C} \ \overset{\mathcal{D}^5}{F}}{F}\cdot E, \cdot E}{\underset{\mathcal{D}^6}{F}}\vee E$$

Assume $\mathcal{D}$ with two cut segments $C \vee D$ and $E \vee F$ and $\mathcal{D} \succ \mathcal{D}'$ and $\mathcal{D} \succ \mathcal{D}''$ as below by permutations. There holds further by permutation:
$\mathcal{D}' \succ \mathcal{D}''''$ and $\mathcal{D}'' \succ \mathcal{D}'''$. Finally two permutations give: $\mathcal{D}'''' \succ^2 \mathcal{D}'''$.
So, at a glance: $\mathcal{D} \succ \mathcal{D}'$ and $\mathcal{D} \succ \mathcal{D}''$ and $\mathcal{D}' \succ^3 \mathcal{D}'''$ and $\mathcal{D}'' \succ \mathcal{D}'''$.

$$\dfrac{G \vee H \quad \dfrac{E \vee F \quad \dfrac{C \vee D \quad A \vee B \quad C \vee D}{C \vee D} \quad E \vee F}{E \vee F} \quad G \vee H}{G \vee H} = \mathcal{D}$$

$$\dfrac{4 \quad \dfrac{3 \quad \dfrac{2\ 1\ 2}{2} \quad 3}{3} \quad 4}{4} = \mathcal{D}$$

$$\dfrac{4 \quad \dfrac{\dfrac{3\ 2\ 3}{3} \quad 1 \quad \dfrac{3\ 2\ 3}{3}}{3} \quad 4}{4} = \mathcal{D}'$$

$$\dfrac{\dfrac{4\ 3\ 4}{4} \quad \dfrac{2\ 1\ 2}{2} \quad \dfrac{4\ 3\ 4}{4}}{4} = \mathcal{D}''$$

$$\dfrac{\dfrac{\dfrac{4\ 3\ 4}{4} \quad 2 \quad \dfrac{4\ 3\ 4}{4}}{4} \quad 1 \quad \dfrac{\dfrac{4\ 3\ 4}{4} \quad 2 \quad \dfrac{4\ 3\ 4}{4}}{4}}{4} = \mathcal{D}'''$$

$$\dfrac{\dfrac{4 \quad \dfrac{3\ 2\ 3}{3} \quad 4}{4} \quad 1 \quad \dfrac{4 \quad \dfrac{3\ 2\ 3}{3} \quad 4}{4}}{4} = \mathcal{D}''''$$

The argument is similar but simpler if $E \vee F$ is a conjunction, say $E \wedge F$, on which a $\wedge E$ is applied with conclusion F, or if $E \vee F$ is an implication, say $E \to F$, on which a $\to E$ is applied with conclusion F.

The last case of two cut segments touching each other concluded the proof of strict local confluence for intuitionistic logic.

Let us now contemplate on strict local confluence for Intuitionistic Linear Logic in the language $\to, \bullet, \cap, \cup$, i.e. multiplicative implication, multiplicative conjunction and additive conjunction and disjunction. Again, it may be surprising, this set of operators does not add new complications, but the sad story is, that the complexity of the cases does not decrease either. Multiplicative implications are treated completely similar to intuitionistic implication: now duplications of subdeduction do not occur due to arbitrary large sets of assumptions but due to repeated pairs of context formulas. Additive conjunction is treated completely similar to intuitionistic conjunction. And, finally, multiplicative conjunction and additive disjunction are treated similar to intuitionistic disjunction. Here permutations are to be taken into account and, again, duplication of subdeduction may occur due to repeated pairs of context formulas, not due to arbitrary large sets of assumptions.

4 Intuitionistic Logic in Natural Deduction

Definition 10 For the natural numbers with their natural order $\langle \omega, > \rangle$, i.e. $\omega = \{0, 1, \dots\}$ and $> \subseteq \omega \times \omega$, a nested multiset M is a element of ω or a multiset with elements from ω or a multiset with elements from ω and elements of multisets of ω and so on. A nested multiset order $\blacktriangleright$ on nested multisets M, M' of natural numbers is defined recursively: $M \blacktriangleright M'$ iff
$M, M' \in \omega$ and $M > M'$; or
$M' \in \omega$ and $M \notin \omega$; or
$\exists X \subseteq M$ and $X \neq \emptyset$ s.t. $M' = M \setminus X \cup Y$ and $\forall y \in Y \exists x \in X\ x \blacktriangleright y$.
$d(M)$ is the *depth* of a nested multiset M: $d(M) = 0$ if $M \in \omega$; $d(M) = k + 1$ if $m \in M$ and $d(m) = k$ and k is the largest depth of elements of M. In fact, in the sequel nested multisets of depth 2 are regarded only. So, multisets of pairs $\{\{a\}, b\}$ are regarded, for $a, b \in \omega$, which designates a cut segment with degree a and length b. $S(\mathcal{D})$ designates the nested multiset of cut segments of deduction $\mathcal{D}$; an example being $S(\mathcal{D}) = \{\{\{7\}, 2\}, \{\{4\}, 3\}, \{\{4\}, 2\}, \{\{4\}, 2\}\}$, so deduction $\mathcal{D}$ has a cut segment of degree 7 and length 2, an other of degree 4 and length 3 and so on. Nested Multisets of pairs $\{\{a\}, b\}$ are measures for deductions and pairs $\{\{a\}, b\}$ are measures for cut segments. Since pairs $\{\{a\}, b\}$ are themselves nested multisets $\blacktriangleright$ is an order even on pairs $\{\{a\}, b\}$.

Theorem 11 *Well-Foundedness of* ▶
$<$ is well-founded iff ▶ *is well-founded.*

Proof by Dershowitz and Manna [2].

Definition 12 Reduction $\succ$ in IL and its decomposition $\dot{\succ}$.

$\to$-conversions and their decomposition:

$$\begin{array}{c}\dfrac{\dfrac{\begin{array}{c}[A]^u\\ \vdots\\ B\end{array}}{A\to B}{\scriptstyle \to Iu}\quad\begin{array}{c}\vdots\\ A\end{array}}{B}{\scriptstyle \to E}\\ \vdots\end{array} \quad \succ \quad \begin{array}{c}\vdots\\ [A]\\ \vdots\\ B\\ \vdots\end{array}$$

$\dot{\succ}$: $\|[A]\| \leq 1$ or in $\begin{array}{c}\vdots\\ A\end{array}$ are only cuts of smaller measure than $A \to B$.

$\wedge$-conversions and their decomposition: $\dot{\succ} = \succ$:

$$\begin{array}{c}\dfrac{\dfrac{\begin{array}{cc}\vdots & \vdots\\ A_1 & A_2\end{array}}{A_1 \wedge A_2}{\scriptstyle \wedge I}}{A_i}{\scriptstyle \wedge E}\\ \vdots\end{array} \quad \succ \quad \begin{array}{c}\vdots\\ A_i\\ \vdots\end{array}$$

$\vee$-conversions and their decomposition:

$$\begin{array}{c}\dfrac{\begin{array}{c}[A_1]^u\\ \vdots\\ C\end{array}\quad \dfrac{\begin{array}{c}\vdots\\ A_i\end{array}}{A_1 \vee A_2}{\scriptstyle \vee I}\quad \begin{array}{c}[A_2]^u\\ \vdots\\ C\end{array}}{C}{\scriptstyle \vee Eu}\\ \vdots\end{array} \quad \succ \quad \begin{array}{c}\vdots\\ [A_i]\\ \vdots\\ C\\ \vdots\end{array}$$

$\dot{\succ}$: $\|[A_i]\| \leq 1$ or in $\begin{array}{c}\vdots\\ A_i\end{array}$ are only cut segments of smaller measure than $A_1 \vee A_2$.

$(\vee E, \vee E)$ permutations and their decomposition:

$$\cfrac{\begin{array}{c}[C]^v\\ \vdots\\ F\end{array} \quad \cfrac{\begin{array}{c}[A]^u\\ \vdots\\ C\vee D\end{array}\quad \begin{array}{c}\vdots\\ A\vee B\end{array}\quad \begin{array}{c}[B]^u\\ \vdots\\ C\vee D\end{array}}{C\vee D}\,\vee Eu \quad \begin{array}{c}[D]^v\\ \vdots\\ F\end{array}}{\begin{array}{c}F\\ \vdots\end{array}}\,\vee Ev \ \succ$$

$$\succ \cfrac{\cfrac{\begin{array}{c}[C]^x\\ \vdots\\ F\end{array}\quad \begin{array}{c}[A]^v\\ \vdots\\ C\vee D\end{array}\quad \begin{array}{c}[D]^x\\ \vdots\\ F\end{array}}{F} \quad \begin{array}{c}\vdots\\ A\vee B\end{array} \quad \cfrac{\begin{array}{c}[C]^y\\ \vdots\\ F\end{array}\quad \begin{array}{c}[B]^v\\ \vdots\\ C\vee D\end{array}\quad \begin{array}{c}[D]^y\\ \vdots\\ F\end{array}}{F}\,\vee Ex, \vee Ey}{\begin{array}{c}F\\ \vdots\end{array}}\,\vee Ev$$

$\dot{\succ}$: cut segments in $\begin{array}{c}\vdots\\ F\end{array}$, F are of smaller measure than $C \vee D$.

$(\vee E, \rightarrow E)$ permutations and their decomposition:

$$\cfrac{\cfrac{\begin{array}{c}[A]^u\\ \vdots\\ C\rightarrow D\end{array}\quad \begin{array}{c}\vdots\\ A\vee B\end{array}\quad \begin{array}{c}[B]^u\\ \vdots\\ C\rightarrow D\end{array}}{C\rightarrow D}\,\vee Eu \quad \begin{array}{c}\vdots\\ C\end{array}}{\begin{array}{c}D\\ \vdots\end{array}}\rightarrow E \ \succ \ \cfrac{\cfrac{\begin{array}{c}[A]^u\\ \vdots\\ C\rightarrow D\end{array}\quad \begin{array}{c}\vdots\\ C\end{array}}{D}\quad \begin{array}{c}\vdots\\ A\vee B\end{array}\quad \cfrac{\begin{array}{c}[B]^u\\ \vdots\\ C\rightarrow D\end{array}\quad \begin{array}{c}\vdots\\ C\end{array}}{D}\rightarrow E, \rightarrow E}{\begin{array}{c}D\\ \vdots\end{array}}\,\vee Eu$$

$\dot{\succ}$: cut segments in $\begin{array}{c}\vdots\\ C\end{array}$ are of smaller measure than $C \rightarrow D$.

$(\vee E, \wedge E)$ permutations and their decomposition: $\dot{\succ} = \succ$:

$$\dfrac{\dfrac{\begin{matrix}[A]^u \\ \vdots \\ C \wedge D\end{matrix} \quad \begin{matrix}\vdots \\ A \vee B\end{matrix} \quad \begin{matrix}[B]^u \\ \vdots \\ C \wedge D\end{matrix}}{C \wedge D}\vee Eu}{\begin{matrix}C \\ \vdots\end{matrix}}\wedge E \quad \succ \quad \dfrac{\dfrac{\begin{matrix}[A]^u \\ \vdots \\ C \wedge D\end{matrix}}{C} \quad \begin{matrix}\vdots \\ A \vee B\end{matrix} \quad \dfrac{\begin{matrix}[B]^u \\ \vdots \\ C \wedge D\end{matrix}}{C}\wedge E, \wedge E}{\begin{matrix}C \\ \vdots\end{matrix}}\vee Eu$$

Lemma 13 *Termination of* $\dot{\succ}$ *in IL*
If $\mathcal{D}\dot{\succ}\mathcal{D}'$, *then* $S(\mathcal{D}) \blacktriangleright S(\mathcal{D}')$.
$\dot{\succ}$ *terminates in IL.*

Proof For the first statement assume a cut segment of degree a with length 1, i.e. a cut in $\mathcal{D}$ and $S(\mathcal{D}) = \Gamma \cup \{\{\{a\}, 1\}\}$. A conversion of this cut s.t. $\mathcal{D}\dot{\succ}\mathcal{D}'$ has the effect that in $\mathcal{D}'$ this cut is removed, but there might exist r additional cut segments of degree $a - k_r$ and length n_r, so $S(\mathcal{D}') = \Gamma \cup \{\{\{a - k_1\}, n_1\}, \ldots, \{\{a - k_r\}, n_r\}\}$. Of course $S(\mathcal{D}) \blacktriangleright S(\mathcal{D}')$ in the nested multiset order. Assume further r cut segments of degree a and length n_r ending in one and the same elimination rule and $S(\mathcal{D}) = \Gamma \cup \{\{\{a\}, n_1\}, \ldots, \{\{a\}, n_r\}\}$. A permutation of this elimination rule s.t. $\mathcal{D}\dot{\succ}\mathcal{D}'$ has the effect that in $\mathcal{D}'$ these cut segments are shorted to $n_r - 1$; but there might exist s additional cut segments of degree b_s and length q_s, s.t. either $b_s < a$ or $b_s = a$ and $\exists r$ s.t. $q_s < n_r$; so $S(\mathcal{D}') = \Gamma \cup \{\{\{a\}, n_1 - 1\}, .., \{\{a\}, n_r - 1\}\} \cup \{\{\{b_1\}, q_1\}, .., \{\{b_s\}, q_s\}\}$. Of course $S(\mathcal{D}) \blacktriangleright S(\mathcal{D}')$ in the nested multiset order.

For the proof of the second statement assume an infinite descending sequence $\mathcal{D}^1\dot{\succ}\mathcal{D}^2\dot{\succ}\mathcal{D}^3\dot{\succ}\ldots$; with the first statement this would amount to a further infinite descending sequence $S(\mathcal{D}^1) \blacktriangleright S(\mathcal{D}^2) \blacktriangleright S(\mathcal{D}^3) \blacktriangleright \ldots$, a contradiction, since $\blacktriangleright$ is terminating. So $\dot{\succ}$ terminates for IL.

Lemma 14 *Confluence of* $\dot{\succ}$ *in IL*
$\dot{\succ}$ *is locally confluent in IL:* $\forall \alpha, \beta, \gamma \ : \ (\alpha\dot{\succ}\beta \wedge \alpha\dot{\succ}\gamma \ \rightarrow \ \exists \delta, m, n(\beta\dot{\succ}^m\delta \wedge \gamma\dot{\succ}^n\delta))$.
$\dot{\succ}$ *is confluent in IL.*

Proof The second statement, confluence of $\dot{\succ}$ in IL, is an immediate consequence of the first statement, locally confluence and termination of $\dot{\succ}$, known as Newman's Lemma and quoted as pure rewriting result in the first section of this chapter. So the proof of local confluence of $\dot{\succ}$ rests to be shown. This proof is according the same lines as the proof of strict local confluence of $\succ$ lemma 9, as a proof by cases. But of course, strict local confluence is lost for $\dot{\succ}$ due to interference of reductions, i.e. due to interference of discharged assumptions, which are going to be composed by reduction. This can be shown best by example. In many cases termination of $\dot{\succ}$ is used. So lets assume a deduction $\mathcal{D}$ with two cuts $A \rightarrow B$ and $C \rightarrow D$ as below and decomposed reductions of these cuts giving $\mathcal{D}\dot{\succ}\mathcal{D}^1$ and $\mathcal{D}\dot{\succ}\mathcal{D}^2$.

$$\mathcal{D}^1 = \begin{array}{c} \dfrac{\dfrac{\begin{array}{c}[A]\\ \vdots\\ [C]\\ \vdots\\ D\\ \vdots\\ B\end{array}}{A \to B} \quad \begin{array}{c}\vdots\\ A\end{array}}{B} \\ \vdots \end{array} \quad \dot{\prec}\, \mathcal{D} = \begin{array}{c} \dfrac{\dfrac{\begin{array}{c}\dfrac{\dfrac{\begin{array}{c}[C]\\ \vdots\\ D\end{array}}{C\to D} \quad \begin{array}{c}[A]\\ \vdots\\ C\end{array}}{D}\\ \vdots\\ B\end{array}}{A \to B} \quad \begin{array}{c}\vdots\\ A\end{array}}{B} \\ \vdots \end{array} \quad \dot{\succ}\, \mathcal{D}^2 = \begin{array}{c} \dfrac{\dfrac{\begin{array}{c}[C]\\ \vdots\\ D\end{array}}{C\to D} \quad \begin{array}{c}\vdots\\ [A]\\ \vdots\\ C\end{array}}{D}\\ \vdots\\ B\\ \vdots \end{array}$$

Given $\mathcal{D}\dot{\succ}\mathcal{D}^1$ and $\mathcal{D}\dot{\succ}\mathcal{D}^2$ it has to be shown that there are $\mathcal{D}^*, m, n$ s.t. $\mathcal{D}^1\dot{\succ}^m\mathcal{D}^*$ and $\mathcal{D}^2\dot{\succ}^n\mathcal{D}^*$. If full reduction $\succ$ were available cuts $A \to B$ would be converted in $\mathcal{D}^1$ and $C \to D$ in $\mathcal{D}^2$ to yield a converging deduction. But, since decomposition $\dot{\succ}$ obeys some restrictions on $\succ$, these conversions often can be done only after some preparatory $\dot{\succ}$ steps. In order to apply $\dot{\succ}$ $\begin{array}{c}\vdots\\ A\end{array}$ in $\mathcal{D}^1$ is normalised to $\begin{array}{c}\vdots\!\vdots\\ A\end{array}$ s.t. $\mathcal{D}^1\dot{\succ}^*\mathcal{D}^{1a}$. This is copied for $\mathcal{D}^2$ s.t. $\mathcal{D}^2\dot{\succ}^*\mathcal{D}^{2a}$. Further $\begin{array}{c}\vdots\!\vdots\\ [A]\\ \vdots\\ C\end{array}$ in $\mathcal{D}^{2a}$ is normalised to $\begin{array}{c}\vdots\!\vdots\\ [A]\\ \vdots\!\vdots\\ C\end{array}$ s.t. $\mathcal{D}^{2a}\dot{\succ}^*\mathcal{D}^{2b}$.

$$\mathcal{D}^{1a} = \begin{array}{c} \dfrac{\dfrac{\begin{array}{c}[A]\\ \vdots\\ [C]\\ \vdots\\ D\\ \vdots\\ B\end{array}}{A \to B} \quad \begin{array}{c}\vdots\!\vdots\\ A\end{array}}{B} \\ \vdots \end{array} \quad \dot{\prec}^*\, \mathcal{D} = \begin{array}{c} \dfrac{\dfrac{\begin{array}{c}\dfrac{\dfrac{\begin{array}{c}[C]\\ \vdots\\ D\end{array}}{C\to D} \quad \begin{array}{c}[A]\\ \vdots\\ C\end{array}}{D}\\ \vdots\\ B\end{array}}{A \to B} \quad \begin{array}{c}\vdots\\ A\end{array}}{B} \\ \vdots \end{array} \quad \dot{\succ}^*\, \mathcal{D}^{2b} = \begin{array}{c} \dfrac{\dfrac{\begin{array}{c}[C]\\ \vdots\\ D\end{array}}{C\to D} \quad \begin{array}{c}\vdots\!\vdots\\ [A]\\ \vdots\!\vdots\\ C\end{array}}{D}\\ \vdots\\ B\\ \vdots \end{array}$$

Now conversions of cuts $A \to B, C \to D$ can be carried out and give $\mathcal{D}^{1a} \dot{\succ} \mathcal{D}^{1b}$ and $\mathcal{D}^{2b} \dot{\succ} \mathcal{D}^{2c}$. Finally normalised $\begin{matrix} \vdots\vdots \\ [A] \\ \vdots\vdots \\ C \end{matrix}$ of $\mathcal{D}^{2c}$ is copied for $\mathcal{D}^{1b}$ s.t. $\mathcal{D}^{1b} \dot{\succ}^* \mathcal{D}^{1c}$, as below. Finally $\mathcal{D}^{1b} = \mathcal{D}^{2c}$ is the desired deduction, s.t. $\mathcal{D} \dot{\succ} \mathcal{D}^1$ and $\mathcal{D} \dot{\succ} \mathcal{D}^2$ and $\mathcal{D}^1 \dot{\succ}^* \mathcal{D}^{1b}$ and $\mathcal{D}^2 \dot{\succ}^* \mathcal{D}^{2c}$.

$$\mathcal{D}^{1c} = \begin{matrix} \vdots\vdots \\ [A] \\ \vdots\vdots \\ [C] \\ \vdots \\ D \\ \vdots \\ B \\ \vdots \end{matrix} \quad \dot{\prec}^* \; \mathcal{D} = \begin{matrix} \dfrac{\dfrac{\begin{matrix} \dfrac{\dfrac{\begin{matrix}[C] \\ \vdots \\ D\end{matrix}}{C \to D} \quad \begin{matrix}[A] \\ \vdots \\ C\end{matrix}}{D} \\ \vdots \\ B \end{matrix}}{A \to B} \quad \begin{matrix}\vdots \\ A\end{matrix}}{B} \\ \vdots \end{matrix} \quad \dot{\succ}^* \; \mathcal{D}^{2c} = \begin{matrix} \vdots\vdots \\ [A] \\ \vdots\vdots \\ C \\ \vdots \\ D \\ \vdots \\ B \\ \vdots \end{matrix}$$

Lemma 15 *Commutation of* $\succ$ *with* $\dot{\succ}$ *in IL*
$\forall \alpha, \beta(\alpha \succ \beta \to \exists \delta, m, n(\alpha \dot{\succ}^m \delta \wedge \beta \dot{\succ}^n \delta))$.

Proof The proof proceeds by investigating all reduction cases of $\alpha \succ \beta$.

For commutation property assume $\mathcal{D} \succ \mathcal{D}'$ to be a $\to$ conversion.

$$\mathcal{D} = \begin{matrix} \dfrac{\dfrac{\begin{matrix}[A]^u \\ \vdots \\ B\end{matrix}}{A \to B} \to Iu \quad \begin{matrix}\vdots \\ A\end{matrix}}{B} \to E \\ \vdots \end{matrix} \quad \succ \quad \begin{matrix} \vdots \\ [A] \\ \vdots \\ B \\ \vdots \end{matrix} = \mathcal{D}'$$

$\mathcal{D}\dot{\succ}^m\mathcal{D}''$ indicates that $\begin{array}{c}\vdots\\ A\end{array}$ is normalised to $\begin{array}{c}m\\ \vdots\\ A\end{array}$.

$$\mathcal{D} = \dfrac{\dfrac{\begin{array}{c}[A]^u\\ \vdots\\ B\end{array}}{A \to B}{\to}Iu \quad \begin{array}{c}\vdots\\ A\end{array}}{\begin{array}{c}B\\ \vdots\end{array}}{\to}E \;\dot{\succ}^m\; \dfrac{\dfrac{\begin{array}{c}[A]^u\\ \vdots\\ B\end{array}}{A \to B}{\to}Iu \quad \begin{array}{c}m\\ \vdots\\ A\end{array}}{\begin{array}{c}B\\ \vdots\end{array}}{\to}E = \mathcal{D}''$$

Now there are $\mathcal{D}'''$ and n s.t. $\mathcal{D}'\dot{\succ}^n\mathcal{D}'''$ and $\mathcal{D}''\dot{\succ}\mathcal{D}'''$.

For commutation property assume $\mathcal{D} \succ \mathcal{D}'$ to be a $\vee$ conversion.

$$\mathcal{D} = \dfrac{\begin{array}{c}[A_1]^u\\ \vdots\\ C\end{array} \quad \dfrac{\begin{array}{c}\vdots\\ A_i\end{array}}{A_1 \vee A_2}{\vee}I \quad \begin{array}{c}[A_2]^u\\ \vdots\\ C\end{array}}{\begin{array}{c}C\\ \vdots\end{array}}{\vee}Eu \;\succ\; \begin{array}{c}\vdots\\ [A_i]\\ \vdots\\ C\\ \vdots\end{array} = \mathcal{D}'$$

$\mathcal{D}\dot{\succ}^m\mathcal{D}''$ indicates that $\begin{array}{c}\vdots\\ A\end{array}$ is normalised to $\begin{array}{c}m\\ \vdots\\ A\end{array}$.

$$\mathcal{D} = \dfrac{\begin{array}{c}[A_1]^u\\ \vdots\\ C\end{array} \quad \dfrac{\begin{array}{c}\vdots\\ A_i\end{array}}{A_1 \vee A_2}{\vee}I \quad \begin{array}{c}[A_2]^u\\ \vdots\\ C\end{array}}{\begin{array}{c}C\\ \vdots\end{array}}{\vee}Eu \;\dot{\succ}^m\; \dfrac{\begin{array}{c}[A_1]^u\\ \vdots\\ C\end{array} \quad \dfrac{\begin{array}{c}m\\ \vdots\\ A_i\end{array}}{A_1 \vee A_2}{\vee}I \quad \begin{array}{c}[A_2]^u\\ \vdots\\ C\end{array}}{\begin{array}{c}C\\ \vdots\end{array}}{\vee}Eu = \mathcal{D}''$$

Now there are $\mathcal{D}'''$, n s.t. $\mathcal{D}'\dot{\succ}^n\mathcal{D}'''$ and $\mathcal{D}''\dot{\succ}\mathcal{D}'''$.

For commutation property assume $\mathcal{D} \succ \mathcal{D}'$ to be a $(\vee E, \vee E)$ permutation and F not to be a cut formula.

$$\mathcal{D} = \dfrac{\begin{matrix}[C]^v\\ \vdots\\ F\end{matrix} \qquad \dfrac{\begin{matrix}[A]^u\\ \vdots\\ C\vee D\end{matrix} \quad \begin{matrix}\\ \vdots\\ A\vee B\end{matrix} \quad \begin{matrix}[B]^u\\ \vdots\\ C\vee D\end{matrix}}{C\vee D}\vee Eu \qquad \begin{matrix}[D]^v\\ \vdots\\ F\end{matrix}}{\begin{matrix}F\\ \vdots\end{matrix}}\vee Ev$$

$$\mathcal{D}' = \dfrac{\dfrac{\begin{matrix}[C]^x\\ \vdots\\ F\end{matrix} \quad \begin{matrix}[A]^v\\ \vdots\\ C\vee D\end{matrix} \quad \begin{matrix}[D]^x\\ \vdots\\ F\end{matrix}}{F} \qquad \begin{matrix}\vdots\\ A\vee B\end{matrix} \qquad \dfrac{\begin{matrix}[C]^y\\ \vdots\\ F\end{matrix} \quad \begin{matrix}[B]^v\\ \vdots\\ C\vee D\end{matrix} \quad \begin{matrix}[D]^y\\ \vdots\\ F\end{matrix}}{F}\vee Ex, \vee Ey}{\begin{matrix}F\\ \vdots\end{matrix}}\vee Ev$$

$\mathcal{D} \mathrel{\dot{\succ}}^{m+n} \mathcal{D}''$ indicates $\begin{matrix}[C]^v\\ \vdots\\ F\end{matrix}$ and $\begin{matrix}[D]^v\\ \vdots\\ F\end{matrix}$ normalising to $\begin{matrix}m\\ [C]^v\\ \vdots\\ F\end{matrix}$ resp. $\begin{matrix}n\\ [D]^v\\ \vdots\\ F\end{matrix}$.

$$\mathcal{D}'' = \dfrac{\begin{matrix}m\\ [C]^v\\ \vdots\\ F\end{matrix} \qquad \dfrac{\begin{matrix}[A]^u\\ \vdots\\ C\vee D\end{matrix} \quad \begin{matrix}\\ \vdots\\ A\vee B\end{matrix} \quad \begin{matrix}[B]^u\\ \vdots\\ C\vee D\end{matrix}}{C\vee D}\vee Eu \qquad \begin{matrix}n\\ [D]^v\\ \vdots\\ F\end{matrix}}{\begin{matrix}F\\ \vdots\end{matrix}}\vee Ev$$

Now there are $\mathcal{D}'''$, k s.t. $\mathcal{D}' \mathrel{\dot{\succ}}^k \mathcal{D}'''$ and $\mathcal{D}'' \mathrel{\dot{\succ}} \mathcal{D}'''$.

For commutation property assume $\mathcal{D} \succ \mathcal{D}'$ to be a $(\vee E, \vee E)$ permutation on a cut segment $C_k \vee D_k$ with $k-1$ further cut segments $C_{k-1} \vee D_{k-1}$ below and F not to be a cut formula. All rules displayed in the deductions below are $\vee E$, and for simplification discharged assumptions are left out.

$$\mathcal{D} = \dfrac{\overset{\vdots}{C_{k-1} \vee D_{k-1}} \quad \dfrac{\overset{\vdots}{C_k \vee D_k} \quad \overset{\vdots}{A \vee B} \quad \overset{\vdots}{C_k \vee D_k}}{C_k \vee D_k} \quad \overset{\vdots}{C_{k-1} \vee D_{k-1}}}{C_{k-1} \vee D_{k-1}}$$

$$\vdots \qquad\qquad \vdots \qquad\qquad \vdots$$

$$\dfrac{\overset{\vdots}{F} \quad \dfrac{C_1 \vee D_1 \quad C_2 \vee D_2 \quad C_1 \vee D_1}{C_1 \vee D_1} \quad \overset{\vdots}{F}}{F}$$

$$\vdots$$

Deduction $\mathcal{D}$ is abbreviated by the convention that disjunctions $C_j \vee D_j$ are designated simply by j.

$$\mathcal{D} = \dfrac{\overset{\vdots}{k-1} \quad \dfrac{\overset{\vdots}{k} \quad \overset{\vdots}{A \vee B} \quad \overset{\vdots}{k}}{k} \quad \overset{\vdots}{k-1}}{k-1}$$

$$\vdots \qquad\qquad \vdots \qquad\qquad \vdots$$

$$\dfrac{\overset{\vdots}{F} \quad \dfrac{1 \quad 2 \quad 1}{1} \quad \overset{\vdots}{F}}{F}$$

$$\vdots$$

$$\mathcal{D}' = \dfrac{\dfrac{\overset{\vdots}{k-1} \ \overset{\vdots}{k} \ \overset{\vdots}{k-1}}{k-1} \quad \overset{\vdots}{A \vee B} \quad \dfrac{\overset{\vdots}{k-1} \ \overset{\vdots}{k} \ \overset{\vdots}{k-1}}{k-1}}{k-1}$$

$$\vdots \qquad\qquad \vdots \qquad\qquad \vdots$$

$$\dfrac{\overset{\vdots}{F} \quad \dfrac{1 \quad 2 \quad 1}{1} \quad \overset{\vdots}{F}}{F}$$

$$\vdots$$

Now there are $\mathcal{D}''$, $\mathcal{D}'''$, k as below s.t. $\mathcal{D} \dot{\succ}^k \mathcal{D}''$ and $\mathcal{D}'' \dot{\succ} \mathcal{D}'''$. $\mathcal{D}''$ is approached by permutation from below, as indicated by $\mathcal{D}^a$.

In the sequel $\overset{n}{\underset{F}{\vdots}}$ indicates that the subdeduction with conclusion F is normal.

$\mathcal{D}'' =$

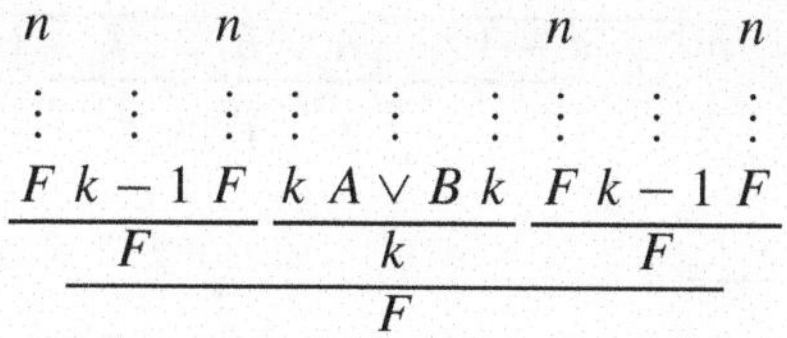

$\mathcal{D}^a =$

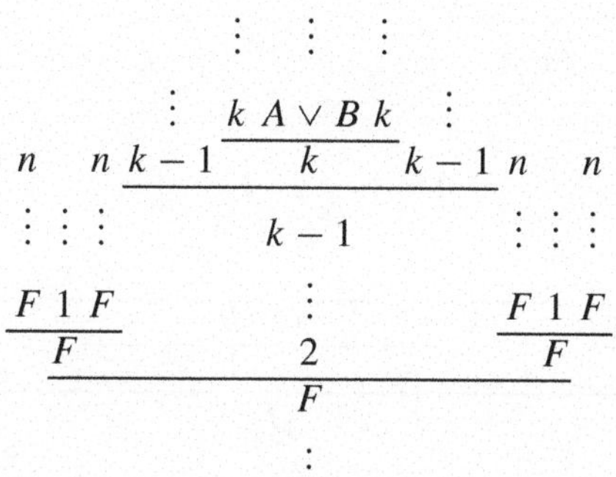

$\mathcal{D}''' =$

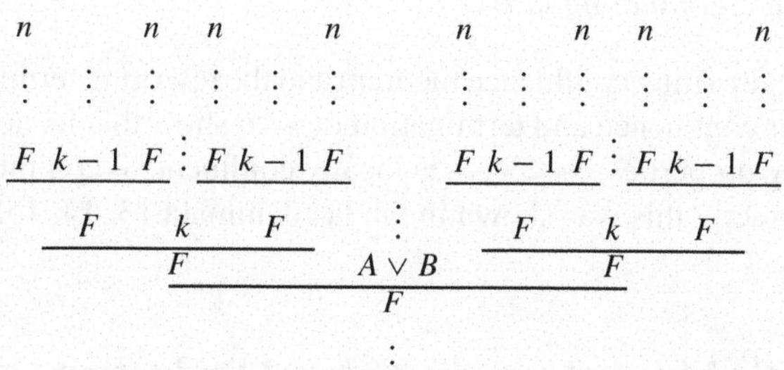

But it holds even that $\mathcal{D}' \dot{\succ}^m \mathcal{D}'''$ for some m, and this starts with $\mathcal{D}' \dot{\succ} \mathcal{D}^b$ and goes on to $\mathcal{D}^b \dot{\succ}^n \mathcal{D}^c$ for some n and $\mathcal{D}^c \dot{\succ}^o \mathcal{D}'''$ for some o.

$\mathcal{D}^b =$

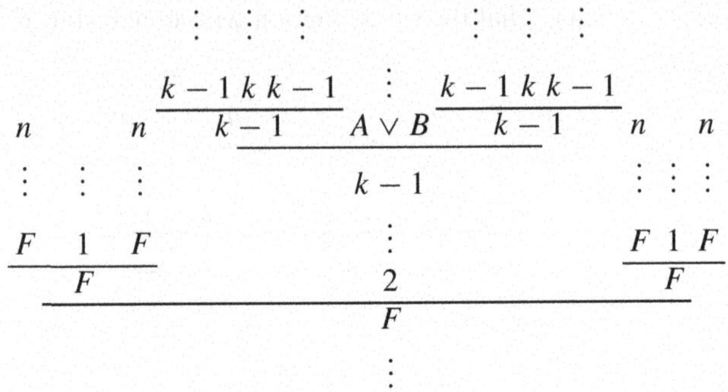

$\mathcal{D}^c =$

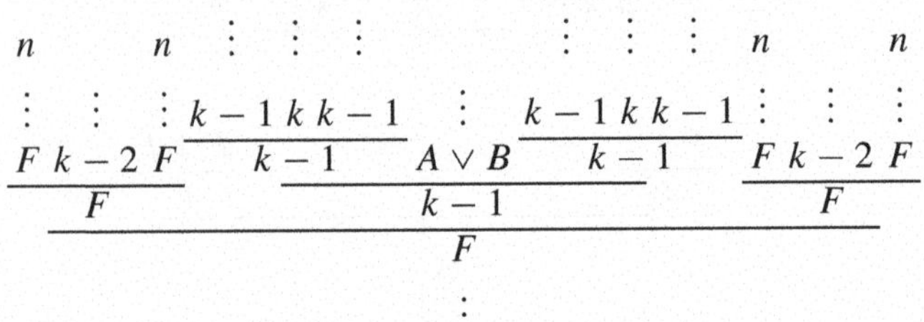

Theorem 16 *Confluence and Termination of* $\succ$ *for IL*
$\succ$ *is confluent and terminating in IL.*

Proof Due to the rewriting confluence theorem and the rewriting termination theorem it is sufficient for confluence and termination of $\succ$ to show that its decomposition $\dot{\succ}$ has the following properties: $\dot{\succ} \subseteq \succ$, $\dot{\succ}$ is locally confluent, $\dot{\succ}$ commutes with $\succ$ and $\dot{\succ}$ terminates. Exactly this was shown in the last lemmata 13, 14, 15.

5 Intuitionistic Linear Logic in Natural Deduction

Definition 17 Reduction $\succ$ in ILL and its decomposition $\dot{\succ}$.

$\rightarrow$-conversions and their decomposition:

$$\dfrac{\begin{matrix}\vdots\\ A\end{matrix}\quad \dfrac{\begin{matrix}[A]^v\\ \vdots\\ B\end{matrix}}{A\to B}{\to I}}{B}{\to Ev}\quad \begin{matrix}\\ \vdots\end{matrix} \qquad \succ \qquad \begin{matrix}\vdots\\ A\\ \vdots\\ B\\ \vdots\end{matrix}$$

$\dot{\succ}$: cut segments in $\begin{matrix}\vdots\\ A\end{matrix}$ are of lower measure than $A \to B$ or $[A]^v$ is no context.

$\bullet$-conversions and their decomposition:

$$\dfrac{\dfrac{\begin{matrix}\vdots\\ A\end{matrix}\quad\begin{matrix}\vdots\\ B\end{matrix}}{A\bullet B}{\bullet I}\quad \begin{matrix}[AB]^v\\ \vdots\\ C\end{matrix}}{C}{\bullet Ev}\quad \begin{matrix}\\ \vdots\end{matrix} \qquad \succ \qquad \begin{matrix}\vdots\ \vdots\\ A\ B\\ \vdots\\ C\\ \vdots\end{matrix}$$

$\dot{\succ}$: cut segments in $\begin{matrix}\vdots\\ A\end{matrix}$ and in $\begin{matrix}\vdots\\ B\end{matrix}$ are of lower measure than $A \bullet B$ or $[A]^v$ and $[B]^v$ are no context.

$\cup$-conversions and their decomposition:

$$\dfrac{\begin{matrix}[A_1\Delta]^v\\ \vdots\\ C\end{matrix}\quad \dfrac{\begin{matrix}\vdots\\ A_j\end{matrix}}{A_1\cup A_2}{\cup I}\quad \begin{matrix}[A_2]^v\Delta\\ \vdots\\ C\end{matrix}}{C}{\cup Ev}\quad \begin{matrix}\\ \vdots\end{matrix} \qquad \succ \qquad \begin{matrix}\vdots\\ A_j\ \Delta\\ \vdots\\ C\\ \vdots\end{matrix}$$

$\dot{\succ}$: cut segments in $\begin{matrix}\vdots\\ A_j\end{matrix}$ are of lower measure than $A \cup B$ or $[A_j]^v$ is no context.

$\cap$-conversions and their decomposition: $\dot{\succ} = \succ$:

$$
\dfrac{\dfrac{\begin{matrix}[\Gamma]^v & \Gamma \\ \vdots & \vdots \\ A_1 & A_2\end{matrix}}{A_1 \cap A_2}\cap I v}{A_j}\cap E \quad \text{and} \quad \dfrac{\dfrac{\begin{matrix}\Gamma & [\Gamma]^v \\ \vdots & \vdots \\ A_1 & A_2\end{matrix}}{A_1 \cap A_2}\cap I v}{A_j}\cap E \;\succ\; \begin{matrix}\Gamma \\ \vdots \\ A_j \\ \vdots\end{matrix}
$$

$$
\begin{matrix}\vdots & & \vdots\end{matrix}
$$

$(\cup E, \cup E)$-permutations and their decomposition:

$$
\dfrac{\begin{matrix}[C\Sigma]^v \\ \vdots \\ F\end{matrix} \quad \dfrac{\begin{matrix}[A\Delta]^u & & [B]^u\Delta \\ \vdots & \vdots & \vdots \\ C \cup D & A \cup B & C \cup D\end{matrix}}{C \cup D}\cup E u \quad \begin{matrix}[D]^v\Sigma \\ \vdots \\ F\end{matrix}}{\begin{matrix}F \\ \vdots\end{matrix}}\cup E v \;\succ
$$

$$
\dfrac{\dfrac{\begin{matrix}[C\Sigma]^x & [A\Delta]^v & [D]^x[\Sigma]^v \\ \vdots & \vdots & \vdots \\ F & C \cup D & F\end{matrix}}{F} \quad \begin{matrix}\vdots \\ A \cup B\end{matrix} \quad \dfrac{\begin{matrix}[C\Sigma]^y & [B]^v\Delta & [D]^y\Sigma \\ \vdots & \vdots & \vdots \\ F & C \cup D & F\end{matrix}}{F}\cup E x, \cup E y}{\begin{matrix}F \\ \vdots\end{matrix}}\cup E v
$$

$\dot{\succ}$: cut segments in $\begin{matrix}\vdots \\ F\end{matrix}$, F are of smaller measure than $C \cup D$.

$(\cup E, \bullet E)$-permutations and their decomposition:

$$
\dfrac{\dfrac{\begin{matrix}[A\Delta]^u & & [B]^u\Delta \\ \vdots & \vdots & \vdots \\ C \bullet D & A \cup B & C \bullet D\end{matrix}}{C \bullet D}\cup E u \quad \begin{matrix}[CD]^v\Sigma \\ \vdots \\ F\end{matrix}}{\begin{matrix}F \\ \vdots\end{matrix}}\bullet E v \;\succ
$$

$$\dfrac{\dfrac{\begin{matrix}[A\Delta]^v & [CD]^x[\Sigma]^v \\ \vdots & \vdots \\ C \bullet D & F\end{matrix}}{F} \quad \begin{matrix}\vdots \\ A \cup B\end{matrix} \quad \dfrac{\begin{matrix}[B]^v\Delta & [CD]^y\Sigma \\ \vdots & \vdots \\ C \bullet D & F\end{matrix}}{F}\bullet Ex, \bullet Ey}{\begin{matrix}F \\ \vdots\end{matrix}}\cup Ev$$

$\dot{\succ}$: cut segments in $\begin{matrix}\vdots \\ F\end{matrix}$, F are of smaller measure than $C \bullet D$.

$(\cup E, \rightarrow E)$-permutations and their decomposition:

$$\dfrac{\begin{matrix}\vdots \\ C\end{matrix} \quad \dfrac{\begin{matrix}[A\Delta]^u & & [B]^u\Delta \\ \vdots & \vdots & \vdots \\ C \rightarrow D & A \cup B & C \rightarrow D\end{matrix}}{C \rightarrow D}\cup Eu}{\begin{matrix}D \\ \vdots\end{matrix}}\rightarrow E \;\succ\; \dfrac{\dfrac{\begin{matrix} & [A\Delta]^u \\ \vdots & \vdots \\ C & C \rightarrow D\end{matrix}}{D} \quad \begin{matrix}\vdots \\ A \cup B\end{matrix} \quad \dfrac{\begin{matrix} & [B]^u\Delta \\ \vdots & \vdots \\ C & C \rightarrow D\end{matrix}}{D}\rightarrow E, \rightarrow E}{\begin{matrix}D \\ \vdots\end{matrix}}\cup Eu$$

$\dot{\succ}$: cut segments in $\begin{matrix}\vdots \\ C\end{matrix}$ are of smaller measure than $C \rightarrow D$.

$(\cup E, \cap E)$-permutations and their decomposition: $\succ = \dot{\succ}$:

$$\dfrac{\dfrac{\begin{matrix}[A\Delta]^u & & [B]^u\Delta \\ \vdots & \vdots & \vdots \\ C \cap D & A \cup B & C \cap D\end{matrix}}{C \cap D}\cup Eu}{\begin{matrix}C \\ \vdots\end{matrix}}\cap E \;\succ\; \dfrac{\dfrac{\begin{matrix}[A\Delta]^u \\ \vdots \\ C \cap D\end{matrix}}{C} \quad \begin{matrix}\vdots \\ A \cup B\end{matrix} \quad \dfrac{\begin{matrix}[B]^u\Delta \\ \vdots \\ C \cap D\end{matrix}}{C}\cap E, \cap E}{\begin{matrix}C \\ \vdots\end{matrix}}\cup Eu$$

Permutations for $\bullet E$, i.e. $(\bullet E, \rightarrow E)$-, $(\bullet E, \bullet E)$-, $(\bullet E, \cup E)$- and $(\bullet E, \cap E)$-permutations are defined similar to the permutations for $\cup E$.

Lemma 18 *Termination of* $\dot{\succ}$ *in ILL*
If $\mathcal{D} \dot{\succ} \mathcal{D}'$, *then* $S(\mathcal{D}) \blacktriangleright S(\mathcal{D}')$.
$\dot{\succ}$ *terminates in IL.*

Proof The first statement is proved for ILL similar to IL with one difference. In IL multiplication of subdeductions by conversions do occur due to discharged assumption classes with cardinality larger 1. In ILL such assumption classes do not occur, but in ILL assumption singletons may occur as context formulas and it is the contexts which may cause multiplication of subdeductions. Nevertheless, conversions are defined such that multiplication of subdeductions of multiple occurring contexts and cut segments therein due to do not increase $\blacktriangleright$. Permutations have the same effects in ILL and IL. So $S(\mathcal{D}) \blacktriangleright S(\mathcal{D}')$ in the nested multiset order whenever $\mathcal{D} \dot{\succ} \mathcal{D}'$. As a consequence $\dot{\succ}$ terminates for IL.

Lemma 19 *Confluence of* $\dot{\succ}$ *in ILL*
$\dot{\succ}$ *is locally confluent in ILL:* $\forall \alpha, \beta, \gamma : (\alpha \dot{\succ} \beta \wedge \alpha \dot{\succ} \gamma \rightarrow \exists \delta, m, n(\beta \dot{\succ}^m \delta \wedge \gamma \dot{\succ}^n \delta))$.
$\dot{\succ}$ *is confluent in ILL.*

Proof The first statement is proved by going through all cases of combinations of reductions, as done in the proof of strict local confluence of $\succ$ in ILL, lemma 9. However, strict local confluence for $\dot{\succ}$ is lost, since $\dot{\succ}$ conversions may generate new cut segments violating the restrictions imposed on $\dot{\succ}$. But normalising respective subdeductions, which is always possible due to termination of $\dot{\succ}$, gives deductions on which $\dot{\succ}$ can be applied, finally. The second statement is a consequence of the first and termination - Newman's Lemma.

Lemma 20 *Commutation of* $\succ$ *with* $\dot{\succ}$ *in ILL*
$\forall \alpha, \beta(\alpha \succ \beta \rightarrow \exists \delta, m, n(\alpha \dot{\succ}^m \delta \wedge \beta \dot{\succ}^n \delta))$.

Proof Given the definition of $\dot{\succ}$ in ILL the proof of this lemma for ILL is similar to the proof in IL.

Theorem 21 *Confluence and Termination of* $\succ$ *for ILL*
$\succ$ *is confluent and terminating in ILL.*

Proof Again, due to the rewriting confluence and termination theorems it is sufficient for confluence and termination of $\succ$ in ILL to show that its decomposition $\dot{\succ}$ has the following properties: $\dot{\succ} \subseteq \succ$, $\dot{\succ}$ is locally confluent, $\dot{\succ}$ commutes with $\succ$ and $\dot{\succ}$ terminates; what was done in the lemmata 18, 19, 20 above.

References

1. Baader, F., & Nipkow, T. (1998). *Term rewriting and all that*. Cambridge: Cambridge University Press.
2. Dershowitz, N., & Manna, Z. (1979). Proving termination with multiset orderings. *Communications of the ACM*, *22*(8), 465–476.
3. Gandy, R. O. (1980). *Proofs of strong normalisation*. To H.B. Curry: Essays on combinatory logic, λ-calculus and formalism (pp. 457–477). Boston: Academic Press.
4. Gentzen, G. (1934/1935). Untersuchungen über das logische Schließen. *Mathematische Zeitschrift, 39*, 176–210, 405–431.

5. Girard, J.-Y. (1986). *Proof theory and logical complexity*. Napoli: Bibliopolis.
6. Joachimsky, F., & Matthes, R. (2003). Short proofs of normalization for the simple typed λ-calculus, permutative conversions and Gödel's T. *Archive for Mathematical Logic*, *42*, 59–87.
7. Prawitz, D. (1965). *Natural deduction*. Stockholm: Almqvist and Wiksell.
8. Prawitz, D. (1971). Ideas and results in proof theory. *Proceedings of the 2-nd Scand. Logic Symposium*, Amsterdam, pp. 235–307.
9. Schwichtenberg, H., & Troelstra, A. S. (2000). *Basic proof theory*, 2nd edn. Cambridge: Cambridge University Press.
10. Zimmermann, E. (2010). Full Lambek calculus in natural deduction. *Mathematical Logic Quarterly*, *56*, 85–88.

An Approach to General Proof Theory and a Conjecture of a Kind of Completeness of Intuitionistic Logic Revisited

Dag Prawitz

Abstract Thirty years ago I formulated a conjecture about a kind of completeness of intuitionistic logic. The framework in which the conjecture was formulated had the form of a semantic approach to a general proof theory (presented at the 4th World Congress of Logic, Methodology and Philosophy of Science at Bucharest 1971 [6]). In the present chapter, I shall reconsider this 30-year old conjecture, which still remains unsettled, but which I continue to think of as a plausible and important supposition. Reconsidering the conjecture, I shall also reconsider and revise the semantic approach in which the conjecture was formulated.

1 Main Ideas Behind the Conjecture

The question that the conjecture was intended to answer is roughly whether the elimination rules of Gentzen's system of natural deduction for intuitionistic logic are the strongest possible ones. This question already arises against the background of Gentzen's own understanding of his intuitionistic system of natural deduction, in particular the significance of his classification of the inference rules into introduction and elimination rules. Gentzen's idea was that the introduction rule for a logical constant gives the meaning of the constant while the corresponding elimination rule becomes justified by this very meaning. The rules for conjunction constitute a simple illustration of this idea:

$$\frac{A \quad B}{A\&B}\,\&I \qquad \frac{A\&B}{A}\,\&E \qquad \frac{A\&B}{B}\,\&E$$

D. Prawitz (✉)
Stockholm, Sweden
e-mail: dag.prawitz@philosophy.su.se

L. C. Pereira et al. (eds.), *Advances in Natural Deduction*,
Trends in Logic 39, DOI: 10.1007/978-94-007-7548-0_12,

Understanding the introduction rule for conjunction ($\&I$) as giving the meaning of conjunction by telling how a conjunction is proved, we see that the elimination rule ($\&E$) in its two forms becomes justified: according to the meaning of conjunction, a proof of the premiss of an application of the rule &E already contains a proof of the conclusion.

It seems obvious that there can be no stronger elimination rule for conjunction that can be justified in this way in terms of the introduction rule. A similar remark can be made concerning all the other logical constants of predicate logic. To illustrate the idea with a slightly more complicated example, consider the introduction and elimination rules for disjunction:

$$\frac{A}{A \vee B} \vee I \qquad \frac{B}{A \vee B} \vee I \qquad \frac{A \vee B \quad \begin{matrix}[A]\\ \vdots \\ C\end{matrix} \quad \begin{matrix}[B]\\ \vdots \\ C\end{matrix}}{C} \vee E$$

The elimination rule for disjunction ($\vee E$) is justified by the meaning given to disjunction by the introduction rule ($\vee I$) in view of the following consideration. According to the meaning of disjunction, a proof of the major premiss $A \vee B$ of an application of $\vee E$ must contain either a proof of A or of B. In the first case we may substitute this proof of A for the hypothesis A in the proof of the first occurrence of the minor premiss C, and in the second case we may instead substitute the proof of B for the hypothesis B in the proof of the second occurrence of the minor premiss C. The result is in either case a proof of C that does not depend on the hypotheses discharged by the application of $\vee E$, without making use of this instance of $\vee E$. In other words, the proofs of the premisses of an inference by $\vee E$ are seen to contain already elements that combined appropriately yield a proof of the conclusion of the inference.

Can we imagine a stronger elimination rule for disjunction that is possible to justify in terms of the introduction rule in this way? Martin-Löf's type theory for intuitionistic logic does contain such a stronger rule, but its formulation requires the richer language of type theory. The question should therefore be put a little more carefully: Is there an elimination rule for disjunction that can be formulated in the language of predicate logic and is stronger than $\vee E$ but can nevertheless be justified in terms of $\vee I$? My conjecture is that the answer is no and that the corresponding thing holds for all the other logical constants in the language of predicate logic.

It seems obvious that the elimination rules of Gentzen's system are *the* elimination rules that correspond to his introduction rules. Or, again to put it more carefully: although there are of course weaker elimination rules and even elimination rules that are deductively equivalent with the ones formulated by Gentzen, there are no stronger rules that can be formulated in the language of predicate logic and are justifiable in terms of the introduction rules. The problem is to formulate this obvious idea more precisely in the form of a conjecture and then to prove it.

2 The Notion of Validity

The ideas sketched above that are in need of a more precise formulation are in particular the two ideas that an introduction rule gives the meaning of the logical constant in question and that this meaning justifies the corresponding elimination rule. As for the second idea concerning the justification of elimination rules, anyone familiar with the reduction steps in the normalization of natural deductions sees immediately that they are the main elements in these justifications. When I introduced the normalization procedure for natural deductions, I saw it as something that brought out Gentzen's idea concerning how introduction and elimination rules are related to each other.[1] But, admittedly, it does not bring out his idea that the introduction rules are meaning constitutive and that the meanings that they assign to the logical constants justify the elimination rules. To bring out these ideas explicitly requires a semantics in which one can speak about the validity of inferences. Analogously to how meaning is explained in classical semantics by reference to truth conditions, we should now explain the meaning of a sentence by telling what is required of a valid proof of the sentence. The notion of convertibility introduced by [11] for terms and adapted to natural deductions under the name of computability by [3] seemed to me to offer a suitable means for such a semantics. Modifying Martin-Löf's notion, I used it to prove strong normalization for various systems of natural deductions. Furthermore, I argued in an appendix to the chapter "Ideas and results in proof theory" [6] that when slightly modified in another direction, the notion could be used to give a semantic explication of the justification of inference rules. Because of this use, I called it *validity*.

I shall now give a condensed account of that notion. To do so it is convenient to introduce a few other notions. I shall say that an inference such as $\rightarrow$-introduction or $\vee$-elimination, by which occurrences of a hypothesis become discharged, *binds* the occurrences in question and that an inference such as $\forall$-introduction or $\exists$-elimination, which puts restriction on a variable (so-called 'Eigenvariabel') and which would lose its correctness if the variable were replaced by a constant, *binds* the variable in question. An occurrence of a hypothesis in a deduction D or of a variable in a formula in D is said to be *free* in D if it is not bound, and a deduction that contains free hypotheses or free variables is said to be *open*. A deduction that is not open is *closed*. A deduction D is said to reduce to the deduction D^*, when D^* is obtained from D by successively replacing subdeductions by their reductions (as defined in the context of normalizations).

A triple consisting of set $\mathcal{T}$ of closed individual terms, a set $\mathcal{R}$ of relational symbols, and a set $\mathcal{P}$ of profs of atomic sentences built up from $\mathcal{R}$ and $\mathcal{T}$ (thus not containing any proof of the constant for falsehood, $\perp$) will be called a *base*. We shall be concerned with formulas in a first-order language determined by such a base; the set $\mathcal{T}$ is supposed to fix the range of the individual variables (and hence the domain of the quantifiers), and the set $\mathcal{P}$ the meanings of the atomic sentences. A deduction D

[1] Prawitz [4]. The elimination rules are there said to be the inverse of the corresponding introduction rules.

of a sentence A is said to be in canonical form relative to such a base $\mathcal{B} = (\mathcal{T}, \mathcal{R}, \mathcal{P})$, if D is closed and either A is an atomic sentence and D is in $\mathcal{P}$ or A is compound and the last inference of D is an introduction.

We can now formulate two *principles of validity* (relative to a base $\mathcal{B} = (\mathcal{T}, \mathcal{R}, \mathcal{P})$) for closed and open deductions respectively:

(I) A closed deduction D of a sentence A is *valid* (relative to a base $\mathcal{B}$) if and only if either (a) D is in canonical form (relative to $\mathcal{B}$) and, in case A is compound, the immediate subdeductions of D are *valid* (relative to $\mathcal{B}$), or (b) D is not in canonical form but reduces to a deduction that is valid according to clause (a).

(II) An open deduction D is *valid* (relative to a base $\mathcal{B}$) if and only if each closed instance D' of D is valid (relative to $\mathcal{B}$) when D' is obtained by first replacing all free individual variables in D by individual terms (in $\mathcal{T}$) and then replacing all free occurrences of each hypothesis in D by a *valid* closed deduction (relative to $\mathcal{B}$) of that hypothesis.

Principle (I) can be seen as a condensed semantic expression of Gentzen's two basic ideas concerning introduction and elimination rules. His first idea, that an introduction rule for a logical constant gives the meaning of the constant, is here understood as implying that the rule is self-justifying in the sense that inferences conforming to the rule are valid simply in virtue of what the constant in question means. This is expressed in (Ia) by saying that applications of introduction rules preserve the validity of a deduction. However, this states only a sufficient condition for validity, half of the meaning of the logical constant so to say: a deduction is valid, *if* its ends with an introduction and its immediate subdeductions are valid. To take this to be also a necessary condition would result in an obviously too restrictive notion of validity. But we could demand that a deduction is valid *only if* it can be *reduced* to a deduction that ends with an introduction. This requirement is stated in (Ib), which also expresses Gentzen's second idea about justifications of elimination inferences: inferring a conclusion A by such an inference yields a valid deduction if and only if it reduces to a deduction of A that does not use the inference in question.

Principle (II) in turn is an expression of the simple idea that an open deduction is to be seen as a schema in which free individual variables are blanks for (closed) individual terms and free hypotheses A are blanks for closed proofs of A. It is hence valid if and only if all results of appropriately filling in the blanks are valid.

Because of the fact that the premisses of an introduction and the hypotheses that it binds are of lower complexity than that of the conclusion, principles (I) and (II) can be seen as a recursive definition (over the complexity of the end formulas) of the notion of validity of natural deductions.

An *inference rule* R can now be defined as *valid* when it preserves validity relative to an arbitrary base $\mathcal{B}$, that is, when a deduction that ends with an application of R is always valid relative to $\mathcal{B}$ given that its immediate subdeductions are valid relative to $\mathcal{B}$. To prove the validity of the inference rules of Gentzen's system of natural deduction for intuitionistic logic is now an easy exercise.

3 Generalizing the Notion of Validity

A limitation of the notion of validity as described above is that it is defined only for deductions in a given formal system. In contrast, a notion like truth is defined for sentences in general but singles out a subset of them. Similarly, one would like to have a notion of validity defined for a broader range of reasoning, singling out a subdomain of correct reasoning that could properly be called proofs. Such a generalization is especially important in the context of the conjecture discussed here, since in that context we want to consider not only inference rules in a given system but other possible elimination rules that no one has proposed so far.

The goal of the chapter in which the conjecture was formulated was precisely to extend the notion of validity to a broader class of reasoning, what I called *arguments*. The simple idea is to consider not only deductions that proceed by applications of a given set of rules of inferences but trees of formulas built up of arbitrary inferences. As in a natural deduction, the top-formulas of such a tree are to represent axioms or hypotheses, and the inferences are allowed to bind hypotheses and variables. It is supposed to be determined for each inference which occurrences of hypotheses and variables become bound by the inference. Such an arbitrary tree of formulas (in the language of first-order logic) with indications of how variables and hypotheses are bound was called an *argument skeleton*. The notion of being an *open* or *closed* deduction can thus be carried over to argument skeletons.

Of course, most argument skeletons do not represent valid forms of reasoning. But some of them do and can be justified by procedures similar to the ones used when justifying the elimination rules in Gentzen's system. To generalize the idea of such justifying procedures, I define a *justification* of an inference rule R different from the introduction rules to be an operation that is defined on some argument skeletons whose last inference is an application of R and that yields as value, when applied to an argument skeleton S, another argument skeleton with the same end formula and with no more free hypotheses or variables than S. I consider sets J of such justifications of inference rules. They are assumed to be consistent in the sense that they are not to contain two operations j and j' defined on the same skeleton and yielding different values. An argument skeleton S is said to *reduce* to another argument skeleton S^* relative to such a set J of justifications, if there is a sequence $S_1, S_2, \ldots, S_n$ $(n > 0)$ where $S_1 = S$, $S_n = S^*$, and, for each i $(i = 1, 2, \ldots, n-1)$, there is a j in J such that S_{i+1} is obtained form S_i by replacing some subskeleton S' of S_i by $j(S')$.

An argument skeleton together with a set of justifications is called an *argument*. It is still the case that most arguments do not represent valid forms of reasoning. What is called justifications above are in fact only alleged justifications. It is the notion of validity generalized to arguments that is now to lay down the conditions that the alleged justifications have to satisfy in order to be real justifications. This generalization of validity is easily obtained by simply carrying over the two principles of validity from deductions to arguments. The notion of a base $\mathcal{B} = (\mathcal{T}, \mathcal{R}, \mathcal{P})$ and the notion of a canonical form is to be kept as before. Principle (I) now becomes

(I') A closed argument (S, J) for a sentence A is *valid* relative to a base $\mathcal{B}$ if and only if either (a) S is in canonical form relative to $\mathcal{B}$ and, in case A is compound, for each immediate subskeleton S' of S it holds that (S', J) is *valid* relative to $\mathcal{B}$, or (b) S reduces relative to J to a skeleton S^* in canonical form such that (S^*, J) is valid relative to $\mathcal{B}$ according to clause (a).

To generalize principle (II) we need to speak about consistent extensions of a set of justifications, where again we require that there is no conflicting overlap between the different justifying operations. Principle (II) now becomes

(II') An open argument (S, J) is *valid* relative to a base $\mathcal{B} = (\mathcal{T}, \mathcal{R}, \mathcal{P})$ if and only if for each consistent extension J' of J and for each instance S^* of S it holds that (S^*, J') is *valid* relative to $\mathcal{B}$, when S^* is obtained from S by first replacing all free occurrences of individual variables by terms in $\mathcal{T}$ and then replacing all free occurrences of each hypothesis A in S by closed skeletons S' ending with A such that (S', J') is valid relative to $\mathcal{B}$.

An argument that is valid in the sense defined may proceed by quite different inferences than those obtained by applying the inference rules of Gentzen's system, but it may claim to represent a proof with the same right as the deductions in Gentzen's system.

That an inference rule is valid is to mean as before that it preserves validity, now validity of an argument, relative to an arbitrary base. In case R is an inference rule different from the introduction rules, this is to mean that there is a justification j of R such that for all bases $\mathcal{B}$ and for all consistent extension J of $\{j\}$, if S is an argument skeleton whose last inference is an application of R and is such that for each immediate subskeleton S' of S, (S', J) is valid relative to $\mathcal{B}$, then (S, J) is valid, too, relative to $\mathcal{B}$.

Having arrived at this notion of valid inference rule, the conjecture that there are no stronger justifiable elimination rules within first-order logic than the ones formulated by Gentzen can now naturally be formulated as follows:

Every valid inference rule that can be formulated within first-order languages holds as a derivable inference rule within the system of natural deduction for intuitionistic logic.

Is this notion of validity rightly explicating a constructive reading of the logical constants? Besides my own discussions of this question in several papers dealing with constructive approaches to the notion of logical consequence (for a relatively recent paper, see [8]), there is a particularly comprehensive presentation and discussion of essentially my notion of validity in Michael Dummett's book *The Logical Basis of Metaphysics* [1] (for a comparison of my and Dummett's notions of validity, see [9]). (*Added in proof:* After this paper was composed, Peter Schroeder-Heister has presented in [10] a very detailed and thoughtful study of the notion of validity, proposing among other things some changes in my notions of justification and validity. If I had written my paper today, I would have adopted some of them, and would have written a different paper also in many other respects.) In the rest of this paper,

I shall take up some doubts that one may have about the notion of validity presented here, and shall consider a quite radical revision of the notion.

4 A Modified Approach

A shortcoming of the notion of argument as described above is that the justifications are operations defined on argument skeletons rather than on arguments, i.e. skeletons together with justifications, and furthermore that the value of the justifying operations consist of just argument skeletons instead of skeletons with justifications. It is the skeletons with justifications that represent arguments, valid or invalid ones, and when an argument step is to be justified it is conceivable that one wants the justification to depend on the entire arguments for the premisses and not only on their skeletons. As for the value of a justifying operation, it is a shortcoming that it consists of just a skeleton if one wants it to contain new inferences that were not present in the skeleton to which the operation is applied.[2] This last limitation may be taken care of without changing the framework very much. However, in order to achieve that the justifications operate not just on argument skeletons, but, as it were, on skeletons with justifications, we must make a more radical change in the approach. We need then to conceive of the valid arguments, i.e. proofs, as built up of operations defined on proofs and yielding proofs as values.

To outline an approach of that kind, let us as before start with a base $\mathcal{B}$, now determined by a set $\mathcal{T}$ of closed individual terms, a set $\mathcal{R}$ of relational symbols, and a set $\mathcal{C}$ of constants standing for proofs of atomic formulas with relational symbols in $\mathcal{R}$ and terms in $\mathcal{T}$. Each constant in $\mathcal{C}$ is to be typed by the atomic formula A that it is a proof of; we may write such a constant c^A. The base determines as before a first-order language $\mathcal{L}_\mathcal{B}$, whose formulas are built up as usual with the symbols given by $\mathcal{R}$ and $\mathcal{T}$, individual variables, and the logical constants $\bot$, &, $\vee$, $\rightarrow$, $\forall$, and $\exists$.

For each formula A in the language $\mathcal{L}_\mathcal{B}$, we introduce (proof) variables of type A, written α^A. The variables of type A are thought of as ranging over proofs of A. From the constants in $\mathcal{C}$ and the proof variables, we build up what I shall call *proof terms*, using operators that are to be thought of as standing for operations on proofs that yield proofs as vales. The proof terms are to be typed by the formulas of $\mathcal{L}_\mathcal{B}$; a proof form of type A is to be thought of as standing for a proof of A. Constants c^A in $\mathcal{C}$ and variables α^A are thus proof terms of type A. For all the logical constants except $\bot$, we introduce primitive operators, which we may call $\&I$, $\vee I_1$, $\vee I_2$, $\rightarrow I$, $\forall I$, and $\exists I$ (for disjunction there are thus two operators $\vee I_1$ and $\vee I_2$), naming them in the same way as the introduction rules. Some of the operators are variable binding, and the variables that they bind are as usual listed after an occurrence of the operator. The rules for forming proof terms with the help of the primitive operators are as follows:

[2] One may argue that to allow the value to contain new inferences is too liberal, since the justifying operation then produces an argument that goes beyond what was present in the arguments for the premisses. However, this is an angle that is not taken up here.

(1) $\&I(P_1, P_2; A_1, A_2/A_1\&A_2)$ is a proof term of type $A_1\&A_2$, if P_i is a proof form of type $A_i (i = 1, 2)$—the formulas written after the semi-colon and before the slash indicate the types of the terms on which the operator is applied, and the one written after the slash indicates the type of the resulting term,
(2) $\vee I_i(P; A_i/A_1 \vee A_2)$ is a proof term of type $A_1 \vee A_2$, if P is a proof term of type $A_i (i = 1, 2)$,
(3) $\rightarrow I\alpha^A(P; B/A \rightarrow B)$ is a proof term of type $A \rightarrow B$ and binds free occurrences of α^A in P, if P is a proof term of type B,
(4) $\forall Ix(P; A(x)/\forall x A(x))$ is a proof term of type $\forall x A(x)$, if P is a proof term of type $A(x)$—it binds free occurrences of the variable x that occur (in type indices) in P, and there is the restriction that x is not to occur free in the indices of free variables α^A in P,
(5) $\exists I(P, t; A(t)/\exists x A(x))$ is a proof term of type $\exists x A(x)$, if P is a proof term of type $A(t)$ (obtained from $A(x)$ in $\exists x A(x)$ by replacing free occurrences of x by the individual term t).

In addition to these primitive operators, we introduce *operational parameters* for all possible elimination rules. For instance, corresponding to a rule of the form

$$\frac{A_1 \quad A_2 \ldots A_n}{B} C, x$$

that binds occurrences of the hypothesis C and the individual variable x, we introduce a parameter Φ with the formation rule:

> $\Phi\alpha^C x(P_1, P_2, \ldots, P_n; A_1, A_2, \ldots, A_n/B)$ is a proof term of type B, if P_i is a proof term of type $A_i (i = 1, 2, \ldots, n)$—it binds free occurrences of α^C and of x in P_i with the restriction that x is not to occur free in type indices of free variables $\alpha^{C'}$ in P_i except when C' is C.

Particular such operation parameters are those that correspond to the usual elimination rules, which we may write $\&E_1$, $\&E_2$, $\vee E$, $\rightarrow E$, $\forall E$, and $\exists E$. For instance,

(1) $\&E_i(P; A_1 \& A_2/A_i)$ is a proof term of type A_i, if P is a proof term of type $A_1\&A_2 (i = 1, 2)$;
(2) $\vee E\alpha^A\beta^B(P, Q, R; A \vee B, C, C/C)$ is a proof term of type C, if P, Q, and R are proof terms of type $A \vee B$, C, and C, respectively—it binds free occurrences of the variables α^A and β^B in Q and R; and
(3) $\exists E\alpha^{A(x)}x(P, Q; \exists x A(x), B/B)$ is a proof term of type B, if P is a proof term of type $\exists x A(x)$ and Q is a proof term of type B—it binds free occurrences of the variables $\alpha^{A(x)}$ and x in Q with the usual variable restrictions.

To the operation parameters we assign *definitions* that tell how the parameters are to be interpreted as standing for operations on proofs that yield proofs as values. Examples of such definitions are:

$$\&E_2(\&I(p_1, p_2; A_1, A_2/A_1\&A_2); A_1\&A_2/A_2) = p_2$$
$$\vee E\alpha^A\beta^B(\vee I_1(P; A/A \vee B), Q(\alpha^A), R(\beta^B); A \vee B, C, C/C) = Q(P)$$
$$\rightarrow E(\rightarrow I\alpha^A(P(\alpha^A); B/A \rightarrow B), Q; A \rightarrow B, A/B) = P(Q)$$
$$\exists E\alpha^{A(x)}x(\exists I(P, t; A(t)/\exists x A(x)), Q(\alpha^{A(x)}, x); \exists x A(x), B/B) = Q(P, t)$$

A proof term P (of type A) together with a set Δ of definitions of this kind, I shall call an interpreted proof term (of type A). A proof term P is said to reduce to another proof term P^* relative to a set of definitions Δ, if P^* is obtained from P by successively replacing subterms of P that appear as definienda of definitions in Δ by their definientia.

We may now define a notion of validity for interpreted proof terms in essentially the same way as for arguments. We have thus the two principles:

(I") A closed interpreted proof term (P, Δ) of type A is *valid* relative to a base $\mathcal{B} = (\mathcal{T}, \mathcal{R}, \mathcal{C})$ if and only if either (a) A is atomic and P is a constant c^A in C, or A is compound with χ as its principal constant and P is a proof term with the primitive operation χ I as its outer operator such that for each immediate subterm Q of P it holds that (Q, Δ) is *valid*, or (b) P reduces to a proof term P^* relative to Δ such that (P^*, Δ) is *valid* according to (a).

(II") An open proof term (P, Δ) is *valid* relative to a base $\mathcal{B} = (\mathcal{T}, \mathcal{R}, \mathcal{P})$ if and only if each (P^*, Δ^*) is valid relative to B when Δ^* is an extension of Δ and P^* is obtained from P by first replacing all occurrences of free individual variables by terms in T and then in the result got after this substitution replacing all free occurrences of each variable α^A by closed proof terms Q of type A such that (Q, Δ^*) is valid relative to B.

Valid closed interpreted proof terms accord well with proofs as usually described in intuitionism. If (P, Δ) is such an interpreted proof term, then either P has introductory form, i.e. its outer operation is one of the five primitive operators, or is definitionally equal to such a proof term, i.e. reduces effectively to such a term by replacing a definienda in a definition in Δ by its definientia. For instance, if P is of type $A \rightarrow B$, then P has the form $\rightarrow I\alpha^A(P(\alpha^A); B/A \rightarrow B)$ or reduces relative to Δ to such a form. Furthermore, the validity of $(\rightarrow I\alpha^A(P(\alpha^A); B/A \rightarrow B), \Delta)$ implies that $(P(\alpha^A), \Delta)$ is valid, which according to principle (II") means that for any closed proof term Q of type A such that (Q, Δ') is valid, it holds that $(P(Q), \Delta \cup \Delta')$ is valid. Valid closed interpreted proof terms may thus be thought of as representing intuitionistic proofs.

Comparing an interpreted proof term (P, Δ) with arguments as defined earlier, we find that one obtains what was called an argument skeleton when one attends to just the types in the subterms P and that the operators occurring in P when interpreted by the definitions in Δ can be seen as names of operations that appear as justifications of the inference steps in the argument skeleton. The difference between arguments and interpreted proof terms is the one discussed at the beginning of this section, namely that the justifying operations in an argument are applied to argument skeletons while the operations in an interpreted proof term are defined for objects built up from this very kind of operations.

The system of interpreted proof terms may be seen as an extension of a typed lambda calculus or a fragment of Martin-Löf's intuitionistic type theory. Of course, $\rightarrow I$ corresponds to λ-abstraction, and $\rightarrow E$, when defined as above, to application of a lambda term, while the definition of $\rightarrow E$ corresponds to the rule of β-conversion. Also, $\forall I$ is a kind of λ-abstraction, while $\& I$ corresponds to pairing. To this kind of well-known operations, $\vee I$ and $\vee E$ add other specific operations. Haskell Curry seems to have been the first to note the relevance of lambda calculus to the semantic interpretation of a fragment of intuitionistic sentential logic. Extensions of Curry's observation to deal with full intuitionistic predicate logic occur in [2] (and [5] - what is there called construction term corresponds to what is now called proof term).

In the approach outlined above, we may now define the validity of an inference rule in essentially the same way as before. For instance, if R is an inference rule

$$\frac{A_1 \quad A_2}{B} C, x$$

that binds a hypothesis C and an individual variable x, we say that R is valid if there is a definition d of the operational parameter $\Phi\alpha^C x(\pi_1, \pi_2; A_1, A_2/B)$ such that for any base $\mathcal{B}$, if (P, Δ) and (Q, Δ) are two interpreted proof terms valid relative to $\mathcal{B}$ where P and Q are of type A_1 and A_2 respectively, then applying Φ to P and Q we get a proof term $\Phi\alpha^C x(P, Q; A_1, A_2/B)$ such that $(\Phi\alpha^C x(P, Q; A_1, A_2/B), \Delta)$ is valid relative to $\mathcal{B}$.

All the elimination rules of Gentzen's system of natural deduction for intuitionistic logic are easily seen to be valid in the sense now defined. The conjecture is that conversely all inference rules that are valid in this sense hold as derived rules in Gentzen's system of natural deduction for intuitionistic logic.

Acknowledgments Work on this chapter was done within the project *Interpretation and Meaning*, funded by Bank of Sweden Tercentenary Foundation.

References

1. Dummett, M. (1991). *The Logical Basis of Metaphysics*. London: Duckworth.
2. Howard, W. (1980). The formulae-as-types notion of construction. In J. R. Hindley & J. Seldin (Eds.), *To H. B. Curry: Essays on combinatory logic, lambda calculus, and formalism* (pp. 479–490). San Diego: Academic Press.
3. Martin-Löf, P. (1971). Hauptsatz for the intuitionistic theory of iterated inductive definitions. In J. E. Fenstad (Ed.), *Proceedings of the Second Scandinavian Logic Symposium* (pp. 179–216). Amsterdam: North-Holland.
4. Prawitz, D. (1965). *Natural deduction: a proof-theoretical study*. Stockholm: Almqvist & Wicksell. (Reprinted 2006. Mineola, New York: Dover Publications.)
5. Prawitz, D. (1970). Constructive semantics. In *Proceedings of the First Scandinavian Logic Symposium, Åbo 1968, Filosofiska studier 8* (pp. 96–114) Uppsala: Filosofiska Föreningen och Filosofiska institutionen vid Uppsala Universitet.

6. Prawitz, D. (1971). Ideas and results in proof theory. In J. E. Fenstad (Ed.), *Proceedings of the Second Scandinavian Logic Symposium* (pp. 225–250). Amsterdam: North-Holland.
7. Prawitz, D. (1973). Towards a foundation of a general proof theory. In P. Suppes, et al. (Eds.), *Logic, Methodology and Philosophy of Science IV* (pp. 225–250). Amsterdam: North Holland.
8. Prawitz, D. (2005). Logical consequence from a constructivist point of view. In S. Shapiro (Ed.), *The Oxford Handbook of Philosophy of Mathematics and Logic* (pp. 671-695). Oxford: Oxford University Press.
9. Prawitz, D. (2006). Meaning approached via proofs. *Synthese*, 148, 507–524.
10. Schroeder-Heister, P. (2006). Validity concepts in proof-theoretic semantics. *Synthese*, 148, 525–571.
11. Tait, W. (1967). Intentional interpretation of functionals of finite type I. *The Journal of Symbolic Logic*, 32, 198–212.

The manufacturer's authorised representative in the EU is Springer Nature Customer Service Centre GmbH, Europaplatz 3, 69115 Heidelberg, Germany. If you have any concerns regarding our products, please contact ProductSafety@springernature.com

Printed and bound by CPI Group (UK) Ltd, Croydon, CR0 4YY
25/09/2026
02222869-0013